CHEMISTRY RESEARCH AND APPLICATIONS

BORON HYDRIDES, HIGH POTENTIAL HYDROGEN STORAGE MATERIALS

CHEMISTRY RESEARCH AND APPLICATIONS

Additional books in this series can be found on Nova's website under the Series tab.

Additional E-books in this series can be found on Nova's website under the E-books tab.

CHEMISTRY RESEARCH AND APPLICATIONS

BORON HYDRIDES, HIGH POTENTIAL HYDROGEN STORAGE MATERIALS

UMIT B. DEMIRCI
AND
PHILIPPE MIELE
EDITORS

Nova Science Publishers, Inc.
New York

For permission to use material from this book please contact us:
Telephone 631-231-7269; Fax 631-231-8175
Web Site: http://www.novapublishers.com

Additional color graphics may be available in the e-book version of this book.

LIBRARY OF CONGRESS CATALOGING-IN-PUBLICATION DATA

Boron hydrides, high potential hydrogen storage materials / editors, Umit B. Demirci and Philippe Miele.
p. cm.
Includes index.
ISBN 978-1-61668-361-0 (hardcover)
1. Hydrogen--Storage--Materials. 2. Boranes. I. Demirci, Umit B. II. Miele, Philippe.
TP245.H9B67 2009
665.8'1--dc22
2010016084

Published by Nova Science Publishers, Inc. + New York

CONTENTS

PREFACE

Boron hydrides are hydrogen storage materials which are the object of intensive investigation because they pose tangible solution to the hydrogen storage issue. Today many boron hydrides are intensively investigated: lithium borohydride LiBH4, sodium borohydride NaBH4, potassium borohydride KBH4, new borohydrides like e.g. Ca(BH4)2, Zn(BH4)2 or Zr(BH4)4, combination of hydrides like e.g. LiBH4 and LiNH2 or LiBH4 and MgH2, ammoniaborane NH3BH3, alkali amidoboranes like e.g. LiNH2BH3 or NaNH2BH3, and so on. This new book reviews research on boron hydrides and gives a general view of the perspectives of application.

Chapter 1- The emergence of a (near-future) hydrogen economy implies the development of hydrogen as energy carrier, which is especially intended for fuel cell applications. However, this is today hindered by many issues related to the hydrogen production, storage, distribution and utilization. The present chapter briefly discusses all of these aspects of the hydrogen economy as well as the most important issues. A special attention is paid on the hydrogen storage issue, which is actually a key issue. The various storage ways that are currently under investigation are briefly reported whereas the solid-state hydrogen storage is more widely discussed. Solid-state hydrogen storage is attractive because it offers a safe and efficient way to store hydrogen, and the highest hydrogen storage capacities. It is reported that complex hydrides are promising hydrogen storage materials by virtue of their high theoretical hydrogen capacities. Foremost among complex hydrides are boron compounds, and especially boron hydrides, which are surveyed throughout this book.

Chapter 2- Boranes (or boron hydrides) are known for their capacity of hydrogen storage since the 1950s, at least for the borohydrides, but it is since the early 2000s, that is 50 years after the discovery of the most known hydrogen storage boron hydride, namely sodium borohydride, that these compounds have attracted much attention with respect to the potential as hydrogen storage materials they have for technological applications. Today, boron hydrides seem to have a major role to play in the hydrogen economy we are intensely trying to develop. The present chapter is devoted to these hydrides and more generally to the boron compounds, which are listed and briefly discussed in a first section. It is shown that only the boron hydrides are promising candidates for storing hydrogen. The boron hydrides, and especially the borohydrides, ammoniaborane and the newly developed metal amidoboranes, are more largely discussed in terms of their gravimetric hydrogen storage capacities. The safety aspect is also, besides, tackled. Actually, this chapter is an introduction to the next chapters, dealing more in details with the presently discussed boron hydrides.

Chapter 3- Hydrogen is currently regarded as one of the most suitable energy carriers for the future sustainable economy, but its full implementation requires paradigm shifts in its production, distribution and storage. It is beyond the scope of this chapter to cover these three sectors; we only want to focus on the safe storage, which stands as a great challenge for various applications. While attractive, the implementation of fuel cell technology in EV transportation is, for instance, beset by our inability to come out with efficient means to safely store large amounts of hydrogen. Such applications require hydrogen absorbing materials having a gravimetric capacity of at least 6 wt.% and a volumetric capacity higher than 70 g/l. Metal-hydrides have long been considered as potential candidates, but their performances have been deceiving, so that a lot of attention has been recently focused on lightweight materials like alanates or borohydrides, which are frequently termed as complex hydrides. Among them, LiBH4 has been the subject of intense studies, which have led to attractive and numerous findings; so this material deserves in itself an entire chapter reuniting the main methods of synthesizing LiBH4 and discussions about its physico-chemical properties, e.g. crystal structure, ionic conduction, thermodynamic stability. LiBH4 undergoes an orthorhombic to hexagonal structural transition at 116°C leading to a compound with a high ionic conductivity. This compound is thermally decomposed via a complex mechanism of hydrogen release forming a new Li2B12H12 salt. Moving towards applications, different approaches to enhance the hydrogen release kinetic are reviewed; the most promising one being the inclusion of LiBH4 into mesoporous carbons, which leads to materials capable of releasing 12 wt.% of hydrogen at high rates and at temperatures as low as 200°C, making this material somewhat unique for hydrogen desorption. Nevertheless, the practical future of this material is in jeopardy if we remain unable to improve the feasibility of its re-hydrogenation process, which is presently very poor. Owing to late developments, optimism must, however, prevail. Besides hydrogen storage applications, LiBH4 could have a great impact within the field of Li-batteries, due to its high ionic conduction and its ability to lead to a Li-based electrolyte.

Chapter 4- In this chapter, production of NaBH4 as hydrogen storage material is discussed. Linking this with other processes, which constitutes a system to recycle hydrogen is shown. Difficulties and advantages of NaBH4 synthesis by natural available boron source kolemanite and borax minerals are discussed in details. It is presented that basic chemicals suffice for NaBH4 production in a procedure designed to work below 275˚C. That proves the feasibility of NaBH4 production. Offered procedure may be used to compensate for the loss of materials during the recycle. In this procedure, hydrogen including substances has been used to remove oxygen and bond remaining hydrogen to the by-product NaBO2 of dehydrogenation reaction. The wet and dry applications of producing NaBH4 have been also discussed. Moreover a shorter procedure can be proposed to directly obtain NaBH4 from borax mineral. Among these processes, dehydrogenation is critical but it requires plenty of water. Hence, new types of catalysts are important to reduce the amount of water required during dehydrogenation. Catalyst types and their production techniques are also given in this chapter. Performance of the catalysts during the hydrolysis of NaBH4 is tried to explain according to the dehydrogenation conditions such as temperature, concentration of the reactants and by-product effect.

Chapter 5- Direct borohydride fuel cells (DBFC) exhibit some potential for the powering of small nomad electronic devices, thanks to their high energy density as well as the facile and safe storage of borohydride salts. Since their development a decade ago, numerous groups have focused on the cell geometry, operating parameters and materials, yielding to appreciable performance improvements. However, from an electrocatalytic point of view, the DBFC remain hindered because (i) the borohydride oxidation reaction (BOR) is complex, (ii) its mechanism is imperfectly determined yet and (iii) no practical electrocatalyst exhibits both fast BOR kinetics and high faradic efficiency. The existence of these three major drawbacks renders mandatory more fundamental research on the BOR. In this context, the BOR mechanism and kinetics was characterized for one of the most studied model BOR electrocatalyst: polycristalline bulk gold. Particular attention was focused on the reaction pathway, using both electrochemistry and coupled physicochemical techniques: Fourier transformed infrared (FTIR) spectroscopy and on-line mass spectrometry (OLEMS).

While one usually admits that gold promotes the complete 8-electron BOR, the present chapter, thanks to the coupling between on-line mass spectrometry and electrochemistry (OLEMS), reveals that H2 is produced when the BOR proceeds on gold electrodes, as a result of some extent of heterogeneous hydrolysis of BH4-. Thus, gold cannot be considered as a perfect faradaic-efficient BOR electrocatalyst.

From this knowledge, modeling cyclic voltammetry (CV) and electrochemical impedance spectroscopy (EIS) data, a simplified reaction pathway was proposed, the theoretical behavior of which agrees with the experimental findings, including those from in situ FTIR spectroscopy. This pathway includes at least two irreversible electrochemical steps (EE) for BH4- oxidation, which compete with the electrochemical adsorption reaction (EAR) of OH- anions at high potentials. This shows that the BOR mechanism on gold is not as simple as usually admitted, and that gold promotes two pathways as a function of the electrode potential (one ca. 4-electron pathway at low potentials, likely resulting from concomitant direct borohydride oxidation and heterogeneous hydrolysis reactions, whereas one quantitative ca. 8-electrons pathway proceeds at higher potentials).

As a result, due to its slow BOR kinetics and incomplete BH4- electrooxidation, one may conclude that Au is not an ideal electrocatalyst for DBFC anodes.

Chapter 6- Crystal chemistry of $M(BH_4)_n$, where M is a 2nd-4th period element, is reviewed here. It is shown that except certain cases, the BH_4 group has a nearly ideal tetrahedral geometry. Corrections of the experimentally- determined H-positions, accounting for the displacement of the electron cloud relative to an average nuclear position and for a libration of the BH_4 group, are considered. Recent studies of structural evolution with temperature and pressure are reviewed. Some borohydrides involving less electropositive metals (e.g. Mg and Zn) reveal porous structures and dense interpenetrated frameworks, thus resembling metal-organic frameworks (MOFs). Analysis of phase transitions, and the related changes of the coordination geometries for M atoms and BH_4 groups, suggests that the directional BH_4...M interaction is at the origin of the structural complexity of borohydrides. The ways to influence their stability by chemical modification are discussed.

Chapter 7- Ammonia borane Borazane (NH_3BH_3) also called borazane ammonia borane has been considered, since its synthesis in 1955, a promising compound for hydrogen storage due to its remarkably high hydrogen content (19.6 wt%) and volumetric density (146 $g(H_2)\ l^{-1}$); moreover it is non-toxic and stable at ambient temperature. A large amount of scientific literature has, therefore, been produced on ammonia borane dehydrogenation processes.

Hydrogen can be produced by hydrolysis or thermolysis of ammonia borane. In the present chapter we focus our attention on thermal decomposition of ammonia borane. Some hints are also given about ammonia borane synthesis and regeneration. Upon Heating heatingammonia borane, a structural modification occurs in ammonia borane at about 70 °C before hydrogen evolution. About 2 moles of hydrogen per mole of ammonia borane (15 wt%) are thermally released in the 110-150 °C temperature range through a two-steps exothermic reaction. However, the decomposition rate is extremely low (several hours). Complete decomposition of ammonia borane takes place at temperatures above 500°C. The decomposition reaction steps have been extensively studied by complementary experimental techniques: differential scanning calorimetry (DSC), thermogravimetric analyses (TGA) and differential thermo-analysis (DTA) at various temperature scan rates followed by isothermal desorption, thermogravimetric analyses (TGA), volumetric and spectroscopic methods. The study of thermodynamic and of the structural and thermodynamic modifications occurring at each step and the identification of the corresponding reaction products allowed a detailed description of the involved chemical reactions. Doping and or dispersing ammonia borane into nanostructures have showned beneficial effects on hydrogen ammonia borane desorption rate. Ammonia borane metal composites have also been investigated as high capacity hydrogen storage materials. Some hints are also given about ammonia borane synthesis and regeneration.

Chapter 8- Ammonia borane NH3BH3 combines low molecular weight (30.7 g mol-1) and high gravimetric hydrogen storage capacity (19.6 wt%), what makes it more attractive than both sodium borohydride (37.8 g mol-1 and 10.8 wt%) and lithium borohydride (21.8 g mol-1 and 18.3 wt%). Like sodium borohydride, ammonia borane is a hydrogen source but ammonia borane is maybe the most promising boron hydride for hydrogen storage applications. Stored hydrogen can be released through either thermolysis or hydrolysis. The present chapter deals especially with hydrolysis of ammonia borane. Its basics, state-of-the-art, and potential are largely discussed on the basis of theoretical data and data available in the open literature. To summarize, it is showned that ammonia borane hydrolysis faces the same issues than those encountered by sodium borohydride hydrolysis and that the latter is superior to the former.

Still like sodium borohydride, ammonia borane is an electron source. Both have attracted attention for their own potential to be electrooxidized because they are able to release up to eight and six electrons, respectively, for electricity generation. Electrooxidation of ammonia borane, i.e. direct ammonia borane fuel cell, is the second purpose of this chapter. Its state-of-the-art, advantages and issues are tackled. Finally direct ammonia borane fuel cell is compared to direct borohydride fuel cell. It is stressed that today research devoted to direct borohydride fuel cell develops even if issues hinder its development but research devoted to direct ammonia borane is at its beginning. However it is showed that the former fuel cell is superior to the latter one. And a question is asked: what can be the advantage of developing direct ammonia borane fuel cell when it is less effective, less practical (because of inevitable formation of ammonium ion), and more complex than direct borohydride fuel cell?

The present chapter discusses thus ammonia borane hydrolysis and ammonia borane electrooxidation and can be seen as a continuation of chapter 7 dealing with ammonia borane thermolysis.

Chapter 9- Metal amidoboranes, consisting of $[NH_2BH_3]^-$ anion and various metal cations, are a new class of compounds that have recently attracted great attentions due to their intriguing crystal structures and novel chemical and physical properties. Similar to ammonia borane (AB), metal amidoboranes have great potential to meet the stringent gravimetric and volumetric hydrogen storage capacity targets needed for transportation applications. Moreover, compared to AB, metal amidoboranes have been shown to release hydrogen at lower temperatures, and at higher rates and with the release of fewer volatile contaminants. Particularly, several metal amidoboarnes also appear to release hydrogen less exothermically than AB. In this chapter, we review recent research progress on metal amidoborane compounds, including their syntheses, structures, crystal chemistry, and hydrogen storage properties.

Chapter 10- Since about ten years, boron hydrides, and especially sodium borohydride, lithium borohydride and ammonia borane (and derivatives such as amidoboranes), are hydrogen storage materials that attract much attention by virtue of their high theoretical gravimetric hydrogen storage capacities. This has been the main subject of the present book. From Chapter 1 to Chapter 9, their state-of-the-art and their potential (especially for on-board automotive applications) have been reviewed, discussed and analyzed. The present chapter aims to conclude all of these chapters and tries to answer, or at least to give some elements of an answer, to the following question: which future for boron hydrides?

Boron hydrides are widely recognized as being promising hydrogen storage materials but none has met the technical criteria set for automotive applications yet. They all suffer from low effective (gravimetric and volumetric) hydrogen storage capacities (i.e. < 9 wt% at 85 °C) and storage irreversibility (at best they are off-board regenerable). These are the main issues. There are other important issues to be addressed, which are e.g. the hydrogen release kinetics and the hydride cost. The issues list is, unfortunately, not exhaustive. However, it is also widely recognized that research on these hydrides is at its early beginnings. In other words, there is still an ample room for improvement, especially in the case of materials like ammonia borane.

In: Boron Hydrides High Potential Hydrogen Storage Material ISBN:978-1-61668-361-0
Editors: Umit B. Demirci and Philippe Miele

Chapter 1

SOLID-STATE HYDROGEN STORAGE

Umit B. Demirci[1] and Philippe Miele[2]
Laboratoire des Multimatériaux et Interfaces, UMR 5615 CNRS Université Lyon 1, Université de Lyon, 43 boulevard du 11 Novembre 1918, F-69622 Villeurbanne, France

ABSTRACT

The emergence of a (near-future) hydrogen economy implies the development of hydrogen as energy carrier, which is especially intended for fuel cell applications. However, this is today hindered by many issues related to the hydrogen production, storage, distribution and utilization. The present chapter briefly discusses all of these aspects of the hydrogen economy as well as the most important issues. A special attention is paid on the hydrogen storage issue, which is actually a key issue. The various storage ways that are currently under investigation are briefly reported whereas the solid-state hydrogen storage is more widely discussed. Solid-state hydrogen storage is attractive because it offers a safe and efficient way to store hydrogen, and the highest hydrogen storage capacities. It is reported that complex hydrides are promising hydrogen storage materials by virtue of their high theoretical hydrogen capacities. Foremost among complex hydrides are boron compounds, and especially boron hydrides, which are surveyed throughout this book.

INTRODUCTION

"Water! cried Pencroft, water as a fuel for steamers and engines! Water to heat water!

Yes, but water decomposed into its primitive elements, replied Cyrus Harding, and decomposed doubtless, by electricity [...] Yes, my friends, I believe that water will one day be employed as a fuel, that hydrogen and oxygen which constitute it, used singly or together, will furnish an inexhaustible source of heat and light [...]. Some day the coalrooms of

[1] E-mails: Umit.Demirci@univ-lyon1.fr;, Tel.: +33 (0)4 72 44 84 03; +33 (0)4 72 43 12 05, Fax: +33 (0)4 72 44 06 18.
[2] Philippe.Miele@univ-lyon1.fr.

steamers and the tenders of locomotives will, instead of coal, be stored with these two condensed gases, which will burn in the furnaces with enormous calorific power. There is, therefore, nothing to fear. As long as the earth is inhabited it will supply the wants of its inhabitants […] I believe, then, that when the deposits of coal are exhausted we shall heat and warm ourselves with water. Water will be the coal of the future.

I should like to see that, observed the sailor.

You were born too soon, Pencroft, returned Neb, who only took part in the discussion by these words." From Jules Verne, *The Mysterious Island*, 1874.

In a recent feature article, van den Berg and Areán [1] quoted this discussion to emphasize a well known but sometimes overlooked fact, which is that hydrogen is not available on Earth as a primary energy source.

Hydrogen is indeed not available as a primary fuel and also is not an energy source. Nevertheless it has a high potential for meeting our future energy needs. At least it is often accepted as such, considering the historical development of the energy carriers towards hydrogen-rich fuels and the necessity to avoid the carbon dioxide emission (Figure 1) [2].

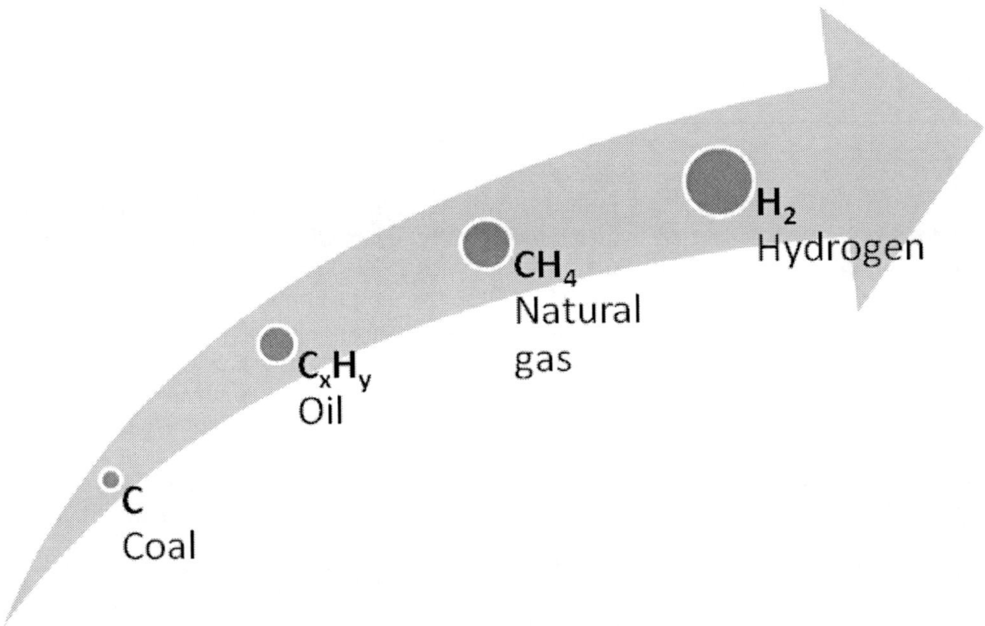

Figure 1. Towards hydrogen.

Actually, the current context favors that. Today, we are completely energy-dependent and our huge energy needs are significantly met by fossil fuels, namely coal, natural gas and hydrocarbons. The most pessimistic opinions announce a depletion of the fossil fuels in a near-future. The 2050s are often cited. Besides the fossil fuels depletion, these energy sources have turned undesirable from an environmental point of view. These carbon-based materials inevitably emit carbon dioxide, one of the most harmful greenhouse gases. Carbon dioxide is one of the main compounds to blame because of global warming and so climate change. Both contexts, namely energy and environmental, have catalyzed emergence of new concepts and behaviors. One of them is sustainability [3]. Another is green chemistry [4].

Sustainability is an emerging concept that explicitly suggests that it should be possible to achieve economic growth without environmental negative impacts. Actually, several definitions of the sustainability have been given over the past decades. Basically, sustainability is engaged in developing the needs and addressing the environmental issues of the present while it is preparing a sustainable future in terms of economic growth, environment preservation and poverty alleviation. The core of mainstream sustainability thinking has become the idea of three dimensions, namely environmental, social, and economic growth [5]. Sustainability is also engaged in developing sustainable energy technologies, being stem from sun, directly or indirectly. Those energies are renewable. Renewable energy technologies directly stem from sun are e.g. solar heating, concentrating solar power and photovoltaic. Renewable energy technologies indirectly stem from sun are e.g. biomass, biofuels, hydropower, ocean energy, geothermal power, and wind power [6].

Green chemistry is a concept that addresses hazards, whether physical (flammability, explosivity), toxicological (carcinogenicity, endocrine disruption), or global (ozone depletion, climate change) as an inherent property of a molecule. Note that global warming is one of the most important issues addressed by green chemistry. Green chemistry is also engaged in addressing energy needs through the development of more sustainable energy technologies like hydrogen and fuel cells [4]. "A central goal of green chemistry is not only to ensure that energy efficiency is ingrained from the molecular level […], but also to ensure that the nature of that energy is sustainable to both humans and the biosphere" [7].

It is thus in such a context that hydrogen has become a potential, promising candidate for meeting our future energy needs. Hydrogen can be basically compared to fossil fuels (Table 1).

It is obvious that hydrogen has several advantages over fossil fuels (for example, abundance, not depleting hydrogen sources, high energy density, and no emission of additional anthropogenic greenhouse gas like carbon dioxide) but in the meantime it suffers from several drawbacks that are at the origin of significant issues hindering the development of an hydrogen economy. The main issues are as follows:

- Hydrogen has to be produced;
- It has to be safely stored and distributed; and,
- It has to be converted into electrical energy in new energy devices (namely, fuel cells).

Each of these steps is under intense investigations and many efforts in research and development are devoted to solve these issues.

The present chapter deals with the issues hindering the development of a hydrogen economy. A special attention is paid on the hydrogen storage issue. The various storage solutions are briefly tackled while the solid-state hydrogen storage is much more largely discussed because this kind of storage is the main purpose of the present book.

Table 1. Hydrogen versus fossil fuels.

	Hydrogen	Fossil fuels
Production		
Availability	Combined with other elements	In their final forms
Sources	Water, biomass [a]	n.a. [b]
Resources	Abundant / Not depleting	Limited / depletion
Renewability	Yes	No
Production routes	Various inefficient processes [c]	Extraction and refining
Is production an issue?	Yes	No
Storage and distribution		
Fuel state	The lightest gas	Gases, liquids, solids
Storage	Various inefficient ways [c]	n.a. [b]
Is storage an issue?	Yes	n.a. [b]
Distribution	Related to storage	n.a. [b]
Oxidation / Combustion		
Energy nature	Energy carrier	Energy source
Energy density (MJ kg^{-1})	141.8	43 [d]
Energy density (kJ m^{-3})	13	40 [e]
Utilization	Fuel cells [c]	Combustion
Emissions	Water	Carbon dioxide
Safety		
Hazards	Explosive	Flammable
Contribution to global warming	No	Yes

[a] Fossil fuels are also hydrogen sources but they are not renewable.
[b] This does not apply to fossil fuels.
[c] Under research and development yet.
[d] For petroleum (coal: 15-27 MJ kg^{-1}; petrol: 48 MJ kg^{-1}; gasoline: 47 MJ kg^{-1}).
[e] For methane (town gas: 18 kJ m^{-3}; butane: 133 kJ m^{-3}).

2. ISSUES ENCOUNTERING HYDROGEN ECONOMY

The development of a hydrogen economy presupposes that hydrogen is efficiently, cost-effectively produced, stored, distributed and converted. Today, hydrogen is in a phase of research, either research is fundamental or applied, which depends on the studied technology. However that may be, the fact that hydrogen is still in this phase is understood by the following properties (and so problems):

- Hydrogen does not exist alone in nature; it is combined with other elements like in water and it has to be produced;
- Hydrogen is a light gas, even the lightest, and this property confronts it with serious technical problems in terms of storage and distribution;
- Hydrogen is an energy carrier and thus it has to be converted in specific convertors that are none others than fuel cells (devices converting chemical energy stored in e.g. hydrogen into electrical energy).

To summarize, hydrogen production, storage, distribution and conversion, which are the links of the hydrogen economy, encounter one or more obstacles in the way of their development. Hereafter, each of the links is briefly discussed while emphasizing the issues relating to efficiency, environment and cost.

Hydrogen can be produced from fossil fuels (coal and oil), natural gas (mainly composed of methane), biomass (green wastes), and water (Figure 2). Hydrogen is currently generated mainly from fossil fuels (96 % of world hydrogen production) with release of carbon dioxide (13.7 kg per kilogram of net hydrogen) [8-10]. The hydrogen production routes aim to separate hydrogen from these various hydrogen-containing sources.

Figure 2. Hydrogen sources.

They are based on thermal, electrochemical, or biological processes. Actually, various hydrogen generation systems can be distinguished [11]: steam reforming (natural gas), partial oxidation (natural gas, hydrocarbons), autothermal reforming (hydrocarbons), plasma reforming (hydrocarbons), aqueous phase reforming (carbohydrates), ammonia reforming, thermolysis or gasification (coal, biomass), photolysis (water), anaerobic digestion (biomass), photo fermentation (biomass), microbial electrolysis cells (biomass), alkaline electrolyzer, polymer electrolyte membrane electrolyzer, solid oxide electrolysis cell, high temperature electrolysis, thermochemical water splitting, and photoelectrochemical water splitting. The most mature technologies are reforming and gasification while electrolysis coupled with renewable energy is near term low emission technology [11]. Each of these processes entails a specific hydrogen source (given in brackets when required) and energy needs, and each supposes emissions of by-products. For example, the thermal and electrochemical processes are energy-guzzling, implying high costs for the hydrogen produced and, the fossil fuels are carbon dioxide-emitting feedstocks, implying emissions of an anthropogenic greenhouse gas. Note that today none of the hydrogen production technologies (involving a renewable energy) under investigation is efficient enough for large-scale productions whereas the efficient and sustainable generation of hydrogen is necessary for the development of an ultimately sustainable hydrogen economy [12].

Once hydrogen has been produced it has to be stored in order to be transported, distributed and on-board stored in vehicular, portable applications. Due to the fact that hydrogen is a gas, the most evident storage way may appear being high pressure. The highest the hydrogen pressure in a tank, the lowest the storage volume. However the high pressure storage suffers from high cost, low storage capacities, and safety problems (risks of hydrogen leakage). An alternative storage way is the low temperature hydrogen storage (or cryogenic storage). Hydrogen is liquid at -253 °C and that implies storage under this temperature. High costs, manipulation and safety problems hinder its development. Accordingly several alternative storage ways have been proposed: namely, physisorption in porous materials like for example carbon nanostructures, metal hydrides like for example magnesium hydride MgH_2, complex hydrides like for example sodium borohydride $NaBH_4$, and storage via chemical reactions such as the reaction between sodium Na and water [13]. Complex hydrides are generally showed as the most promising storage solution [14] but all of the storage solutions suffer from high costs, lack of safety, and inefficiency in terms of storage capacities [12].

Although hydrogen is highly energetic (Figure 3), it is not considered as an energy source, except in few specific cases such as aerospace applications. Actually hydrogen can be a possible fuel of internal combustion engine but efficiency is limited by the physics of the Carnot cycle to about 25 % [8]. Hydrogen is an energy carrier that has to be converted in a fuel cell, given fuel cells escape from the limitations of the Carnot cycle [8], having efficiencies up to 65 % [12]. Actually, the issue relating to the hydrogen utilization is rather dependent on the fuel cell technology. The fuel cell is a device intended to convert chemical energy into electrical energy. It is a technological solution for storing electricity indirectly. Hydrogen may then be seen as an electricity carrier. The fuel cell technology is still in a development phase but it is not in a commercialization phase yet. Indeed, various major technological hurdles (cost, durability, performance of materials and components) must still be overcome before fuel cells can compete effectively with conventional energy conversion technologies [12]. Hence, the hydrogen utilization issue is more an issue of the fuel cell (which suffers from limited efficiency and high costs).

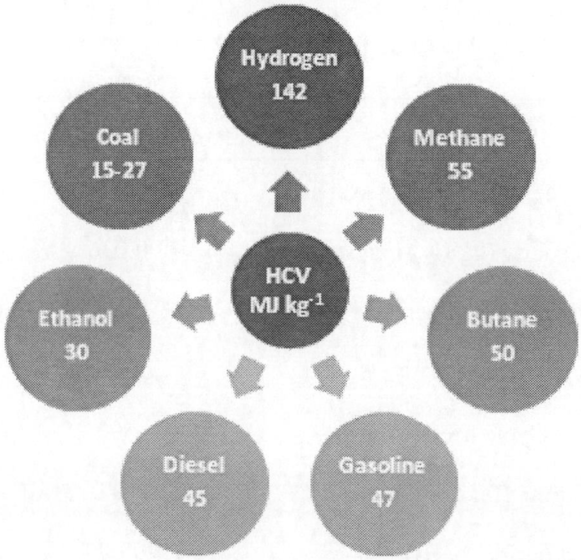

Figure 3. Higher calorific values (HCV) of some fuels.

One has to admit that we are still far from taking advantage of a hydrogen economy. Nevertheless huge scientific, technical, economic, and political efforts are today devoted to solve each of the issues discussed above. The hydrogen storage issue is a good example of that and it is the purpose of the present chapter. This issue and more specifically the materials investigated for the solid-state hydrogen storage are discussed in the following sections. Prior to that, it is important to briefly tackle the hydrogen safety issue.

Hydrogen safety is a justified concern since hydrogen can combust, but like any flammable fuel, and can be explosive, but at concentrations of 18.3-59 % in air. Nevertheless hydrogen is lighter than air and has a rapid diffusivity, diluting quickly into a non-flammable concentration. Compared to other fuels such as gasoline vapor and natural gas, hydrogen is not more or less dangerous. It simply is different and education of the differences is the key enabler to making hydrogen a consumer-handled fuel [15]. Furthermore, hydrogen is non-toxic and non-poisonous, that is it is environmentally safe.

3. Hydrogen Storage Issue

3.1 Technical Targets

Foremost among hydrogen storage systems is on-board vehicular hydrogen storage. Vehicular applications are the most discussed applications throughout the open literature. The main current drive to use hydrogen as an energy carrier comes from its potential to replace petrol derived fuels in cars and other vehicles [1]. Technical targets for these applications have been set by the U.S. Department of Energy [16] while, to the best of our knowledge, no technical target has been proposed for portable or mobile applications up to now. Some of the technical targets for on-board hydrogen storage systems [16] are reported in Table 2.

Table 2. Some technical targets for on-board hydrogen storage systems [16].

Storage parameter	2007	2010	2015
Gravimetric capacity (kg H_2 / kg system)	0.045	0.060	0.090
Volumetric capacity (kg H_2 / L system)	0.036	0.045	0.081
Hydrogen desorption temperature (°C)	-	up to 85	up to 85
Storage system cost ($ / kg H_2)	200	133	67
Durability / Cycle life [a] (cycles)	500	1000	1500
System fill time (for 5 kg; in min)	10.0	3.0	2.5
Fuel purity (% H_2)	99.99	99.99	99.99

[a] 1/4 tank to full; equivalent to 62000, 124000, and 186000 km respectively (current gasoline tank spec).

More often than not the gravimetric hydrogen storage capacity (in kg H_2 per kg of system, or in wt%) is the target utilized for discussing the storage capacities of new storage systems. The targets 4.5, 6.0 and 9.0 wt% for 2007, 2010 and 2015 are well known. Obviously it is a mistake not to consider much more systematically the volumetric hydrogen storage capacity. Note that the storage capacities given in Table 2 refer to the hydrogen storage system as a whole, which includes all of the hardware (e.g. tank, valves and regulators, cooling or heating systems if needed, insulation, etc.) [1].

Simply speaking, hydrogen has to be packed as closely as possible. This is a first criterion for the hydrogen storage systems. A second important criterion is the reversibility of uptake and release of hydrogen, what excludes all covalent hydrogen-carbon compounds [13].

As introduced in the previous section, there is a tremendous amount of research being pursued in the development of hydrogen storage systems. The major hydrogen storage approaches are physical:

- High pressure tanks;
- Cryogenic methods; or chemical:
- Reversible adsorption on solids;
- Reversible absorption in metal or main groups hydrides;
- Storage in the form of metals that can liberate hydrogen from water [8].

The physical methods, or conventional storage methods, are the most mature of the hydrogen storage technologies [17]. They have reached the engineering prototype stage while the research about the chemical methods is at its beginning [18]. Nevertheless, the conventional physical methods are unlikely to meet the aggressive U.S. Department of Energy targets (Table 2) [19]. It is discussed hereafter.

3.2 Physical Methods for Hydrogen Storage

3.2.1 High Pressure Storage

High pressure is a well known storage way for any gas (helium, argon, nitrogen, oxygen, air, etc). With respect to hydrogen, the storage is technically complex and expensive because of its lightness. A conventional steel hydrogen cylinder achieves about 1 wt% hydrogen [18], which is not enough to meet the technical targets for automotive applications. The recent years have seen notable improvements in the storage capacities with the development of fiber-reinforced (for example, glass or carbon) resin. Pressures of 350 bar have been reached, pressures of 200 bar being become typical, but today reaching pressures of 700 bar (density of 39.6 kg m^{-3}) is the main objective [14]. The 700-bar pressure corresponds to about half the density of liquid hydrogen (namely 70.8 kg m^{-3}) [18]. Pressures of 800 bar (density of 36.0 kg m^{-3}) have even been reported but the gravimetric hydrogen storage capacity decreases with increasing pressure due to the increasing thickness of the cylinder walls [2,13].

Innovative materials and technologies are necessary in order to reach the high pressures expected for the various applications and this has a significant impact in the storage costs. Furthermore, the low storage density together with the very high pressures, the diffusion problems and the cyclic stability in the systems are important drawbacks to the technically simple high pressure storage method[3]. Also, the safety of pressurized cylinders is an issue of concern especially in highly populated regions [2,13].

3.2.2 Cryogenic Storage

Liquid hydrogen, obtained at temperatures below -253 °C at atmospheric pressure, can be stored and this opens opportunities for on-board vehicular storage. The density of liquid hydrogen is 70.8 kg m^{-3} [2,13] which is very attractive as such a value can only be reached at 1820 bar for high pressure storage. Gravimetric hydrogen storage capacities of about 10 wt% can be obtained [18]. Cryogenic storage requires performing high-technology materials and highly severe safety criteria. In other words it is an expensive storage way. Moreover, it has two major disadvantages [18]:

- There is a boil-off rate of about 1 % per day;
- There is an energy loss because the refrigeration process amounts to 30 % of the energy available from burning the hydrogen.

Such disadvantages limit the possible applications for liquid hydrogen storage systems. Hence the challenges of the liquid hydrogen storage[4] are, respectively, the thermal insulation of the cryogenic storage vessel and the energy efficiency liquefaction process [2,13].

---

3 For more details about high pressure storage of hydrogen, the reader can refer to references [2,13].

4 For more details about liquid hydrogen storage, the reader can refer to references [2,13].

3.3 Chemical Methods for Hydrogen Storage

As that has been well remarked by Guo et al. [20], the "grand challenge" in finding an ideal hydrogen storage solid is to meet all the targets for on-board vehicular applications, given that the targets are closely associated with the following parameters: structure (crystalline form, specific surface area, and structural defects), chemistry (components, composition, phases, catalysts, and impurities) and reaction path. These are the key parameters that have to be tuned in order to improve the properties of the hydrogen storage solid materials. The targets for practical applications are as follows [20]:

- Storage capacity, but it is not really a limiting factor, given the large number of solids having capacities well above 6 and 9 wt%;
- Thermodynamic stability, for rapid hydrogen desorption/adsorption (or dehydrogenation/hydrogenation);
- Hydrogen generation kinetics, and also kinetics of the hydrogen adsorption or absorption;
- Storage reversibility: practical reversibility is referred to as the easiness of reversing the hydrogenation and dehydrogenation;
- High thermal conductivity; and,
- Cost.

On the basis of such targets, solid materials for hydrogen storage must be developed which outperform high-pressure and cryogenic storage systems. They are possible candidates, perhaps the only viable methods, for more compact, low-weight, low-cost, safe and efficient storage systems operating at or near room temperatures and low pressures [12].

The chemical methods are generally based on the storage of hydrogen in solid materials, either by adsorption (physisorption) or absorption (chemisorption). The former phenomenon is that taking place with porous materials like carbon nanostructures. The latter one takes into account the hydrides either they are metallic or complex. To these it must be added a different chemical method, which is based on the chemical reaction between a reacting metal and water, the reaction generating hydrogen. The three methods can be seen as the main ones. They are tackled more in details in the next section (4) devoted to the "solids for hydrogen storage". Besides these three chemical methods, other ways of storing hydrogen have also received attention; for example:

- ➢ Encapsulation of hydrogen inside a guest solid structure to form a clathrates, from which hydrogen is released by a pressure and temperature swing [1];
- ➢ Liquid organic heterocycles, such as e.g. N-ethyl carbazole, from which hydrogen is released by the help of catalyst [8];
- ➢ Formic acid or methanol, from which hydrogen is produced using a catalyst [17,21].

These are new avenues of research and not so many efforts have been devoted to their development yet. Nevertheless conceiving formic acid is an interesting possibility since this chemical can be generated from anthropogenic methane stem from natural sources. Accordingly acid formic should be a green, renewable, sustainable hydrogen storage system.

Note that according to Sartbaeva et al. [12], the grand challenge reflects the fact that, at present, no known solid material meets the critical requirements listed above.

4. Solids for Hydrogen Storage or Chemical Storage

According to Mao and Chen [22], solid-state hydrogen storage is potentially the most convenient and the safest method for storing hydrogen, especially for vehicular applications. We agree and it is the reason why the present section is completely devoted to this storage way.

Because adsorption (physisorption) and absorption (chemisorption) are distinguishable chemical phenomena, we are able to distinguish solid porous materials from hydrides, both being solid-state hydrogen storage systems. Both sorptions are largely discussed hereafter in two distinct sections. A third section is devoted to the storage via chemical reactions.

4.1 Adsorption in Porous Materials

The physical adsorption (or physisorption) of a gas like hydrogen in a solid enables the increase of the gas density over the solid surface, the gas being physisorbed over the surface due to the existence of intermolecular attractions. The higher the gas pressure, the higher its density, which is besides heightened as low temperatures. In other words, adsorption decreases when the pressure decreases or when the temperature increases. Storage by physical adsorption is reversible and it is one of the main advantages. The other main advantages are the fast kinetics because physisorption is a non-activated process, the low operating pressures, the relatively low cost of the materials, and the simple design of the storage system [2]. However, physisorption involves weak van der Waals forces and this has for consequence that significant hydrogen adsorption takes place at very low temperatures, up to the cryogenic one [1]. Actually, the amount of hydrogen adsorbed remains rather small even with the low temperatures that are necessary and these are significant drawbacks [2].

The materials of choice for this way of storage are solids with high specific surface area, which are finely divided powders with extremely high porosity. Activated carbons including carbon nanostructures, zeolites, metal-organic frameworks and related microporous solids, and polymers having intrinsic microporosity are the main kinds of adsorbent currently being investigated [1].

4.1.1 Activated Carbons including Carbon Nanostructures

The activation of carbon (natural or synthetic) is a highly efficient process that can be achieved through various procedures [23]. Microporous materials with very large specific surface area and considerable microporous volume can be obtained. Surface areas from 1000 to 3000 m^2 g^{-1} have been obtained [23]. Such materials are then suitable for the hydrogen adsorption in large amounts. Actually it has been reported that adsorption of hydrogen is approximately proportional to specific surface area and micropore volume [24]. For example, Ströbel et al. [25] have suggested, from theoretical considerations that a gravimetric hydrogen storage capacity of 6.0 wt% would be reached at a specific surface area greater than 4000 m^2

g^{-1} even if such area has not been obtained up to now. According to van den Berg et al. [1], interaction energy between hydrogen molecules and carbon adsorbents should be enhanced in narrow pores, knowing that the hydrogen molecule has a kinetic diameter of 0.29 nm. However, from the literature available, it appears that gravimetric hydrogen storage capacities of 3.0-4.0 wt% can be obtained but only at liquid nitrogen temperature [1,26].

Since the publication of Dillon et al.'s paper [27], which has reported a hydrogen desorption experiment and a hydrogen storage capacity of 5-10 wt% with carbon nanotubes, nanostructured carbon has attracted much attention as it is the case for all of the nanostructured materials. Immediately nanostructured carbon has aroused interest and expectations in hydrogen storage because of its high specific surface area. Accordingly hydrogen storage in these materials has been intensely investigated. However, according to van den Berg et al. [1], recent reports do not confirm any substantial advantage of e.g. carbon nanotubes (or even carbon nanofibres) over activated carbon. And it is true that gravimetric hydrogen storage capacities of about 4 wt% at -196 °C or <1 wt% at 20 °C have been reported up to now [1,25,26,28]. The amount of adsorbed hydrogen is proportional to the surface area of the nanostructured carbon. For example it has been reported that the reversibly adsorbed quantity of hydrogen amounts to 1.5 wt% per 1000 m^2 g^{-1} of specific surface area at -196 °C [14,29]. The main difference between carbon nanotubes (graphene sheets rolled up into a cylindrical shape) and high surface area graphite is the curvature of the graphene sheets and the cavity inside the tube [2]. However there is no evidence that the geometric structure of nanostructured carbon influences the amount of adsorbed hydrogen [2,13]. The curvature of nanotubes may only influence the adsorption energy but not the amount of adsorbed hydrogen [2]. Hydrogen storage in carbon nanotubes can occur by physisorption (preferentially at cryogenic temperatures) or chemisorption (at elevated temperatures, hydrogen becoming covalently bonded to carbon) [30]. Carbon nanotubes as hydrogen storage materials continue interesting although there are discrepancies in the published data [30].

Ordered porous carbon can be prepared by using a template such as a zeolite. Such carbon shows then large surface area and regular pore size. Yang et al. [31] has reported the fabrication of an ordered porous carbon obtained by using zeolite beta as template. The as-obtained carbon had a specific surface area of 3200 m^2 g^{-1} and reversibly adsorbed 6.9 wt% hydrogen at liquid nitrogen temperature and 20 bar. However, no data is available regarding hydrogen uptake at room temperature [1].

As a conclusion about carbon materials, one may cite Schlapbach and Züttel [14]. In a review article, the authors have concluded that they consider it is scientifically interesting and challenging to continue research on the interaction of hydrogen with carbon nanostructures. However they were wondering whether a hydrogen-storage material will emerge from it remains an open question. It was in 2001. In 2006, the U.S. Department of Energy Hydrogen Program [32] decided to stop (no-go decision) future applied research and development investment in pure, undoped single-walled carbon nanotubes for vehicular hydrogen storage applications, because these materials have not achieved 6 wt% hydrogen storage at close to room temperature. However, it has been decided to invest in certain areas of carbon nanotube research, such as multi-walled nanotubes and metal-doped hybrid materials. Another possibility is the development of functionalization methods and carbon composite materials [23].

4.1.2 Hydrogen Physisorption in other Materials

As reported by Ross, the search for a viable high surface area storage material has extended from carbon systems to zeolites, metal-organic frameworks (MOFs) and other new porous materials. The present sub-section briefly discusses, successively, each of these new hydrogen storage materials.

Zeolites are crystalline aluminosilicate materials. They are defined by a three-dimensional network of cavities and pores, which give the ability to selectively sort molecules based primarily on a size exclusion process. Hydrogen storage can follow either an encapsulation mechanism (or trapping, at elevated temperature) [33] or an adsorption regime (dominating below room temperature) [30]. Hydrogen adsorption has been showed to correlate with the specific surface area as well as with micropore volume [29]. Zeolites as hydrogen storage materials are attractive because they are cheap, readily available, easy to synthesize, robust, non-flammable in air or a hydrogen atmosphere; furthermore there exists a wide range of framework types and their composition and properties are tunable by ion exchange [30]. However, according to van den Berg and Areán [1], zeolite is unlikely to be a material of choice for hydrogen storage in the transportation sector. The authors supported their conclusion with the following arguments. First, maximum hydrogen storage capacity has been calculated to be lower than 3 wt% [34]. Second, the interaction energies between zeolite and hydrogen can be either too high or too low, leading to non-reversibility of hydrogen uptake or to hydrogen uptake well below the target 6 wt%. In fact, Langmi and McGrady [30] have got to a similar conclusion. While reviewing the effective capacities reported throughout the open literature for various zeolites, it has been concluded that though zeolites are an interesting class of inorganic materials, their hydrogen storage capacities fall short of the 6 wt% target set by the U.S. Department of Energy for automotive applications.

Metal-organic frameworks (MOFs) are potential hydrogen storage solids. They are crystalline. They consist of multidentate organic ligands connecting metal ions or small metal-containing clusters, forming uniform pores that are inter-connected forming an ordered network of channels [1]. Metal-organic frameworks have very large specific surface areas, with values from 1500 to 5000 m^2 g^{-1} [35,36], and their pores are designable. Both features confer considerable potentials to these as materials for hydrogen storage. Metal-organic frameworks offer great possibilities of varying topologies and chemical composition and as remarked by van den Berg and Areán [1] such particularities catalyze their fast development, faster than that for any other type of hydrogen storage materials. Hydrogen uptake is reported to be proportional to surface area as for carbons [36-38]. Effective gravimetric hydrogen storage capacities in the range of 5.2 to 7.5 wt% at -196 °C and moderate pressure were found [1,36,39]. However, at ambient temperature, the effective capacities reported so far do not exceed 1.5 wt% [1,39]. The low effective capacities in ambient conditions made Zhu and Xiao [40] concluding that because the adsorption is very small metal-organic materials are unsuitable for hydrogen storage (but excellent for methane storage). Despite this pessimistic conclusion, intense investigations on metal-organic frameworks are in progress. In order to increase the hydrogen uptake at near ambient temperature, there are two tracks to explore, both being related: namely, narrower pores and increased interaction between the porous solid and adsorbed hydrogen molecules [1].

All of the materials that may have the ability to physisorb gases such as hydrogen (or even methane or carbon dioxide) are envisaged as potential hydrogen storage materials. Here are given some examples of porous materials that would have a potential as hydrogen storage materials:

- Polymers with intrinsic microporosity, with specific surface areas in the range of 500 to 1000 m^2 g^{-1} and hydrogen uptake up to about 3 wt% at -196 °C [1,41]; according to van den Berg and Areán [1] the maximum hydrogen uptake near ambient temperature is probably much lower than 3 wt%;
- Boron nitride-based materials, namely nanotubes and cages, with storage capacities of about 3 [42] and 4 wt% [43];
- Lithium nitride, Li_3N, with a theoretical hydrogen uptake capacity of 11.5 wt%, the hydrogen adsorption occurring at temperatures well above 100 °C [30];
- Oxide nanostructures such as titanium oxide TiO_2 nanotubes, being able to store up to approximately 2 wt% hydrogen and the hydrogen adsorption being a reversible process [22,44,45];
- Titanium-decorated BC_4N nanotubes, with storage capacity up to 5.6 wt% (determined by first-principles calculations) [46];
- Porous silica with a regular hexagonal array of cylindrical pores, with specific surface area of about 1000 m^2 g^{-1} and hydrogen uptake up to 2 wt% at -196 °C [47];
- Titanium-decorated doped silicon fullerene clusters, with a storage capacity of up to about 5 wt% corresponding to full titanium coverage (determined by first-principles calculations) [48];
- Alkali metal-doped B_{80} fullerenes, with gravimetric densities of 10-11 wt% (determined by first-principles calculations) [49]

According to van den Berg and Areán [1], a wide variety of porous solids have been tested but none of them meets the U.S. Department of Energy targets. Furthermore it has been suggested that more experimental measurements are needed in order to assess the potential of the newly developed nanostructured materials. Otherwise, many papers have reported theoretical values determined by first-principles calculations.

4.2 Storage by Absorption

The absorption of hydrogen presupposes a chemical combination of hydrogen with an element different from carbon that can be for example a metal. Such compounds are generally grouped together in one category, that of hydrides. A huge number of hydrides exist and most of them have a theoretical potential as hydrogen storage materials. In the early 2000s, Schlapbach and Züttel [14] have interestingly showed this theoretical potential for some of the most common hydrides. For example, it has been stressed the fact that metal hydrides such as magnesium hydride MgH_2 have high volumetric hydrogen storage capacities. It has besides been showed that complex hydrides such as lithium borohydride $LiBH_4$ intrinsically store 18 wt% of hydrogen. Since then a great interest has been showed to the hydrides by the scientific community involved in solid-state hydrogen storage: namely, metal hydrides, complex hydrides, and storage via chemical reaction [13][5].

4.2.1 Metal Hydrides

Basics of both metal hydrides and hydrogen adsorption-desorption processes have been reported by Züttel in different review papers [2,13,14]. The reader can refer to these papers for detailed information, especially with respect to the fundamentals of the hydrogen absorption. Here is only reported the main aspects.

Hydrides exist as ionic (namely, elements of the lithium and beryllium columns in the periodic table), polymeric covalent (namely, the copper, zinc and boron columns, except silver), volatile covalent (namely, the carbon, nitrogen, oxygen and fluoride columns) and metal hydrides (namely, scandium, titanium, vanadium, chrome, nickel, yttrium, zirconium, niobium, palladium, lanthanides, hafnium, and actinides) [13]. Polymeric covalent hydrides are also called complex hydrides and they are discussed in the next sub-section. The present sub-section deals with the metal hydrides.

Binary metal hydrides are generally denoted as MH_n. Many of these compounds show deviations from ideal stoichiometry (i.e. n = 1, 2, 3) and can exist in multiphase systems. These hydrides are also called interstitial hydrides because their lattice structure is that of a typical metal with hydrogen atoms on the interstitial sites [13]. Ternary metal hydrides, denoted as AB_xH_n (with A, a rare earth or an alkaline earth metal; B, a transition metal), imply the incorporation of a new element, namely A, that allows to modify the hydride properties. The AB_xH_n hydrides are rather stable compounds while the B element forms unstable hydrides in the absence of A [2]. The most important intermetallic hydrides have the following compositions: AB_5 (e.g. $LaNi_5$), AB_2 (e.g. ZrV_2), AB_3 (e.g. YFe_3), A_2B_7 (Y_2Ni_7), A_6B_{23} (Y_6Fe_{23}), AB (ZrNi), and A_2B (Mg_2Ni) [13,50]. The metal-hydrogen bond offers the advantage of high hydrogen density at moderate pressure and desorption of all stored hydrogen at the same pressure [14]. For example, Mg_2FeH_6 can store up to 150 kg m^{-3} hydrogen [17].

Hydrogen can react with the above listed transition metals and their alloys, forming thus solid metal-hydrogen compounds [50]. Molecular hydrogen reversibly adsorbs over metal by dissociating at the surface and desorbs through a recombination process of two metal adsorbed hydrogen atoms. The hydrogen absorption in metals is generally described as a process implying six distinct states. Figure 4 shows a scheme of the hydrogen adsorption processes.

[5] There is no clear distinction between metal and complex hydrides throughout the open literature. For example, it is sometimes observed that some complex hydrides (like e.g. alanates) are considered as metal hydrides. Here the distinction between metal and complex hydrides that Züttel [13] gave has been chosen. The differences are thus detailed in sections 4.2.1. and 4.2.2.

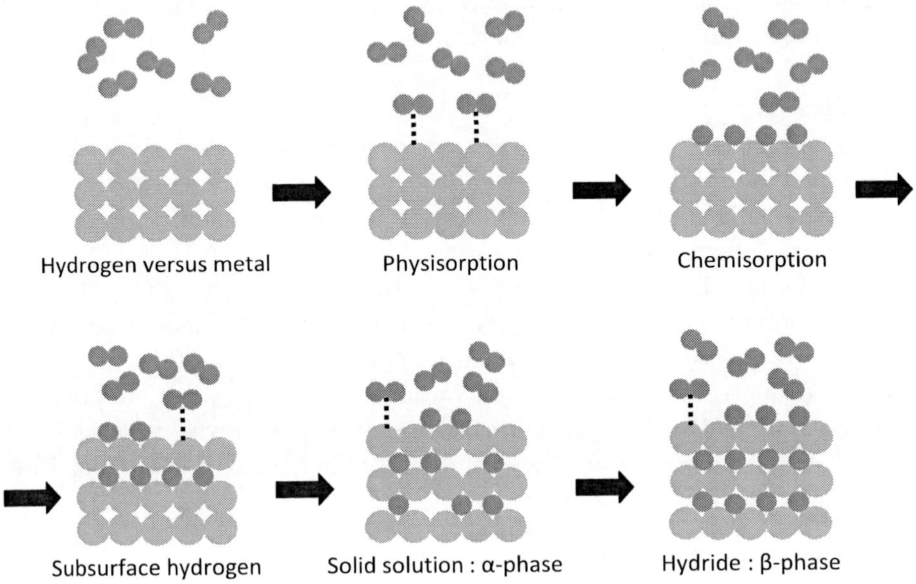

Figure 4. Hydrogen absorption in metals; adapted from [2].

The six states can be summarized as follows [2,18,50]:

- State 1: Molecular hydrogen gas and metal in two separate phases;
- State 2: Physisorption of molecular hydrogen gas over the metal surface at low temperature;
- State 3: Molecular hydrogen splitting and binding to the surface metal atoms, namely chemisorption of hydrogen;
- State 4: Jumping hydrogen between interstitial sites (i.e. in a subsurface metal layer);
- State 5: Hydrogen diffusion into the metal host lattice as a solid solution, namely formation of a phase called α-phase;
- State 6: Strong hydrogen-hydrogen interaction (i.e. lattice expansion at $H/M > 0.1$), namely formation of a phase called β-phase or hydride phase (often $H/M = 1$).

The formations of the α- and β-phases are thermodynamically controlled (Figure 5). When the hydrogen concentration (or H/M) increases, the isotherms follow the following trends. First, the equilibrium pressure increases in the pure α-phase. Second, when the α- and β-phases (i.e. solid solution and hydride phases, respectively) coexist, there is a plateau in the isotherms. Third (and finally), when a critical point is reached, the two-phase region ends, the hydrogen pressure steeply rises with the concentration (pure β-phase) [14]. Note that, as reported by Züttel [2,13], empirical models have been developed for estimating the stability and the hydrogen concentration in metal hydrides. The reader can refer to these articles as well as the references cited.

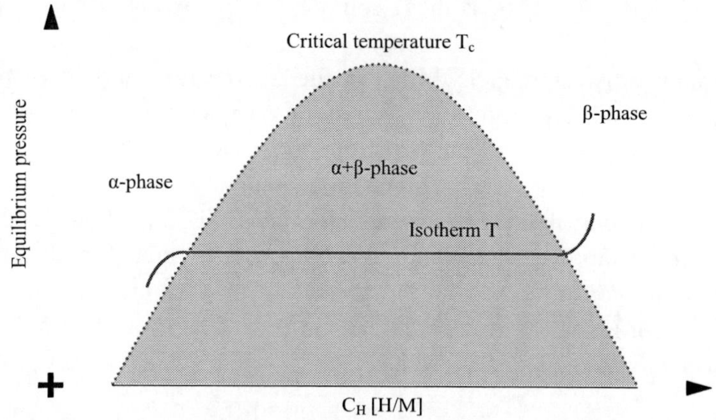

Figure 5. Pressure composition isotherms for hydrogen absorption in metal hydride; adapted from [13].

Metal hydrides have high volumetric hydrogen storage capacities. For example, a capacity of 115 kg m^{-3} in the intermetallic $LaNi_5$ (or $LaNi_5H_6$) has been reported. Most metal hydrides absorb hydrogen up to a H/M ratio of 2 and the hydrides having a H/M ratio above 2 belong to the complex hydrides group (because of more ionic and covalent characters) [13].

Hydrogen storage via metal hydrides is attractive because of mild adsorption/desorption storage, storage reversibility, high volumetric storage capacities, safety and compactness. Nevertheless it suffers from low gravimetric hydrogen storage capacities (< 1.5 wt%) [50] if compared to the targets for vehicular applications, given that much of the current research activity on metal hydrides is focused on developing materials that can meet the criteria of automotive applications [19]. According to Principi et al. [50], none of the up to now studied metal hydrides has the required characteristics for such practical applications. These authors have appeared quite pessimistic in the capacities of the metal hydrides and have not seemed to believe in finding materials having, at the same time, low enthalpy (of hydride formation and desorption), good reaction kinetics and high hydrogen density. The van den Berg and Areán's conclusion [1] is as severe as Principi et al.'s [50]. They have seen the metal hydrides applicable only in some niche applications.

4.2.2 Complex Hydrides

Züttel [13] defined the complex hydrides as being hydrides formed with light metals, such as e.g. Li, Mg, B and Al, which give rise to a large variety of metal-hydrogen complexes. The complex hydrides benefit from both the lightness of the metals and the high H/M ratio (generally of 2 and in few cases higher than 2), which permit to reach gravimetric hydrogen storage capacities among the highest ones. For example, lithium borohydride $LiBH_4$ has a theoretical capacity of 18 wt%. In complex hydrides, the hydrogen atoms are covalently bonded to a central light metal such as e.g. boron or aluminum and stabilized by a cation (alkali, alkaline earth metal or transition metal) [19].

Borohydrides (or tetrahydroborates, $M(BH_4)_y$) and alanates (or tetrahydroaluminates, $M(AlH_4)_z$) are maybe the most studied complex hydrides because they generally offer the following advantages: high gravimetric hydrogen storage capacities, high volumetric hydrogen capacities, stability, and ability to release stored hydrogen through thermal or chemical decomposition [14]. Besides these complex hydrides, magnesium hydride MgH_2 is also of interest.

Foremost among most studied hydrides is the magnesium derivative [18]. It has the advantages to have a relatively large hydrogen storage capacity compared to metal hydrides (i.e. 7.6 wt%) and, above all, a low cost. Nevertheless, its development is mainly hindered by very low hydrogenation and dehydrogenation kinetics [14,20]. For example, the dehydrogenation only occurs at temperatures higher than 300 °C [51], which does not make it suitable for automotive applications. With the objective to facilitate the dissociation and recombination of hydrogen molecules at the surface of magnesium particles [20], that is by decreasing the dehydrogenation temperature and the dehydrogenation kinetics, several solutions have been envisaged and assessed:

- Magnesium alloying with metals such as Cu [51];
- Addition of noble metal catalysts such as Pt [52];
- Destabilization by addition of Si through the formation of Mg_2Si [53,54];
- Nanocrystalline magnesium [50,54,55];
- Nanocrystalline magnesium mixed with transition metals or metal oxides [50,56].

Among these solutions, a breakthrough was achieved with nanocrystalline magnesium mixed with transition metals or metal oxides [20]. Interestingly Mao and Chen [22] have reported, pessimistically, maybe realistically, that the desorption characteristics of magnesium hydride MgH_2 in the bulk are inconvenient for practical applications. But it is true that its theoretical gravimetric hydrogen storage capacity does not meet the criteria set for 2010 (Table 2).

Alanates are well known complex hydrides. According to Graetz [19], they are the group of complex hydrides that have received the most attention since the 1990s. To illustrate that purpose, Bogdanović et al.'s works [57,58] are often cited. Indeed it was showed for the first time adsorption and desorption pressure-concentration isotherms for titanium-catalyzed sodium alanate $NaAlH_4$ at 180 and 210 °C. The storage was showed to be reversible and the effective gravimetric hydrogen storage capacity was of 4.2 wt% (that is about twice the reversible capacity of conventional metal hydrides [19]). This material is also of interest because it is readily available at low cost [1]. Since Bogdanović et al.'s works [57,58], catalytic doping of alanates has been a research area for improving and optimizing the storage reversibility, the effective storage capacities and the dehydrogenation kinetics. Graetz [19] recently commented that the discovery of reversibility in the complex hydrides has been a breakthrough in solid-state hydrogen storage but the capacities of this material are unlikely to meet all of the U.S. Department of Energy targets. Many other alanates exist but those ones are generally criticized for their low-hydrogen content [20], the storage irreversibility [50], their relative instability, and/or the too high temperatures required for dehydrogenating them [50]. For example, Na_2LiAlH_6, $KAlH_4$ and K_3AlH_6 are capable of rehydrogenation at moderate temperatures and pressures but they have lower hydrogen density than $NaAlH_4$; and, $LiAlH_4$, $Mg(AlH_4)_2$ and $Ti(AlH_4)_4$ are less stable than $NaAlH_4$ but exhibit high hydrogen densities [19]. However these materials are not reversible under moderate conditions [19].

Alkali borohydrides are attractive hydrogen material because they have high hydrogen densities. For example, lithium borohydride[6] $LiBH_4$ and sodium borohydride $NaBH_4$ have theoretical gravimetric hydrogen storage capacities of about 18 and 11 wt% and theoretical volumetric hydrogen storage capacities of about 120 kg m^{-3} for both. However, such compounds have large decomposition enthalpy and desorb hydrogen at too high temperatures (for example, 1 wt% released at > 400 °C) for hydrogen release [15,19,59-61]. Thermal decomposition of $LiBH_4$ in the presence of silica permitted to decrease the decomposition temperature from 600 to 400 °C [59], leading to the formation of LiH (much more stable than Li) and B as by-products, which implies an effective gravimetric hydrogen storage capacity of 13.8 wt% [59]. Metal oxides (e.g. TiO_2 and ZrO_2) can also be used to decrease the temperature of dehydrogenation of $LiBH_4$ [59]. Besides metal oxides, other ways for destabilizing the hydride are investigated [54]: for example, metal halides as additives [62], size effects by e.g. infiltrating the hydride into nanoporous scaffolds [63], tuning small ion mobility by preparing binary mixtures such as $LiBH_4/MgH_2$ [64,65], and cation substitution to produce metal-mixed borohydrides such as $LiK(BH_4)_2$ [66]. Recently an interest towards alkaline earth (e.g. $Ca(BH_4)_2$ [67,68]) and transition metal borohydrides (e.g. $Zr(BH_4)_4$ [69]) has been showed[7]. Nakamori et al. [70] investigated the thermal desorption properties of several borohydrides: namely, Mg, Ca, Mn, Zn, Al, Y, Zr, and Hf-based borohydrides. It has been showed that hydrogen desorption temperature decreases with increasing Pauling's electronegativities of metals. Much stable than the alkali borohydrides, they have showed promising hydrogen storage capacities but the reversibility of storage remains a problem. There is no doubt about the fact that improvements should be soon published since studies about these new borohydrides are at the beginning. With respect to sodium borohydride $NaBH_4$, maybe foremost among investigated borohydrides (Figure 6), the thermal hydrogen desorption is not applicable because of temperatures higher than 400 °C even in the presence of catalytic activators [60,61,71].

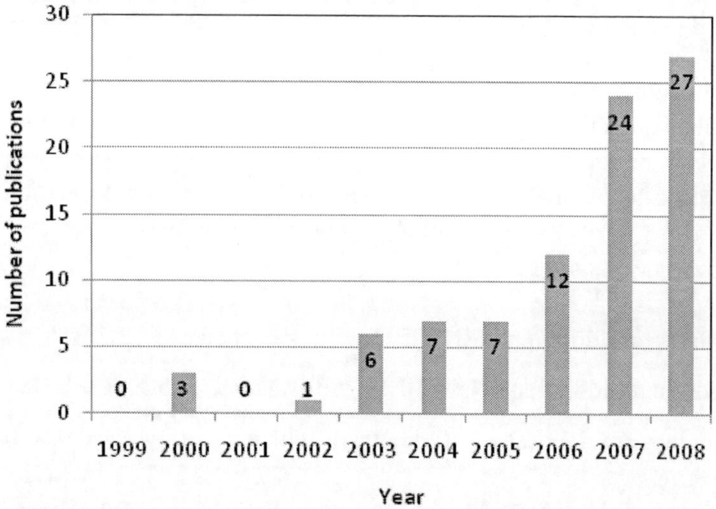

Figure 6. Number of publications dealing with sodium borohydride as hydrogen storage material.

6 Chapter 3 specifically deals with the thermolysis of lithium borohydride $LiBH_4$.

7 Chapter 6 specifically deals with the newly developed borohydrides.

Actually, hydrogen generation from $NaBH_4$ is mainly considered through its catalyzed hydrolysis. However, despite a theoretical gravimetric hydrogen storage capacity of about 11 wt%, its development is hindered by some significant issues[8]: for example, effective hydrogen storage capacities of only 4 to 7 wt%, by-products recycling (i.e. storage irreversibility), and high cost [18,60,61]. However that may be, the U.S. Department of Energy has recommended a no-go for $NaBH_4$ as hydrogen storage system for automotive applications [72]. In other words, this means that $NaBH_4$ can only have applications in portable and niche applications. Actually these last months the review papers dealing with solid-state hydrogen storage does even not consider $NaBH_4$ as a high-potential material anymore (see for example reference [1]).

While the U.S. Department of Energy recommended a no-go for $NaBH_4$, it was suggested to focus on ammonia borane NH_3BH_3 [72], another promising boron-based hydrogen storage material. Net ammonia borane contains 19.6 wt% of hydrogen and such theoretical capacity is higher than that of $LiBH_4$, which has often been presented as the solid storing the highest content of hydrogen. Ammonia borane is potentially consistent with the U.S. Department of Energy targets but some major challenges still remain [60,61,73,74]: hydrogen release kinetics and temperature, pure hydrogen generation, and storage reversibility (regeneration of the by-products by an economically viable process). Ammonia borane releases hydrogen at temperatures higher than 100 °C, two third being generated below 200 °C and the last hydrogen molecule at 400-500 °C [75]. The final product is boron nitride BN, which means that storage via ammonia borane is not reversible [30]. In order to solve the hydrogen release issues, several possibilities have been examined: namely, decomposition in ionic liquid solvents [76], incorporation into a scaffold such as mesoporous silica [77], and utilization of transition metal complex-based catalysts [78]. According to Marder [73], the main issues for future research will be minimizing the formation of volatile by-products (detrimental for fuel cells) and developing economically viable methods for ammonia borane regeneration. Note that the high cost of $NaBH_4$ regeneration was one of the arguments against $NaBH_4$ [72]. Actually all of the researchers involved in developing ammonia borane agree the latter issue [60,74,79].

Another interesting strategy for improving the potential of ammonia borane in hydrogen storage is the chemical activation via substitution of the protonic hydrogen in the hydride by a metal atom. Metal amidoborane MNH_2BH_3 are then obtained. The metal M is used to potentially decrease the enthalpy of hydrogen loss, to eliminate formation of volatile co-products (hydrogen purity), and to help control of hydrogen release. Several metal amidoboranes have been synthesized:

- Lithium amidoborane $LiNH_2BH_3$ with a capacity of 13.6 wt % [80];
- Sodium amidoborane $NaNH_2BH_3$ with a capacity of 9.5 wt% [81]; and,
- Calcium amidoborane $Ca(NH_2BH_3)_2$ with a capacity of 10.0 wt% [80].

According to Hamilton et al. [74], the range of metals used in metal amidoborane complexes can be expanded. Xiong et al. [81] have confirmed and added that the anion

[8] Chapter 4 specifically deals with the hydrolysis of sodium borohydride $NaBH_4$.

$[NH_2BH_3]^-$ is likely to bond with other metals or metalloid elements, which will not only open up significant opportunities for hydrogen storage materials development, but will also bring considerable insights to B–N chemistry. Amidoboranes are new compounds in hydrogen storage and thus research is at the beginning. Compared to ammonia borane, the first results reported for amidoboranes so far are quite promising with decreased hydrogen release temperatures, less exothermic dehydrogenation and no formation of volatile by-products [60].

Amides are also of interest as solid hydrogen storage materials. This is a new research area and the first results reported do not show applicability potential in terms of temperature of hydrogen release yet. For example, lithium amide $LiNH_2$ in the presence of lithium hydride LiH decomposed at temperatures higher than 400 °C and showed a theoretical gravimetric hydrogen storage capacity of 10.3 wt% [82]. Another example may be cited. Magnesium amide $Mg(NH_2)_2$ and lithium hydride LiH generated 5.6 wt% hydrogen at temperatures much lower than those achieved with $LiNH_2$ [83]. Nevertheless the hydrogen release temperatures remain too high for automotive fuel cell applications. Investigations for optimizing the hydrogen release and the storage reversibility are in progress. Efforts are especially devoted to promote the dehydrogenation by using additives and to hinder emission of ammonia NH_3, which is one of the drawbacks with the utilization of amides [19,84]. One of the strategies adopted to make amides suitable for applications is tuning small ion mobility [54]. The utilization of metal oxides to improve the decomposition of amides did not show significant effect on the hydrogen kinetics [20,85].

In addition to the complex hydrides discussed above, the most common ones being both alanates and borohydrides while ammonia borane and its derivatives and amides being new compounds, many hydrides have been, at least, assessed once, or at best, studied in details. Hereafter are reported some of them although we are aware that the list that is given is not exhaustive due to the huge amount of literature:

- Aluminum hydride AlH_3 with a theoretical volumetric hydrogen storage capacity of 148 kg m^{-3} and a theoretical gravimetric hydrogen storage capacity exceeding 10 wt% [19,86]; though being a non-reversible hydride, its decomposition rapidly occurs at low temperatures (< 100 °C) [87]; the challenge of regeneration is a major obstacle to its development [19];
- Aluminoborane compounds such as AlB_4H_{11} with a theoretical gravimetric hydrogen storage capacity of 13.5 wt% and attractive hydrogen desorption temperature [88]; it is a reversible hydride;
- Ammonia NH_3 with a theoretical gravimetric hydrogen storage capacity of 17.7 wt% [61,89]; its decomposition into nitrogen N_2 and hydrogen H_2 requires high temperatures and there are currently two possibilities for decreasing this temperature, which are the utilization of a metal (Ru)-based catalyst and the formation of lithium amide $LiNH_2$ from lithium hydride LiH and NH_3;
- Metal ammine complexes [89] such as $Mg(NH_3)_6Cl_2$ with a theoretical gravimetric hydrogen storage capacity of 9.2 wt% and a volumetric hydrogen storage capacity of 109 kg m^{-3}; it has the advantage to store as much hydrogen as ammonia NH_3 while being safer ;
- Hydrazine N_2H_4 with a theoretical gravimetric hydrogen storage capacity of 12.6 wt% [61];

These compounds have a certain potential in solid-state hydrogen storage but as the research devoted to them are at the beginning they still have to provide evidence of their effective potential. There is no doubt that chemists will propose new compounds yet.

The present section can be summarized in two points. First, from the theoretical capacities of some of the discussed complex hydrides it appears to be evident that most of them are attractive compared to high pressure storage although few are promising in relation to cryogenic storage. Nevertheless they still have to put evidence of their effective capacities, knowing besides that other criteria such as the storage reversibility have to be fulfilled. The second point emphasizes the safety of few of the complex hydrides. It has been showed that sodium alanate $NaAlH_4$, lithium borohydride $LiBH_4$, lithium amide $LiNH_2$, and magnesium amide $Mg(NH_2)_2$ are hazardous and have to be much cautiously handled.

4.2.3 Storage via Chemical Reactions

Storage via chemical reactions is in fact based on hydrogen generation from water by making water reacting with a reacting metal. If sodium borohydride $NaBH_4$ was not a complex hydride, capable of liberating its hydrides in the presence of water (two molecules) through the hydrolysis reaction, it could be regarded as a compound allowing storage via a chemical reaction. Actually, for being considered as, the reacting metal has to be hydrogen-free. Several metals have already shown such properties. Sodium, lithium and zinc are the most known examples.

It is well known that sodium Na (specifically, two atoms) reacts with water (namely, two molecules) while producing one hydrogen molecule and sodium hydroxide NaOH (two compounds). The latter product may be recovered and recycled back to metallic sodium Na by reduction in a solar furnace. Hence the reaction is not directly reversible. The gravimetric hydrogen storage capacity of the system sodium-water is only of 3.0 wt% because a third water molecule is required for a further hydrogen generation, one of the required water molecule being stem from hydrogen combustion (or oxidation) process. Lithium Li is another metal that can be used to generate hydrogen by reaction with water. In that case, the gravimetric hydrogen storage capacity is of 6.3 wt% [2,13]. However, sodium Na and lithium Li can pose safety problems because both are water-reactive. The reactivity of both with water is exothermic and this may lead to the flammability of the as-generated hydrogen. Note that the process feasibility has also been showed with zinc Zn [90].

CONCLUSION

Today, a hydrogen economy already exists but this hydrogen economy is much more chemistry-turned (for example hydrogen is currently much used in the petrol industry). It is “the” hydrogen economy or more specifically the hydrogen economy today. However, in our minds when we talk about the hydrogen economy we think to the future hydrogen economy, that is, to a hydrogen economy that will be energy-turned. It is “a” hydrogen economy. Such difference is important because we are not ready, from the technological and commercial points of view, for developing a hydrogen economy yet.

The development of a hydrogen economy is hindered by many issues, one of them being the hydrogen storage. Several solutions are under investigation and much effort is devoted to each of them. The most common hydrogen storage solutions are as follows: high pressure, low temperature (cryogenic), porous materials, metallic hydrides, complex hydrides, and metals reacting with water. The methods are either physical or chemical. They have been discussed throughout the present chapter but the solid-state hydrogen storage has focused more attention. Table 3 summarizes some of their main advantages and issues [17].

Table 3. Some of the main advantages (denoted +) and issues (denoted -) of the hydrogen storage technologies (+/- is given when the property depends on the material nature); adapted from reference [17].

	High pressure 350-800 bar	Cryogenic	Absorption Porous mater.	Adsorption Metal hydrides	Adsorption Complex hydr.
Availability	+	+	+ / -	+ / -	+ / -
GHSC [a]	-	+	-	-	+
VHSC [a]	-	+	-	+	+
Reversibility	+	+	+	+ / -	-
H_2 release	+	+	+	-	-
Safety	-	-	+	+	+ / -
Cost	-	-	+ / -	+ / -	-

[a] GHSC as gravimetric hydrogen storage capacity and VHSC as volumetric hydrogen storage capacity.

Solid-state hydrogen storage is attractive because it offers a safe and efficient way to store hydrogen. It does not have the safety problems encountered with the physical storage methods while it shows the highest hydrogen storage capacities (gravimetric and volumetric). For example, it has been reported that on the basis of theoretical estimations volumetric hydrogen storage capacities higher than 200 kg m^{-3} could be reached [13]. Nevertheless the targets established for automotive fuel cell applications are highly restrictive. They mainly apply to the effective storage capacities, the storage reversibility and the hydrogen release conditions. All of these criteria are in fact highly sensitive since the technologically viable candidate should fulfill all of them. To these criteria the cost issue has to be added because it is highly sensitive in the view of commercialization. Principi et al. [50] are quite severe and pessimistic in their conclusion with respect to the new materials for solid-state hydrogen storage. They have remarked that up to now none of the studied systems including metal and complex hydrides has all the required characteristics for practical applications for large scale. They rather have suggested niche applications. For a few of the hydrogen storage materials, the U.S. Department of Energy seems to agree since recent reports recommending no-go for both carbon nanotubes and sodium borohydride have been recently spread [32,72]. Otherwise, van den Berg and Areán [1] agree Principi et al.'s conclusion [50]. They reported that “we are still short of making Jules Verne's visionary come true”. However, they have added that there is ample room for improvement.

Complex hydrides are especially of interest because they, with few of the metal hydrides, are the compounds showing the highest storage capacities. In fact their development is hindered by issues related to e.g. the storage reversibility, the hydrogen release conditions, the generated hydrogen purity, and the cost. Moreover, as interestingly reported by Graetz [19] the high capacity hydride must not only have the suitable properties as a material but also the suitable properties as a whole storage system. Indeed, we all make the same mistake. We systematically refer to the hydride storage capacities for discussing their suitability for e.g. automotive applications while we should also consider the storage system including all of its components. In the latter case the hydride effective capacities could be divided by, for example, a factor of 2. However that may be, we, that is, most of the scientists involved in hydrogen, are convinced that there is ample room for improvement.

In our opinion, foremost among complex hydrides are boron compounds, and especially boron hydrides. According to Fakioğlu et al. [91], boron hydrides are very important for hydrogen economy concept. We agree and that's why we propose the present book, which is completely devoted to boron hydrides as hydrogen storage materials with high (maybe the highest one? This will be discussed in the next chapters) potential. We hope that this book will provide to the reader all the needed information regarding these very interesting hydrides and that it will show how boron has become a significant element in solid-state hydrogen storage.

We would like to conclude this chapter quoting van den Berg and Areán [1] because we completely support their enthusiasm and optimism. "We are eager to see fruitful developments coming out soon".

REFERENCES

[1] van den Berg, A. W. C.; Areán, C. O. *Chem. Commun.* 2008, 668-681.
[2] Züttel, A. *Mitig. Adapt. Strat. Glob. Change* 2007, *12*, 343-365.
[3] Adams, W. M. The future of sustainability: re-thinking environment and development in the twenty-first century, Publication of IUCN The World Conservation Union. http://cmsdata.iucn.org/downloads/iucn_future_of_sustainability.pdf.
[4] Anastas, P. T. *Green Chem.* 2003, G29-G34.
[5] Adams, W. M. The future of sustainability: re-thinking environment and development in the twenty-first century, Publication of IUCN, The World Conservation Union. http://cmsdata.iucn.org/downloads/iucn_future_of_sustainability.pdf.
[6] Jacobson, M. Z. *Energy Environ. Sci.* 2009, doi:10.1039/b809990c.
[7] Horváth, I. T.; Anastas, P. T. *Chem. Rev.* 2007, *107*, 2169-2173.
[8] Crabtree, R. H. *Energy Environ. Sci.* 2008, *1*, 134-138.
[9] Muradov, N. Z.; Veziroglu, T. N. *Int. J. Hydrogen Energy* 2005, *30*, 225-237.
[10] Dufour, J.; Serrano, D. P.; Gálvez, J. L.; Moreno, J.; García, C. *Int. J. Hydrogen Energy* 2009, *34*, 1370-1376.
[11] Holladay, J. D.; Hu, J.; King, D. L.; Wang, Y. *Catal. Today* 2008, *139*, 244-260.
[12] Sartbaeva, A.; Kuznetsov, V. L.; Wells, S. A.; Edwards, P. P. *Energy Environ. Sci.* 2008, *1*, 79-85.
[13] Züttel, A. *Mater. Today* 2003, *September*, 24-33.
[14] Schlapbach, L.; Züttel, A. *Nature* 2001, *414*, 353-358.

[15] U.S. Department of Energy, Energy Efficiency and Renewable Energy. http://www1.eere.energy.gov/hydrogenandfuelcells; or, Hydrogen Association. http://www.hydrogenassociation.org.
[16] U.S. Department of Energy, Energy Efficiency and Renewable Energy (2007). Hydrogen storage. http://www1.eere.energy.gov/hydrogenandfuelcells/mypp/pdfs/storage.pdf.
[17] Rufford, T. E.; Zhu, Z. H.; Lu, G. Q. *Dev. Chem. Eng. Mineral Process.* 2006, *14*, 85-99.
[18] Ross, D. K. *Vacuum* 2006, *80*, 1084-1089.
[19] Graetz, J. *Chem. Soc. Rev.* 2009, *38*, 73-82.
[20] Guo, Z. X.; Shang, C.; Aguey-Zinsou, K. F. *J. Eur. Ceram. Soc.* 2008, *28*, 1467-1473.
[21] Fellay, C.; Dyson, P. J.; Laurenczy, G. *Angew. Chem. Int. Ed.* 2008, *47*, 1-4.
[22] Mao, S. S.; Chen, X. *Int. J. Energy Res.* 2007, *31*, 619-636.
[23] Stein, A.; Wang, Z.; Fierke, M. *Adv. Mater.* 2008, *20*, 1-29.
[24] Parra, J. B.; Ania, C. O.; Arenillas, A.; Rubiera, F.; Palacios, J. M.; Pis, J. J. *J. Alloys Compd.* 2004, *379*, 280-289.
[25] Ströbel, R.; Garche, J.; Moseley, P. T.; Jörisen, L.; Wolf, G. *J. Power Sources* 2006, *159*, 781-801.
[26] Dash, R.; Chmiola, J.; Yushin, G.; Gogotsi, Y.; Landisio, G.; Singer, J.; Fisher, J.; Kucheyev, S. *Carbon* 2005, *44*, 2489-2497.
[27] Dillon, A. C.; Jones, K. M.; Bekkedahl, D. A.; Kiang, C. H.; Bethune, D. S.; Heben, M. J. *Nature* 1977, *386*, 377-379.
[28] Panella, B.; Hirscher, M.; Ludescher, B. *Microporous Mesoporous Mater.* 2007, *103*, 230-234.
[29] Nijkamp, M. G.; Raaymakers, J. E. M. J.; Van Dillen, A. J.; De Jong, K. P. *Appl. Phys. A* 2001, *72*, 619-623.
[30] Langmi, H. W.; McGrady, G. S. *Coord. Chem. Rev.* 2007, *251*, 925-935.
[31] Yang, Z.; Xia, Y.; Mokaya, R. *J. Am. Chem. Soc.* 2007, *129*, 1673-1674.
[32] U.S. Department of Energy (2006). Go/no-go decision: pure, undoped single walled carbon nanotubes for vehicular hydrogen storage. http://www.hydrogen.energy.gov/pdfs/go_no_go_nanotubes.pdf.
[33] Fraenkel, D.; Shabtai, J. *J. Am. Chem. Soc.* 1977, *99*, 7074.
[34] Vitillo, J. G.; Ricchiardi, G.; Spoto, G.; Zecchina, A. *Phys. Chem. Chem. Phys.* 2005, *7*, 3948-3954.
[35] Roswell, J. L. C.; Millward, A. R.; Park, K. S.; Yaghi, O. M. *J. Am. Chem. Soc.* 2004, *126*, 5666-5667.
[36] Wong-Foy, A. G.; Matzger, A. J.; Yaghi, O. M. *J. Am. Chem. Soc.* 2006, *128*, 3494-3495.
[37] Thomas, K. M. *Catal. Today* 2007, *120*, 389-398.
[38] Lee, J. Y.; Pan, L.; Kelly S. P.; Jagiello, J.; Emge, T. J.; Li, J. *Adv. Mater.* 2005, *17*, 2703-2706.
[39] Collins, D. J.; Zhou, H. C. *J. Mater. Chem.* 2007, *17*, 3154-3160.
[40] Zhu, L. G.; Xiao, H. P. *Z. Anorg. Allg. Chem.* 2008, 845-847.
[41] Ghanem, B. S.; Msayib, K. J.; McKeown, N. B.; Harris, K. D. M.; Pan, Z.; Budd, P. M.; Butler, A.; Selbie, J.; Book, D.; Walton, A. *Chem. Commun.* 2007, 67-69.
[42] Ma, R.; Bando, Y.; Zhu, H.; Sato, T.; Xu, C.; Wu, D.; *J. Am. Chem. Soc.* 2002, *124*, 7672-7673.
[43] Sun, Q.; Wang, Q.; Jena, P. *Nano Lett.* 2005, *5*, 1273-1277.
[44] Lim, S. H.; Luo, J.; Zhong, Z.; Ji, W.; Lin, J. *Inorg. Chem.* 2005, *44*, 4124-4126.

[45] Bavikyn, D. V.; Lapkin, A. A.; Plucinski, P. K.; Friedrich, J. M.; Walsh, F. C. *J. Phys. Chem. B* 2005, *109*, 19422-19427.
[46] Bhattacharya, S.; Majumder, C.; Das, G. P. *J. Phys. Chem. C* 2008, *112*, 17487-17491.
[47] Sheppard, D. A.; Buckley, C. E. *Int. J. Hydrogen Energy* 2008, *33*, 1688-1692.
[48] Barman, S.; Sen, P.; Das, G. P. *J. Phys. Chem. C* 2008, *112*, 19963-19968.
[49] Li, Y.; Zhou, G.; Li, J.; Gu, B. H.; Duan, W. *J. Phys. Chem. C* 2008, *112*, 19268-19271.
[50] Principi, G.; Agresti, F.; Maddalena, A.; Lo Russo, S. *Energy* 2008, doi:10.1016/j.energy.2008.08.027.
[51] Shang, C. X.; Bououdina, M.; Song, Y.; Guo, Z. X. *Int. J. Hydrogen Energy* 2004, *29*, 73-80.
[52] Eisenberg, F. G.; Zagnoli, D. A.; Sheridan, J. J. *J. Less-Common Met.* 1980, *74*, 323-331.
[53] Vajo, J. J.; Mertens, F.; Ahn, C. C.; Bowman Jr., R. C.; Fultz, B. J. *J. Phys. Chem. B* 2004, *108*, 13977-13983.
[54] Wu, H. *Chem. Phys. Chem.* 2008, *9*, 2157-2162.
[55] Huot, J.; Liang, G.; Boily, S.; Van Neste A.; Schulz, R. *J. Alloys Compd.* 1999, *293-295*, 495-500.
[56] Aguey Zinsou, K. F.; Ares-Fernandez, J. R.; Klassen, T.; Bormann, R. *Mater. Res. Bull.* 2006, *41*, 1118-1126.
[57] Bogdanović, B.; Schwickardi, M. *J. Alloys Compd.* 1997, *253-254*, 1-9.
[58] Bogdanović, B.; Brand, R. A.; Marjanovic, A.; Schwickardi, M.; Tolle, J. *J. Alloys Compd.* 2000, *302*, 36-58.
[59] Züttel, A.; Wenger, P.; Rentsch, S.; Sudan, P.; Mauron, Ph.; Ernmenegger, Ch. *J. Power Sources* 2003, *118*, 1-7.
[60] Wang, P.; Kang, X. D. *Dalton Trans.* 2008, 5400-5413.
[61] Umegaki, T.; Yan, J. M.; Zhang, X. B.; Shioyama, H.; Kuriyama, N.; Xu, Q. *Int. J. Hydrogen Energy* 2009, *34*, 2303-2311.
[62] Au, M.; Jurgensen, A. R.; Spencer, W. A.; Anton, D. L.; Pinkerton, F. E.; Hwang, S. J.; Kim, C.; Bowman Jr., R. C. *J. Phys. Chem. C* 2008, *112*, 18661-18671.
[63] Gross, A. F.; Vajo, J. J.; Atta, S. L.; Olsen, G. L. *J. Phys. Chem. C* 2008, *112*, 5651-5657.
[64] Sudik, A.; Yang, J.; Halliday, D.; Wolverton, C. *J. Phys. Chem. C* 2007, *111*, 6568-6573.
[65] Wan, X.; Markmaitree, T.; Osborn, W.; Shaw, L. L. *J. Phys. Chem. C* 2008, *112*, 18232-18243.
[66] Nickels, E. A.; Jones, M. O.; David, W. I. F.; Johnson, S. R.; Lowton, R. L.; Sommariva, M.; Edwards, P. P. *Angew. Chem. Int. Ed.* 2008, *47*, 2817-2819.
[67] Ronnebro, E.; Majzoub, E. H. *J. Phys. Chem. B* 2007, *111*, 12045-12047.
[68] Kim, J. H.; Shi, J. H.; Cho, Y. W. *J. Power Sources* 2008, *181*, 140-143.
[69] Nakamori, Y.; Li, H. W.; Miwa, S.; Towata, S.; Orimo, S. *Mater. Trans.* 2006, *47*, 1898-1901.
[70] Nakamori, Y.; Li, H. W.; Matsuo, M.; Miwa, S.; Towata, S.; Orimo, S. *J. Phys. Chem. Solids* 2008, *69*, 2292-2296.
[71] Çakanyildirim, Ç.; Gürü, M. *Int. J. Hydrogen Energy* 2008, *33*, 4634-4639.
[72] U.S. Department of Energy Hydrogen Program. (2007). Independent review. November 2007. Go/no-go recommendation for sodium borohydride for on-board vehicular hydrogen storage. NREL/MP-150-42220. http://www.hydrogen.energy.gov/.
[73] Marder, T. B. *Angew. Chem. Int. Ed.* 2007, *46*, 8116-8118.
[74] Hamilton, C. W.; Baker, R. T.; Staubitz, A.; Manners, I. *Chem. Soc. Rev.* 2009, *38*, 279-293.

[75] Sit, V.; Geanangel, R. A.; Wendlandt, W. W. *Thermochim. Acta* 1987, *113*, 379-382.
[76] Bluhm, M. E.; Bradley, M. G.; Butterick III, R.; Kusari, U.; Sneddon, L. G. *J. Am. Chem. Soc.* 2006, *128*, 7748-7749.
[77] Gutowska, A.; Li, L.; Shin, Y.; Wang, C. M.; Li, X. S.; Linehan, J. C.; Smith, R. S.; Kay, B. D.; Schmid, B.; Shaw, W.; Gutowski, M.; Autrey, T. *Angew. Chem. Int. Ed.* 2005, *44*, 3578-3582.
[78] Denney, M. C.; Pons, V.; Hebden, T. J.; Heinekey, D. M.; Goldberg, K. I. *J. Am. Chem. Soc.* 2006, *128*, 12048-12049.
[79] Peng, B.; Chen, J. *Energy Environ. Int.* 2008, *1*, 479-483.
[80] Wu, H.; Zhou, W.; Yildirim, T. *J. Am. Chem. Soc.* 2008, *130*, 14834-14839.
[81] Xiong, Z.; Wu, G.; Chua, Y. S.; Hu, J.; He, T.; Xu, W.; Chen, P. *Energy Environ. Sci.* 2008, *1*, 360-363.
[82] Chen, P.; Xiong, Z. T.; Luo, J. Z.; Lin, J. Y.; Tan, K. L. *Nature* 2002, *420*, 302-304.
[83] Luo, W.; Sickafoose, S. *J. Alloys Compd.* 2006, *407*, 274-281.
[84] Orimo, S.; Nakamori, Y.; Eliseo, J. R.; Züttel, A.; Jensen, C. M. *Chem. Rev.* 2007, *107*, 4111-4132.
[85] Yao, J. H.; Shang, C.; Aguey Zinsou, K. F.; Guo, Z. *J. Alloys Compd.* 2007, *118*, 1-7.
[86] Kato, T.; Nakamori, Y.; Brown, C.; Jensen, C. M. *J. Alloys Compd.* 2007, *446-447*, 276-279.
[87] Graetz, J.; Reilly, J. J.; Kulleck, J. G.; Bowman, R. C. *J. Alloys Compd.* 2007, *446-447*, 271-279.
[88] Zhao, J. C.; Knight, D. A.; Brown, G. M.; Kim, C.; Hwang, S. J.; Reiter, J. W.; Bowman Jr., R. C.; Zan, J. A.; Kulleck, J. G. *J. Phys. Chem. C* 2009, *113*, 2-11.
[89] Klerke, A.; Christensen, C. H.; Nørskov, J. K.; Vegge, T. *J. Mater. Chem.* 2008, *18*, 2304-2310.
[90] Steinfeld, A. *Int. J. Hydrogen Energy* 2002, *27*, 611-619.
[91] Fakioğlu, E.; Yürüm, Y.; Veziroğlu, T. N. *Int. J. Hydrogen Economy* 2004, *29*, 1371-1376.

In: Boron Hydrides High Potential Hydrogen Storage Material ISBN:978-1-61668-361-0
Editors: Umit B. Demirci and Philippe Miele

Chapter 2

BORON HYDRIDES

Umit B. Demirci[1], Ouardia Akdim, Jérôme Andrieux, Rita Chamoun, Julien Hannauer and Philippe Miele[2]

Laboratoire des Multimatériaux et Interfaces, UMR 5615 CNRS Université Lyon 1, Université de Lyon, 43 boulevard du 11 Novembre 1918, F-69622 Villeurbanne, France

ABSTRACT

Boranes (or boron hydrides)[3] are known for their capacity of hydrogen storage since the 1950s, at least for the borohydrides, but it is since the early 2000s, that is 50 years after the discovery of the most known hydrogen storage boron hydride, namely sodium borohydride, that these compounds have attracted much attention with respect to the potential as hydrogen storage materials they have for technological applications. Today, boron hydrides seem to have a major role to play in the hydrogen economy we are intensely trying to develop. The present chapter is devoted to these hydrides and more generally to the boron compounds, which are listed and briefly discussed in a first section. It is showned that only the boron hydrides are promising candidates for storing hydrogen. The boron hydrides, and especially the borohydrides, ammoniaborane and the newly developed metal amidoboranes, are more largely discussed in terms of their gravimetric hydrogen storage capacities. The safety aspect is also, besides, tackled. Actually, this chapter is an introduction to the next chapters, dealing more in details with the presently discussed boron hydrides.

---

[1] E-mails: Umit.Demirci@univ-lyon1.fr, Tel.:+33 (0)4 72 44 84 03; +33 (0)4 72 43 12 05, Fax:+33 (0)4 72 44 06 18

[2] Philippe.Miele@univ-lyon1.fr

[3] The words 'boranes' (mainly used in organic chemistry because having a structure similar to alkanes compounds) and 'boron hydrides' (mainly used in inorganic chemistry because of the presence of a hydride) will be used throughout the present chapter to define the same kinds of compounds, that is, compounds composed of boron and hydrogen, and also other elements.

1. INTRODUCTION

The twentieth century was the oil century. Hydrocarbons have provided a specific, significant comfort to industry, houses, and transport, namely sectors that are insatiable consumers of energy. However, the twenty-first century does not seem to be an oil century. And this has, at least, four distinct reasons.

- First, fossil fuels are stem from depleting resources (which costs critically escalate) and today most of experts are quite alarmist with respect to the remaining feedstocks.
- Second, most of the (industrialized) nations are dependent on oil and the awareness of this has motivated the searching for alternative fuels that would be produced in any places.
- Third, fossil fuels are carbon-based materials and they emit anthropogenic carbon dioxide as product of the combustion process; carbon dioxide is an additional greenhouse gas that contributes to global warming.
- Fourth, the concept of sustainability has emerged in the late twentieth century. Sustainability explicitly suggests that it should be possible to achieve economic growth without environmental negative impacts. It is engaged in developing our needs and addressing the environmental issues of the present while preparing a sustainable future in terms of economic growth, environment preservation and poverty alleviation [1].

Hence, there is an urgent need to find sustainable (i.e. renewable, clean, efficient and cost-effective) energy sources and, above all, ecologically friendly biofuels [2].

Among the various energy sources or carriers that are under investigation, hydrogen as energy carrier is considered as being the most promising [3]. However, the development of a hydrogen economy is hindered by many key issues (Figure 1):

- Issue relating to its production (and purification), because hydrogen exists only combined with other elements like e.g. carbon (hydrocarbons, biomass) or oxygen (water, biomass);
- Issue relating to its storage, because it is the lightest compound, what does not make easy its storage;
- Issue relating to its distribution;
- Issue relating to its consumption, because it is mainly intended to feed fuel cell technologies, which are still under investigation.

There are great efforts that are devoted to solve these issues but one has to recognize that hydrogen is not mature enough for conceiving of a short-term development of a hydrogen economy.

Today, hydrogen storage draws much attention. This issue may be seen as the most critical since, if hydrogen cannot be efficiently, safely stored, it will not be possible to distribute this energy carrier and thus to develop the hydrogen economy [3]. Efforts devoted to find a solution to this key issue has permitted the emergence of several storage solutions that are under intense investigations each: e.g. high pressure (250-700 bars), low temperature (cryogenic storage at -252 °C), porous materials (e.g. porous carbon materials), metal hydrides (e.g. Mg-based materials), and complex hydrides (e.g. alanates and boron hydrides) [4]. Solid state hydrogen storage has represented a breakthrough, especially because of safer storage and higher gravimetric hydrogen storage capacity [5]. Nevertheless, up to now, none of the studied materials for hydrogen storage has reached a scientific, technical maturity for applications [6]. None has taken up all of the following technical challenges yet: effective gravimetric/volumetric hydrogen storage capacity, thermodynamic stability, controllable and effective kinetics of hydrogen desorption/generation, reversibility of storage, efficient thermal conductivity, and low cost [7].

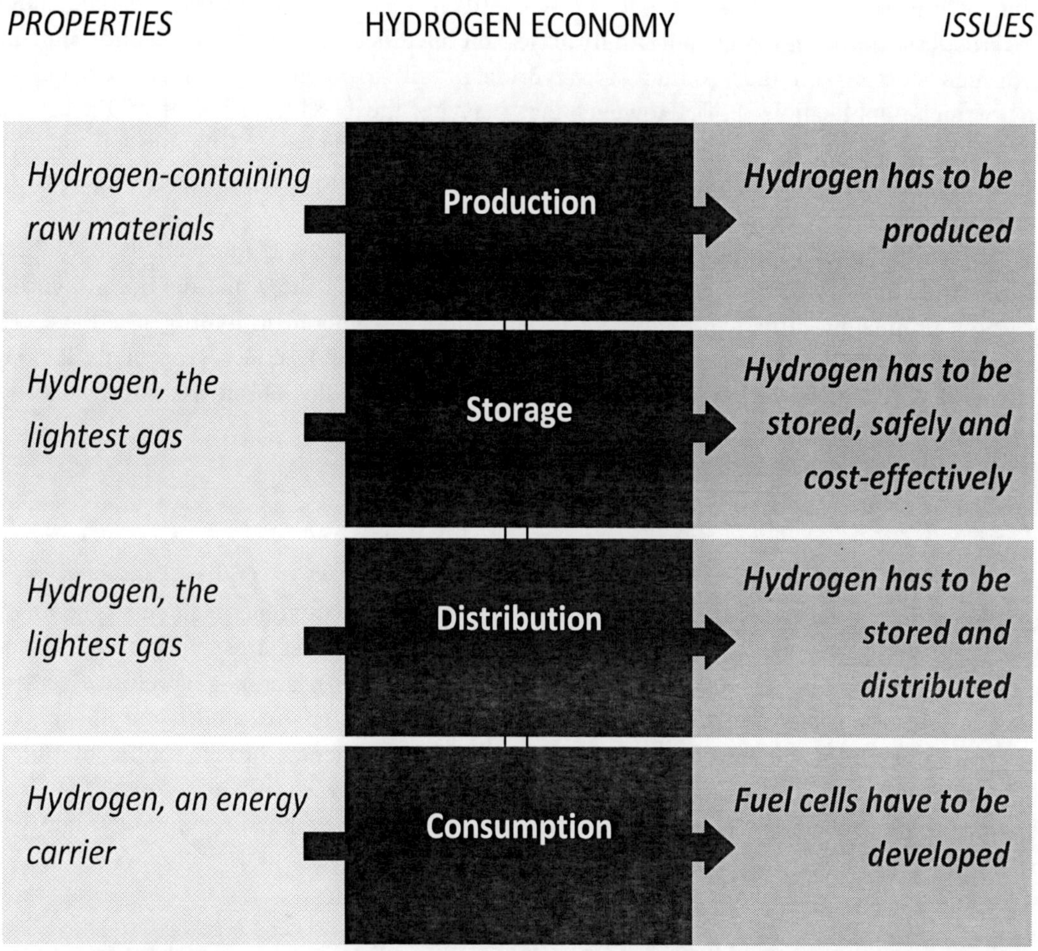

Figure 1. Issues hindering the development of a hydrogen economy.

Complex hydrides are high hydrogen-content materials with gravimetric hydrogen storage capacities above 10 wt%. The most interesting hydrides are those formed by some of the lightest elements in the periodic table like for example lithium, boron, sodium and aluminum [5, 8]. Fakioğlu et al. [9] reported that boron is a promising element for hydrogen storage with its chemical hydrides and nanostructural forms. The hydrogen content reaches values up to 10, 18 and 19 wt% for sodium borohydride ($NaBH_4$), lithium borohydride ($LiBH_4$) and ammonia borane (NH_3BH_3), respectively. Of these three boron hydrides, the oldest (discovered in the middle of the 1940s), the most known and the most investigated is sodium borohydride [10]. It has attracted much attention since the late 1990s [11]. Since then, there has been an eight-year increasing interest towards this boron hydride [12]. However, in November 2007, the U.S. Department of Energy communicated a significant decision [13]. Due to the fact that the hydrogen storage technology considered for the hydrolysis of sodium borohydride for vehicular applications had clearly not met all the 2007 targets (for example, a gravimetric hydrogen storage capacity of 6.00 wt% taking the storage system as a whole), the U.S. Department of Energy has seen no promising path forward for this technology to reach all of the 2010 targets (for example, a gravimetric hydrogen storage capacity of 9.00 wt% taking the storage system as a whole). A no-go decision has therefore been recommended. Nevertheless, the no-go decision is only based on specifications set for on-board vehicular hydrogen storage and thus sodium borohydride is still a potential candidate for portable and/or niche applications [14]. However that may be, the U.S. Department of Energy has acknowledged that the challenge with sodium borohydride was difficult and that the investigations have provided improvements that could be valorized in other applications like for example hydrogen storage via ammonia borane [13].

Efforts devoted to sodium borohydride and ammonia borane (as well as on lithium borohydride and amidoboranes) as hydrogen storage materials show that the boron hydrides are now promising, attractive compounds that cannot be ignored in hydrogen storage. The present chapter is therefore devoted to an introduction to boron hydrides[4], especially the ones having a potential in hydrogen storage. Before discussing the boron hydrides, a section introducing the boron compounds is proposed.

2. Boron Compounds

According to Grimes [15], boron is in many ways unlike any other elements. The unique properties of boron are directly traceable to the following features: only three valence electrons and low electronegativities (lower than that of carbon). Boron is therefore a versatile element because it can ring structures with electron-donor atoms (e.g. nitrogen, phosphorus, oxygen and sulfide) and can compensate for its electron deficiency through multicenter bonds (cluster formation). It is possible to distinguish at least five types of boron compounds, each having its own chemical systematic derived from the type of bonding involved and each resulting in individualistic structures and chemical reactions [16]:

- Borides;
- Boron hydrides, and their derivatives like carboranes and polyhedral borane-metal complexes;
- Boron halides and their derivatives;
- Boron-oxygen compounds (borates, borosilicates…);
- Boron-nitrogen compounds (ammonia borane, boron nitride…).

[4] The present chapter rather focuses on an introduction to the various boron hydrides; the state-of-the-art of the most attractive hydrides as hydrogen storage materials are largely discussed in next chapters devoted to borohydrides (e.g. sodium borohydride and lithium borohydride) and B,N-based hydrides (e.g. ammonia borane and amidoboranes).

These various boron compounds are briefly tackled hereafter whereas a large section is devoted to boron hydrides. Before that, two books can be cited for readers that wish to have more detailed information about boron and all of its compounds. The first one, edited by Bailar et al. [16], proposes a large, detailed chapter while the second one, edited by Cotton et al. [15], suggests a shorter view of these compounds. The preset section largely refers to both references.

2.1 Borides

Borides are M_xB_y-type compounds, M being a less electronegative element (metals and metalloids) [15]. Borides constitute a group of over 200 binary compounds showing a diversity of stoichiometries and structural types and there also ternary and more complex phases in which, for example, more than one metal combine with boron [16].

Boron and most of the elements of the periodic table form binary compounds. The most commonly occurring stoichiometries are M_2B, MB, MB_2, MB_4, MB_6 and MB_{12}. These six stoichiometries account for over 70 % of the binary compounds while examples of well-defined borides are known for the following twenty-four compositions: M_5B, M_4B, M_3B, M_5B_2, M_7B_3, M_2B, M_5B_3, M_3B_2, $M_{11}B_8$, MB, $M_{10}B_{11}$, M_3B_4, M_2B_3, MB_2, M_2B_5, MB_3, MB_4, MB_6, M_2B_{13}, MB_{10}, MB_{12}, MB_{18}, MB_{66}, and MB_{100} [16]. For example, the compounds Be_2B, FeB, TiB_2, LaB_4, CaB_6, and YB_{12} have been prepared. These materials exhibit specific solid-state architectures: isolated boron atoms (e.g. Be_2B), pairs of boron atoms (e.g. V_3B_2), zigzag chains of boron atoms (e.g. FeB), branched chains (e.g. $Ru_{11}B_8$), double chains (e.g. Cr_3B_4), layer networks (e.g. TiB_2), three-dimensional networks (e.g. CaB_6). Further, their physical properties (e.g. extreme hardness, inertness, high melting points, and high thermal conductivity) make them attractive to modern technology [15].

The most important non-metal borides are certainly the boron carbides, which compositions vary from $B_{13}C_2$ to $B_{24}C$. These materials are narrow-band semiconductors and are of interest for their potential in high-temperature thermoelectric energy conversion [15].

2.2 Boron Hydrides

Boranes (boron hydrides), which are binary boron-hydrogen compounds, are highly exciting compounds in molecular cluster chemistry since the early work of Alfred Stock, who is the pioneer of the borane chemistry [15]. Large numbers of chemical compounds related to polyborane chemistry exist: boron hydrides, carboranes, metallaboranes, metallocarboranes, mixed compounds with organic moieties, and others [17].

Neutral boranes are B_xH_y-type compounds with x varying from 2 to 20 and y from 6 to 26. About thirty-five neutral binary boranes (including isomers) are known (many derivatives containing organic or halo substituent have been synthesized). The lightest is diborane B_2H_6 that is known to have an ethylene-like structure (Figure 2) and to be explosively reactive in contact with air (readily oxidized). Molecules larger than diborane are called polyboranes. Among the isolated and characterized species, $B_{20}H_{26}$ is the heaviest borane (stable compound) [15, 17-21]. The gravimetric hydrogen storage capacity of these compounds varies from 21.7 wt% for B_2H_6 to e.g. about 11 wt% for $B_{20}H_{26}$ or $B_{12}H_{16}$. These values should ideally be of interest for hydrogen storage but the dangerousness of B_2H_6 and the high stability of the polyboranes do not make such compounds suitable for such utilization. And, to the best of our knowledge, these materials have never been considered as hydrogen storage compounds.

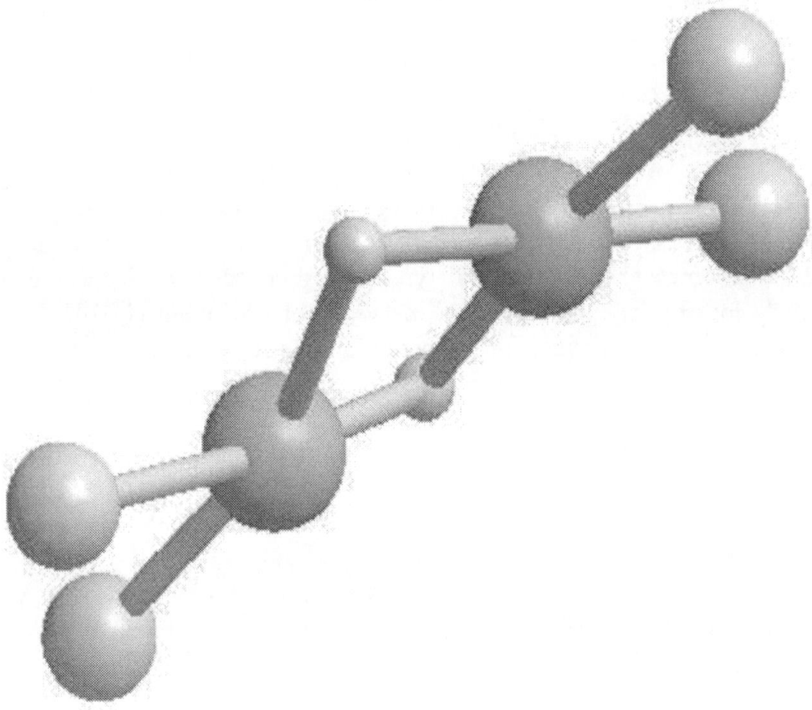

Figure 2. Diborane.

Hydride anions are related to the neutral boranes. There are three classes of these anions: the $B_nH_n^{2-}$ series (with n increasing from 6 to 12); anions derived from boranes by deprotonation or addition of a hydride ion (e.g. BH_4^-, $B_2H_7^-$, $B_3H_8^-$, $B_5H_{10}^-$, $B_{10}H_{12}^{2-}$, $B_{10}H_{15}^-$, $B_{11}H_{13}^{2-}$, etc); and, anions derived from linked boron hydride units ($B_{18}H_{21}^-$, $B_{20}H_{18}^-$, $B_{24}H_{23}^{3-}$, etc) [16, 17, 20, 22]. Borohydrides (or tetrahydroborates, BH_4^-) are well-known compounds in hydrogen storage [9, 23]. The most common ones are lithium borohydride ($LiBH_4$) and sodium borohydride ($NaBH_4$). They have a theoretical gravimetric hydrogen storage capacity of 18.4 and 10.7 wt%, respectively. Regarding the other hydride ions, their gravimetric hydrogen storage capacities are also quite attractive: for example, 20.7 wt% for BeB_2H_8, 14.8 wt% for MgB_2H_8, 12.6 wt% for NaB_3H_8, 4.0 wt% for $K_2B_6H_6$, 10.4 wt% for NaB_9H_{14}, or 7.3 wt% for $Na_2B_{11}H_{13}$. None of these compounds (or any hydride anion different from borohydrides) have been considered for storing hydrogen so far. The reasons are in fact quite evident. Some are unstable (BeB_2H_8, highly reactive air-sensitive solid) while others are too stable (high molecular weight polyboranes). Some have too low hydrogen storage capacities in comparison to the borohydrides and this does not make them suitable for hydrogen storage.

Carboranes are compounds deriving from borane anion clusters, in which one or more carbon atoms replace the isoelectronic B^- ions. Most of the carboranes have two skeletal carbon atoms per cage but one-, three-, four-, and five-carbon compounds are known [15, 16,

20-23]. From boranes and carboranes can be derived metallaboranes and metallacarboranes. The former ones are borane cages containing one or more metal atoms (for example $Cu(B_{11}H_{11})_2^{3-}$). The latter ones contain both carbon and metal atoms in the cluster (for example $Ni(C_2B_9H_{11})_2$). In addition to these compounds, one can cite heteroboranes and heterocarboranes, which are boron cages containing one or more non-metal atoms in the skeletal framework (and carbon atom(s) for the latter ones). Further variation may be achieved by introducing metals to give heterometallaboranes and heterometallacarboranes [15, 21, 22]. None of these boranes is suitable as hydrogen storage material because, being stable, their destabilization and thus their dehydrogenation are unrealizable in mild conditions.

The boranes structure is somewhat complicated. The bonding in boranes considers the concept of the three-centre, two-electron B–H–B bond (Figure 2) [15, 16, 17, 18, 20]. Actually, three types of three-bond find considerable application: B–H–B bridge bonds, central three-centre BBB bonds, and open three-centre BBB bonds (Figure 3). The skeletal structures of the boranes, borane anions, and carboranes can then be deduced. For example, species with n skeletal atoms adopt *closo*- (closed polyhedral geometries), *nido*- (non-closed structure), and *arachno*-structures (non-closed structure), if held by (n+1), (n+2) and (n+3) pairs of skeletal bonding electrons, respectively [15, 16, 20, 21]. For more details about the nomenclature, structures and bonding of boranes, the reader can refer to references [15, 16].

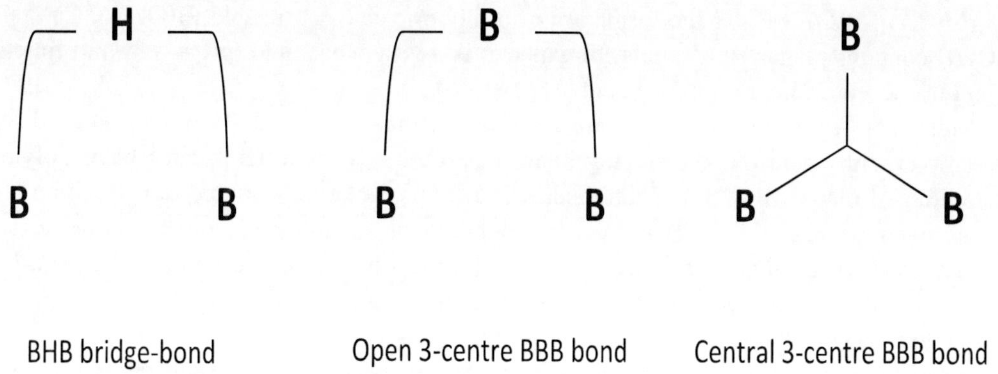

Figure 3. Types of three-bond [16].

2.3 Boron Halides

Among the boron halides, one has to distinguish boron trihalides from boron subhalides. The boron trihalides are monomeric, trigonal, planar molecules [15]. Boron trifluoride (BF_3) has been the first boron halide reported, over 200 years ago. It is a powerful Lewis acid, utilized as an effective reagent in organic chemistry. Boron trichloride (BCl_3), boron tribromide (BBr_3) and boron triiodide (BI_3) are stronger Lewis acids than boron fluoride. The acceptor strength of three of these boron trihalides is as follows: $BF_3 < BCl_3 < BBr_3$. Note that the position of BI_3 in this ranking is complicated by the steric requirement of the iodine atoms [16]. Tetrahalogenoborates (BF_4^-, BCl_4^-, BBr_4^-, BI_4^-) can be derived from these boron trihalides.

The boron subhalides are boron-halogen compounds respecting the following criteria:

- First, they contains B–B bonds:
- Second, the halogen/boron ratio is less than 3.

Boron tetrafluoride (B_2F_4), boron tetrachloride (B_2Cl_4), boron tetrabromide (B_2Br_4) and boron tetraiodide (B_2I_4) are known compounds. They consist of two BX_2 groups joined by a B–B bond [15]. Their thermal stability is very much less than that of the corresponding trihalides [16].

Polyhedral boron halides (B_nX_n) are stable clusters [15]. The other boron subhalides that can be briefly cited are boron monohalides (e.g. BF, BCl, and BBr, stable only at high temperatures and low pressure), triboron pentafluoride (B_3F_5, unstable compound in air and water, and under hydrogen), and polyhedral chloroboranes (for example B_4Cl_4, B_8Cl_8 and B_9Cl_8H) [16].

2.4 Boron-Oxygen Compounds

Boron occurs in nature as oxo-compounds, i.e. directly bonded to oxygen, containing predominantly trigonal planar BO_3, and to lesser degree tetrahedral BO_4 units. Boron-oxygen chemistry is the oldest branch of boron chemistry [15, 16].

The principal boron oxide is boric oxide (B_2O_3), which is very difficult to crystallize. Hydrolysis of B_2O_3 leads to the formation of orthoboric acid (boric acid, $B(OH)_3$). Crystalline $B(OH)_3$ contains planar BO_3^{3-} units. In water, it is a very weak acid (pK = 9.3) and functions as an OH^- acceptor, i.e. as a Lewis acid [15, 16].

Metal borates, found in nature, are for most of them hydrated. They consist of discrete polyanions with some oxygen having attached protons. Anhydrous borates have polymeric assemblies of planar BO_3 and/or tetrahedral BO_4 units, linked via shared oxygen atoms [15]. In hydrogen storage, the most known metal borate is certainly sodium metaborate, a by-product obtained through the hydrolysis of sodium borohydride and ammoniaborane. It has no potential in hydrogen storage but its recyclability back to sodium borohydride is the main challenge it faces [14].

There are many others compounds containing B–O bonds such as e.g. organo-orthoborates, hydroxyl derivatives and their anhydrides, acyloxy derivatives, alkoxyboroxines, peroxy derivatives, alkoxyboranes, and alkoxyhalogenoboranes [16]. None has potential in hydrogen storage.

2.5 Boron-Nitrogen Compounds

Boron-nitrogen compounds have attracted much attention because of the specificity of the B–N moiety:

- First, it (3+5 outer electrons) is isoelectronic with C–C (4+4 outer electrons).
- Second, the electronegativity and the atomic radius of carbon (0.077 nm, 2.5 eV) are almost exactly intermediate between the values of boron (0.088 nm, 2.0 eV) and nitrogen (0.070 nm, 3.0 eV) [15, 16].

Boron nitride (BN) is one of the oldest known boron-nitrogen compounds. Boron nitride in its hexagonal form is an increasingly important ceramic material (used for example as a lubricant). Hexagonal boron nitride is isoelectronic with graphite and has a similar sheet-like structure even if the B and N atoms alternate vertically in successive layers [15]. Each layer consists of a flat or nearly flat network of B_3N_3 hexagons [16]. At high temperature and pressure (e.g. 2000 °C and 60 kbar), it converts to cubic boron nitride. This material has a tetrahedral structure (wurtzite) analogous to diamond [15].

Amine-borane adducts are $R_3NBR'_3$ molecules (with R as H, alkyl or aryl, and R' as H, alkyl, aryl or halide) for which the first adduct H_3NBF_3 was prepared in 1809. The nitrogen-boron bond strength depends on the nature of the groups R and R'. Consequently, the amine-borane adducts can display a wide range of stability. Amine-boranes with composition R_3NBH_3 can derive from boranes [16]. Such materials have today a great potential as solid hydrogen storage materials. Ammoniaborane (H_3NBH_3), isoelectronic and isostructural with ethane, stores 19.5 wt% of hydrogen what makes it attractive. Many investigations on it are in progress, the main objectives being to decrease the hydrogen release temperature and to increase the hydrogen generation kinetics. A variety of amine-boranes as potential materials for hydrogen storage exist: for example, $CH_3NH_2BH_3$ with 17.9 wt% of hydrogen stored, $C_2H_5NH_2BH_3$ with 17.6 wt% of hydrogen, and $(CH_3)_3NBH_3$ with 16.5 wt% of hydrogen. Ammoniaborane-derived alkali amine-boranes (amidoboranes) such as $LiNH_2BH_3$ and $NaNH_2BH_3$ are also candidates for storing hydrogen (13.6 and 9.5 wt% of hydrogen). To summarize, amine-boranes derived from boranes are potential hydrogen storage materials and these kinds of compounds are discussed in details in chapters that follows the present one.

Among the other nitrogen-boron compounds, one may cite the followings: the aminoboranes including the monoaminoboranes ($R_2NBR'_2$ with R as H, alkyl or aryl, and R' as H, alkyl or aryl), the bisaminoboranes ($(R_2N)_2BR'$ with R as H, alkyl or aryl, and R' as H, alkyl, aryl or halide), the trisaminoboranes ($B(NRR')_3$ with R as H, alkyl or aryl); the amino derivatives of diborane(4) (for example, $(CH_3N)_2BB(NCH_3)_2$); the borazines. The borazines represent one of the largest classes of boron-nitrogen compounds. Borazine is a $(–BH–NH–)_3$ cycle. It has no aromatic character, although its physical properties seem to substantiate a benzene-like resonance structure of the molecule. From this compound are derived symmetrically substituted borazines ($(–BR–NR'–)_3$ with R and R' as H, alkyl or aryl), unsymmetrically substituted borazines (for example, $H_3B_3H_2CH_3$), and polyborazines (derivatives containing two or more condensed or linked borazine rings) [16]. The compounds tackled in the present paragraph are also under consideration as hydrogen storage materials.

According to Hamilton et al. [24], B–N compounds are well suited for hydrogen storage because:

- Both boron and nitrogen are lightweight elements capable of bearing multiple hydrogen atoms;
- The B–H and N–H bonds tend to be hydridic and protic respectively, resulting in normally facile hydrogen release.

In a recent critical review, these authors have mentioned various B–N compounds as being suitable for hydrogen storage applications, even though some are not really under consideration in the open literature:

- Hydrazine borane ($N_2H_4{\cdot}BH_3$) and hydrazine bis(borane) ($N_2H_4{\cdot}2BH_3$), with a potential hydrogen storage capacity of 13.1 and 13.4 wt%, respectively; note that the latter is unstable (for example, it is shock-sensitive explosive); to the best of our knowledge, both compounds have never been the object of an experimental study considering them for storing and releasing hydrogen.
- Ammonia-triborane ($H_3N{-}B_3H_7$, 10.6 wt%) and ammonium hydrotriborate ($[NH_4][B_3H_8]$, 13.9 wt%); hydrogen can be generated in specific conditions but these compounds are likely to be explosive.
- Amine compounds of high-order polyborane such as ammonium salts of $(B_{11}H_{14})^-$; such compounds can be hydrolyzed and their effective gravimetric hydrogen storage capacities are below 7 wt%.

2.6 Other Boron-Based Compounds

As an extension of the boron-nitrogen or boron-oxygen bonds, boron is also able to bond to the elements below nitrogen or oxygen in the periodic table.

Boron and sulfur can form an extensive series of molecular and/or solid-state compounds, but there is little resemblance to boron-oxygen chemistry. Boron-selenium chemistry provides much less compounds than the boron-sulfur ones. Of the various compounds, there are binary sulfides and selenides of boron (such as diboron trisulfide B_2S_3 and diboron triselenide B_2Se_3), boron-sulfur ring compounds having alternating S atoms and B(SH) groups (such as borthiin $(HSBS)_3$) and thioborates ($B(SR)_3$) that represent the largest area of boron-sulfur chemistry [15, 16].

The field of boron-phosphorus chemistry is small compared to that of boron-nitrogen chemistry. Among the various compounds with boron-phosphorus bonds, one can cite the binary phosphides (e.g. BP) and the phosphinoboranes (or borophanes i.e. compounds of the type $R_2PBR'_2$). Few boron-arsenic and boron-antimony compounds have also been reported. For example, boron arsenide (BAs) is a binary arsenide that is isomorphous with boron phosphide [16].

None of these materials has any potential as hydrogen storage material.

2.7 Summary

From the previous sub-sections as well as from the open literature, it clearly stands out that the boron compounds having the best potentials as hydrogen storage materials are borohydrides, ammoniaborane and amidoboranes. The following section will focus on listing these materials while briefly tackling their hydrogen storage capacities. Note that among the following chapters some deals with the state-of-the-art of some of these compounds.

3. Borohydrides

Today, it is often noticed that chemical hydrides like the borohydrides offer a safe alternative for the storage of hydrogen [5]. These compounds have a significant advantage that makes them attractive: they all have high hydrogen contents, from about 8 to about 20 wt% [25]. Züttel and co-workers have published several papers about borohydrides as hydrogen storage materials [5, 25, 26] and particularly a viewpoint paper discussing the physical properties of $LiBH_4$, KBH_4, $NaBH_4$, $Mg(BH_4)_2$, $Ca(BH_4)_2$, $Al(BH_4)_3$ and $Be(BH_4)_2$. Filinchuk et al. [27] have reviewed the crystal structure and the structural stability of the same borohydrides. Among the various borohydrides, sodium borohydride ($NaBH_4$) is the most investigated because of the following reasons: non-flammable and stable under anhydrous medium up to 300 °C, commercially available and high hydrogen content (10.7 wt%) [28, 29].

A good hydrogen storing material must reversibly store hydrogen while it must efficiently generate hydrogen (often with the help of an accelerator). For the borohydrides, two routes to generate hydrogen are viewed: thermolysis and hydrolysis [30]. Whereas the former reaction is investigated for a good many borohydrides [26], the latter is mainly applied to sodium borohydride [28]. Unfortunately, these two hydrogen generation routes do not enable the recovering of the whole hydrogen stored (if the reactants are in stoichiometric amounts for the hydrolysis reaction [28, 31]). Furthermore, the reversibility of both reactions is not yet established [5, 28], knowing that reversibility is an important criterion for the hydrogen storage materials from an application point of view.

The search for new materials has an important place in the current research devoted to the hydrogen storage. Since the hydrogen storage capacity of the chemical hydrides is the key criterion, works focus on the synthesis of new borohydrides storing even more hydrogen. The present section reviews the borohydrides that have been suggested as potential hydrogen storage materials and specifically discusses their gravimetric hydrogen storage capacity. The relevance of these compounds in terms of safety is besides tackled.

3.1 All of the Borohydrides

All of the borohydrides that has been described in the open literature as being potential hydrogen storage materials were tested for generating hydrogen by thermolysis. Hence, the literature devoted to this reaction enables to survey exhaustively the borohydrides and this is what is done below. Few interesting results are reported in order to discuss the gravimetric efficiency of these materials.

Züttel and co-workers [26, 32] and Filinchuk et al. [27] have already listed several of these borohydrides. Precisely, they discussed the physical properties of $LiBH_4$, KBH_4, $NaBH_4$, $Mg(BH_4)_2$, $Ca(BH_4)_2$, $Al(BH_4)_3$ and $Be(BH_4)_2$ [26, 27] and the thermodynamical stabilities for $LiBH_4$, KBH_4, $NaBH_4$, $CuBH_4$, $Mg(BH_4)_2$, $Zn(BH_4)_2$, $Sc(BH_4)_3$, $Zr(BH_4)_4$ and $Hf(BH_4)_4$ [32]. It has been reported that the decomposition temperature of the borohydrides depends on the Pauling electronegativity of their cations and that the suitable Pauling electronegativity is expected to be between 1.2 and 1.5, what leaves only the Hf, Mg, Zr, Sc and U-based borohydrides, the cations having electronegativities of 1.30, 1.31, 1.33, 1.36 and 1.38 eV [33].

For both Kong et al. [34] and Au et al. [35], $LiBH_4$ is of interest because it has the largest hydrogen content (18.51 wt%) in the suitable chemical hydrides. Au et al. [35] chose to use an additive (e.g. TiO_2 or $TiCl_3$) to reduce the thermal decomposition temperature of $LiBH_4$. However, the additive has a negative impact on the gravimetric hydrogen storage capacity.

For instance, the capacity of the system $LiBH_4$-TiO_2 (75-25 wt%) [35] is of 11.10 wt% instead of 18.51 wt% for pure $LiBH_4$. The utilization of an additive ($TiCl_3$) was also envisaged by Orimo and co-workers [36] for $Mg(BH_4)_2$. For Züttel and co-workers [37] and Chłopek et al. [38], $Mg(BH_4)_2$ is the promising material with a capacity of 14.93 wt%. Chłopek et al. [38] observed that it is possible to partly rehydrogenate the material, but only up to 3.5 wt% of hydrogen because of the formation of MgH_2. Orimo and co-workers [39] investigated the partial substitution of an electronegative element like Cu for Li to obtain $Li_{0.7}Cu_{0.3}BH_4$. Cho and co-workers [40] remarked the relatively high theoretical gravimetric hydrogen storage capacity of $Ca(BH_4)_2$, namely 11.56 wt%, for which Rönnebro and Majzoub [41] have demonstrated a reversible theoretical gravimetric hydrogen storage capacity of 9.60 wt%. Züttel and co-workers [26] have reported that $Al(BH_4)_3$ (16.92 wt% of hydrogen) is liquid at room temperature (melting point of -65°C). Besides of the hydrocarbons, ammonia and water, it is the only liquid hydride that exists at room temperature. Jeon and Cho [42] synthesized $Zn(BH_4)_2$ but, because it even decomposed at room temperature, it was concluded that this complex is not a suitable material for reversible hydrogen storage for on-board applications. In 1940, Schlesinger and co-workers [43-45], leading researchers in the synthesis of borohydrides in the 1940-50s, suggested that $Be(BH_4)_2$ and $Al(BH_4)_3$ for which melting points are below 200 °C (respectively, 123 and -64.5 °C). Few years later, they attempted the synthesis of $Ga(BH_4)_3$ without being able to isolate it [46]. Besides, ten years later, Schlesinger and Brown [47] prepared $U(BH_4)_4$ and showed that it is fairly stable at temperatures below 70 °C and decomposes at temperatures above 100 °C. It is interesting to note that Prof. H.C. Brown (1912-2004) won the Nobel Prize 1979 in Chemistry [48] "for the development of the use of boron-containing compounds, into important reagents in organic synthesis" [23] as well as because he, with the late Prof. H.I. Schlesinger, is the first researcher having synthesized $NaBH_4$ [49]. Züttel and co-workers [32] were not able to investigate the decomposition of $Hf(BH_4)_4$ because its boiling temperature (118 °C) is lower than its decomposition temperature. Mixed-cation borohydrides [27] have otherwise been proposed: i.e. $LiSc(BH_4)_4$ [50] and $LiK(BH_4)_2$ [51].

More complex boron-based hydrides have been also prepared. Meismer et al. [52] synthesized a quaternary hydride with the approximate composition $Li(BH_4)_{0.33}(NH_2)_{0.67}$. Chater et al. [53] attempted to synthesize $Li_4BH_4(NH_2)_3$ and $Na_4BH_4(NH_2)_3$. The thermal decomposition of these three hydrides led to emissions of NH_3 and NH_2. Züttel and co-workers [26] have suggested, only briefly, the complex $Ca(BH_4)_2Al(BH_4)_3$, which had been first proposed by Titov and Eremin [54]. $Na_4(BH_4)(AlH_4)_3$ [55] and $Mg(BH_4)(AlH_4)$ [56] have been recently reported.

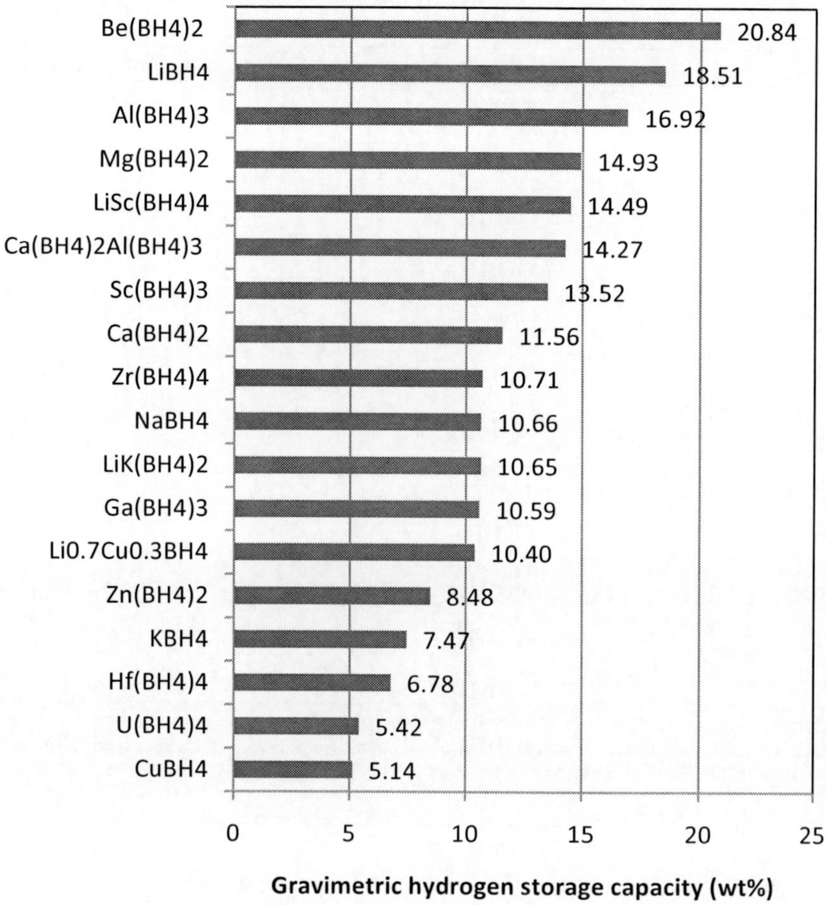

Figure 4.Theoretical gravimetric hydrogen storage capacities of borohydrides.

To summarize, Table 1 shows all of the borohydrides (single-cation and mixed-cations) that have been suggested throughout the literature as materials storing hydrogen: there are 18 compounds, the lighter being $LiBH_4$ with 27.78 g mol^{-1}, the heavier one being $U(BH_4)_4$ with 297.41 g mol^{-1}. Figure 4 compares their theoretical gravimetric hydrogen storage capacities. The capacities varies from 5.14 wt% for $CuBH_4$ to 20.84 wt% for $Be(BH_4)_2$ via 18.51 wt% for $LiBH_4$. The second observation is that the mixed-cations borohydrides (like e.g. $LiK(BH_4)_2$) do not necessarily store more hydrogen than the simple borohydrides (like e.g. $NaBH_4$).

Züttel and co-workers [57] have reported that $LiBH_4$ liberates hydrogen in different reaction steps and temperature regimes according to the following simplified reaction scheme:

$$LiBH_4 \rightarrow LiH + B + 3/2\ H_2 \qquad (1)$$

$$M(BH_4)_n \rightarrow \ldots \rightarrow \ldots \rightarrow MH + B + 3n/2\ H_2 \qquad (2)$$

Table 1. Borohydrides.

Cation	Borohydride	Molecular weight (g mol^{-1})
Li	$LiBH_4$	27.78
Be	$Be(BH_4)_2$	38.70
Na	$NaBH_4$	37.83
Mg	$Mg(BH_4)_2$	54.00
Al	$Al(BH_4)_3$	71.51
K	KBH_4	53.95
Ca	$Ca(BH_4)_2$	69.77
Sc	$Sc(BH_4)_3$	89.49
Cu	$CuBH_4$	78.38
Zn	$Zn(BH_4)_2$	95.06
Ga	$Ga(BH_4)_3$	114.25
Zr	$Zr(BH_4)_4$	150.59
Hf	$Hf(BH_4)_4$	237.86
U	$U(BH_4)_4$	297.41
Li - Cu	$Li_{0.7}Cu_{0.3}BH_4$	38.76
Li - K	$LiK(BH_4)_2$	75.73
Li - Sc	$LiSc(BH_4)_4$	111.27
Al - Ca	$Ca(BH_4)_2Al(BH_4)_3$	141.28

3.2 Thermolysis of Borohydrides

Part of the 4.63 wt% of the hydrogen remains as LiH in the decomposition product and the effective gravimetric hydrogen storage capacity of $LiBH_4$ is then 13.88 wt% [29]. If the reaction is applied to some of the borohydrides reported by Züttel and co-workers in one of their last papers, i.e. $LiBH_4$, KBH_4, $NaBH_4$, $Mg(BH_4)_2$, and $Ca(BH_4)_2$ [26], one can remark that the effective gravimetric hydrogen storage capacity is less attracting than the theoretical one since lowered of a quarter (Figure 5). Apart from KBH_4, the effective values would meet the target set by the U.S. Department of Energy for 2010, namely of 6.00 wt%. In fact, the thermolysis is a quite complex process and experimental evidence is required for each compound.

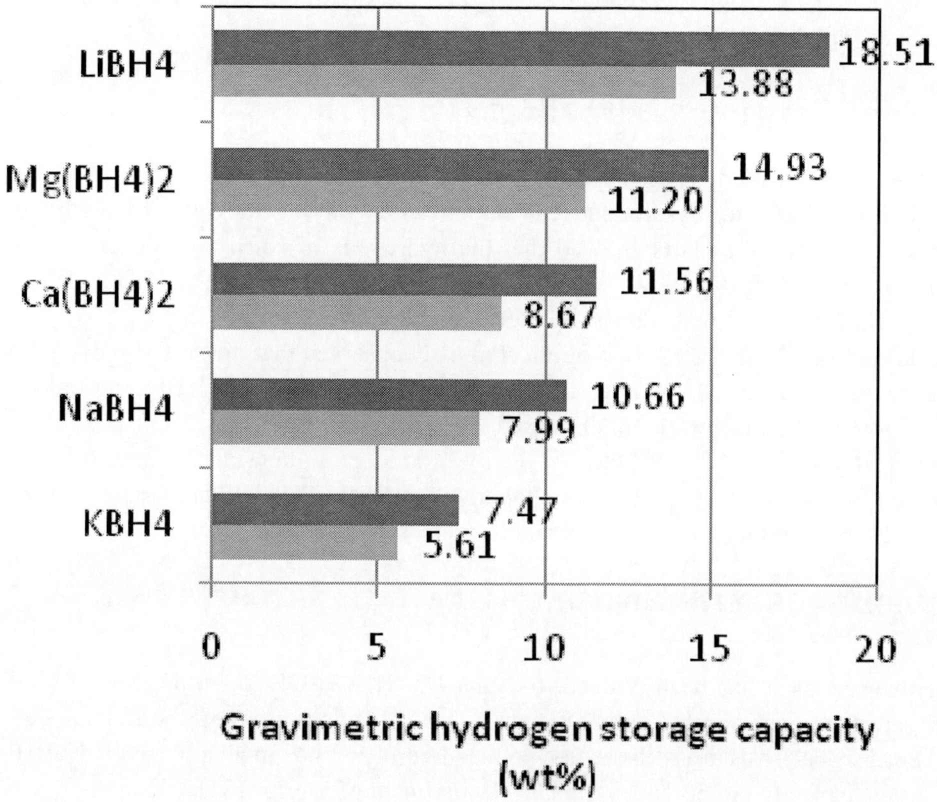

Figure 5. Gravimetric hydrogen storage capacities of some borohydrides if incomplete thermolysis (in light grey; in dark grey the theoretical gravimetric hydrogen storage capacity).

3.3 Hydrolysis of Borohydrides

The hydrolysis is also a possibility for recovering hydrogen from borohydrides. It has the advantage to be realized at low temperatures (below 80 °C, even at room temperature) [9, 58]. It is widely studied for the most common borohydride, $NaBH_4$ [29]. Although few or no study about the hydrolysis of the other borohydrides has been reported in the open literature, the hydrolysis might be envisaged for some of them.

The hydrolysis of sodium borohydride is the subject of an increasing interest since Amendola et al.'s works published in 2000 [11] even though sodium borohydride has been known as a viable hydrogen generator since the 1950s [49]. The sodium borohydride hydrolysis is a follows [58]:

$$NaBH_4 + 2\ H_2O \rightarrow NaBO_2 + 4\ H_2 \qquad (3)$$

Half of hydrogen being provided by water (11.19 wt% of hydrogen stored). The gravimetric hydrogen storage capacity of the system borohydride-water is thus 10.92 wt%, what meets the targets set by the U.S. Department of Energy. However, the reaction by-product $NaBO_2$, sodium metaborate, is usually hydrated because there are the thermodynamically stable compounds. Hence, the hydrolysis involves more water in accordance with the following reaction:

$$NaBH_4 + (2 + x)\ H_2O \rightarrow NaBO_2.xH_2O + 4\ H_2 \quad (4)$$

The term x is called "excess hydration factor" by Marrero-Alfonso et al. [59]. It accounts for the fact that the by-product can exist with varying degrees of hydration (i.e. $NaBO_2.xH_2O$). These authors showed that the hydrolysis of $NaBH_4$ produces $NaBO_2.2H_2O$ (i.e. $x = 2$). In another paper [60], it has been remarked the prevalence of $NaBO_2.2H_2O$ (rather than e.g. $NaBO_2.4H_2O$) that is the stable form at temperatures up to 105 °C.

For the hydrolysis of $LiBH_4$, Kojima et al. [61] observed that, for $H_2O/LiBH_4 = 2$, the by-product was $LiBO_2.2H_2O$. Şahin et al. [62] studied the hydrolysis of KBH_4 but unfortunately the reaction by-products were not discussed.

Interestingly, Lo et al. [31] tackled the by-product hydration-related issue differently. To avoid the formation of the hydrated metaborates, methanol, which has the highest reactivity towards $NaBH_4$ among the primary alcohols, was used instead of water:

$$NaBH_4 + 4\ CH_3OH \rightarrow NaB(OCH_3)_3 + 4\ H_2 \quad (5)$$

However, as it has been noticed, the fuel $NaBH_4$-CH_3OH has a theoretical gravimetric hydrogen storage capacity of only 4.86 wt%. This value does not meet the 2010 target of 6.00 wt% but as that is discussed hereafter neither hydrolysis nor methanolysis of $NaBH_4$ meets the storage capacity targets set by the U.S. Department of Energy [31].

Here, it is important to note two things. First, Kong et al. [34] thought about the suitability of hydrides as candidate hydrolysis hydrides. Using several criteria (e.g. maximum deliverability hydrogen, ambient state, storability, controllability of the reaction, recyclability and economic feasibility), they concluded that the borohydrides are unsuitable because of potential problems in reaction controllability, manufacturing and recycling or disposal. Second, the aqueous solution of $NaBH_4$ is generally stabilized with NaOH in order to avoid its spontaneous hydrolysis (which occurs in the absence of catalyst) [28] and the generation of hydrogen from the NaOH-stabilized fuel requires then a metal catalyst [11].

In Figure 4, one can remark that, except $U(BH_4)_4$ and $CuBH_4$, the borohydrides have theoretical gravimetric hydrogen storage capacities that reach the 2010 target of 6.00 wt%. The capacities of $Zn(BH_4)_2$, KBH_4, and $Hf(BH_4)_4$ are below the 2015 target of 9.00 wt%. The data showed in Figure 4 are theoretical and it is, thus, important to discuss the effective gravimetric hydrogen storage capacities. $Ga(BH_4)_3$ is not regarded anymore here because it has never been isolated [46]. Even if the application of most of the borohydrides discussed above is unlikely because of an excessive reactivity in presence of water (e.g. $Al(BH_4)_3$ and $Mg(BH_4)_2$) or because of the nature of the cation (e.g. $U(BH_4)_4$), they are discussed hereafter for information only. The preliminary point to tackle is the gravimetric hydrogen storage capacity when the amount of water is stoichiometric. Figure 6 compares these values to the theoretical ones. For $Ca(BH_4)_2$, $Zr(BH_4)_4$, $NaBH_4$, $LiK(BH_4)_2$ and $Li_{0.7}Cu_{0.3}BH_4$, both values are quite similar. For $Zn(BH_4)_2$, KBH_4, $Hf(BH_4)_4$, $U(BH_4)_4$ and $CuBH_4$, the presence of water improves slightly the gravimetric hydrogen storage capacity. Henceforth, $Zn(BH_4)_2$ and KBH_4 meet the 2015 target of 9.00 wt% and $Hf(BH_4)_4$, $U(BH_4)_4$ and $CuBH_4$ meet the 2010 target of 6.00 wt%. For the other borohydrides, the capacities decrease of 1 to 6 wt% while they are still above 9.00 wt%. The most important decreases (6.28 and 4.46 wt%) are observed for $Be(BH_4)_2$ and $LiBH_4$.

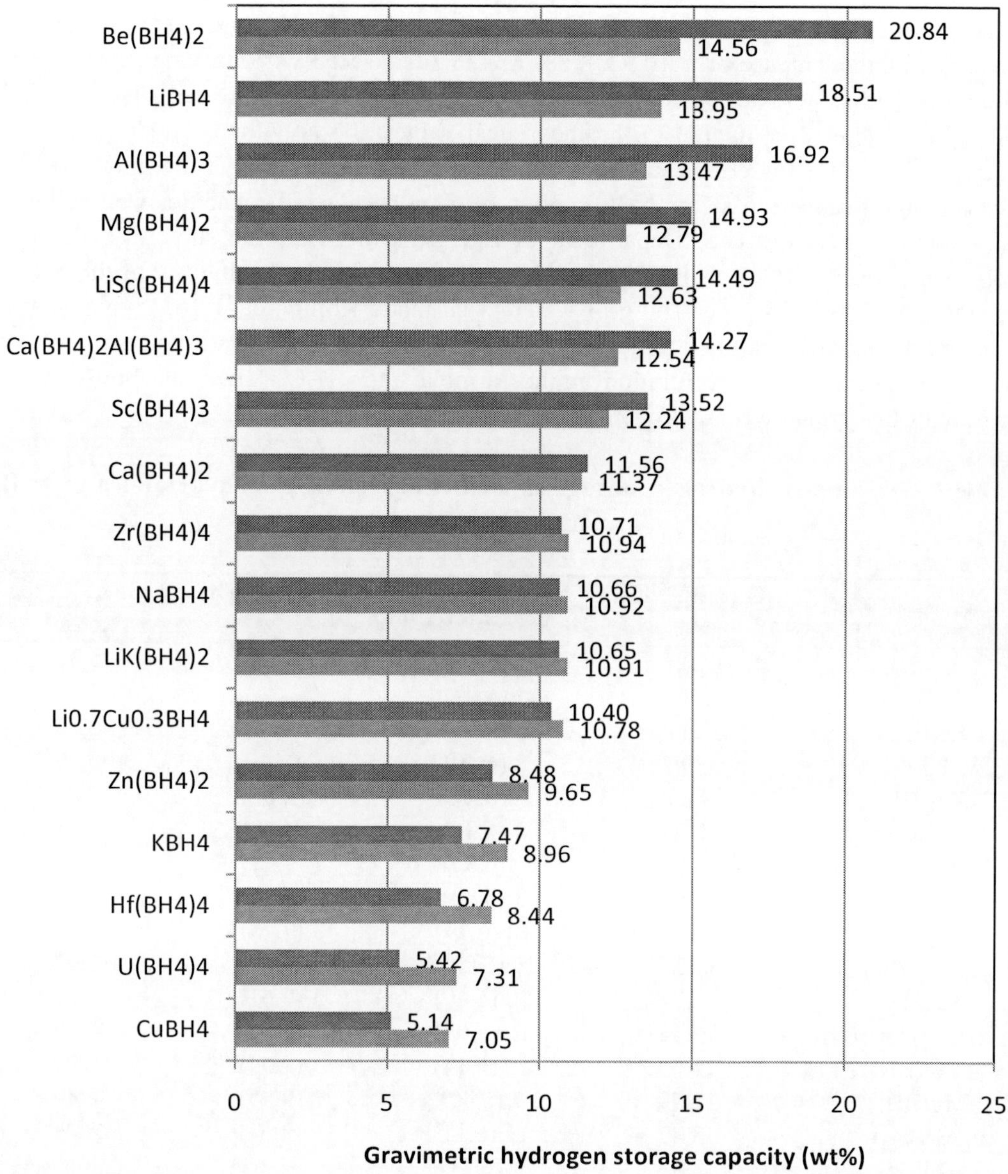

Figure 6. Gravimetric hydrogen storage capacities of systems borohydride-water, with water in stoichiometric amount for hydrolysis (in light grey; in dark grey the theoretical gravimetric hydrogen storage capacity).

However, the reaction by-product $NaBO_2$ is hydrated [59, 60], which reduces the gravimetric hydrogen storage capacity of $NaBH_4$. This is this empirical observation (i.e. excess hydration factor x) that sustains the follow-up to the present discussion. Values of x

from 0 (water in stoichiometric amount) to 10 (excess of 10 molecules of water) are considered for the reaction:

$$M(BH_4)_n + (2n+x)\ H_2O \rightarrow M(BO_2)_n.xH_2O + 4n\ H_2 \text{ (M being the cation)} \quad (6)$$

Let us begin with sodium borohydride. Its gravimetric hydrogen storage capacity decreases with the increase of x: 10.92, 8.78, 7.34, 6.30 and 5.53 wt% for $x = 0$, 1, 2, 3 and 4. If the molar ratio $H_2O/NaBH_4$ is of 6 (excess of $x = 4$ molecules of water), the capacity is divided by a factor 2 compared to the theoretical value. The gravimetric hydrogen storage capacity is 3.17 wt% for $x = 10$, that implies that using sodium borohydride in an excess of water is insignificant from an application point of view because of capacities well below the specifications set by the U.S. Department of Energy. Otherwise, the solubility of sodium borohydride in water at 25 °C is 55 g per 100 g of water while the solubility of the reaction by-product $NaBO_2$ is 28 g per 100 g of water, what made Kojima et al. [63] report that the concentration of $NaBH_4$ at 25 °C should be below 16 g per 100 g of water to keep the liquid state of $NaBO_2$. Such a concentration implies a molar ratio $H_2O/NaBH_4$ of about 13, what corresponds to a gravimetric hydrogen storage capacity of 2.62 wt%.

Table 2. Gravimetric hydrogen storage capacities of systems borohydride-water, with water in excess (x).

$M(BH_4)_n$ [a]	G.H.S.C.[b] for x=0	x for 6.00 wt% [c]	x for 9.00 wt% [d]	G.H.S.C. for x=2n [e]
$CuBH_4$	7.05	1 (6.11)	-	5.36
$U(BH_4)_4$	7.31	5 (6.07)	-	5.51
$Hf(BH_4)_4$	8.44	8 (6.13)	-	6.13
KBH_4	8.96	2 (6.40)	-	6.40
$Zn(BH_4)_2$	9.65	5 (6.27)	0 (9.65)	6.74
$Li_{0.7}Cu_{0.3}BH_4$	10.78	3 (6.26)	0 (10.78)	7.28
$LiK(BH_4)_2$	10.91	6 (6.30)	1 (9.73)	7.34
$NaBH_4$	10.92	3 (6.30)	0 (10.92)	7.34
$Zr(BH_4)_4$	10.94	10 (6.79)	3 (9.25)	7.35
$Ca(BH_4)_2$	11.37	7 (6.02)	2 (9.07)	7.54
$Sc(BH_4)_3$	12.24	10 (6.40)	3 (9.61)	7.91
$Ca(BH_4)_2Al(BH_4)_3$	12.54	19 (6.07)	7 (9.01)	8.04
$LiSc(BH_4)_4$	12.63	15 (6.14)	5 (9.34)	8.07
$Mg(BH_4)_2$	12.79	7 (6.40)	2 (9.95)	8.14
$Al(BH_4)_3$	13.47	12 (6.11)	4 (9.61)	8.41
$LiBH_4$	13.95	4 (6.21)	1 (10.63)	8.59
$Be(BH_4)_2$	14.56	8 (6.33)	3 (9.79)	8.82

[a] $M(BH_4)_n$ with M the cation and n from 1 to 4,
[b] G.H.S.C. as gravimetric hydrogen storage capacity
[c] Lowest value of x to have G.H.S.C. higher than 6.00 wt% (2010 target set by the U.S. Department of Energy) and in brackets the exact G.H.S.C.
[d] Lowest value of x to have G.H.S.C. higher than 9.00 wt% (2015 target set by the U.S. Department of Energy) and in brackets the exact G.H.S.C.
[e] n of $M(BH_4)_n$

All of these calculated data are at variance with the values claimed for $NaBH_4$. In other words, $NaBH_4$ does not appear to be a suitable hydrogen storage material if used in excess of water. Let us continue with the other borohydrides. Table 2 reports the lowest x values in order to have gravimetric hydrogen storage capacities higher than the targets of the U.S. Department of Energy. The 2010 target of 6.00 wt% is met by all of the borohydrides despite water excesses of 1 to 19 (respectively for $CuBH_4$ and $Ca(BH_4)_2Al(BH_4)_3$). The 2015 target of 9.00 wt% is met by $LiK(BH_4)_2$, $Zr(BH_4)_4$, $Ca(BH_4)_2$, $Sc(BH_4)_3$, $Ca(BH_4)_2Al(BH_4)_3$, $LiSc(BH_4)_4$, $Mg(BH_4)_2$, $Al(BH_4)_3$, $LiBH_4$ and $Be(BH_4)_2$ with x from 1 to 7. For $Zn(BH_4)_2$, $Li_{0.7}Cu_{0.3}BH_4$ and $NaBH_4$, any excess of water hinders these compounds from meeting the criterion 9.00 wt%. The best candidate would be $Ca(BH_4)_2Al(BH_4)_3$. It stores 9.01 wt% of hydrogen with a water excess of 7 molecules. The second information given by Table 2 is the gravimetric hydrogen storage capacity reachable if each BO_2 is dihydrated ($x = 2n$). Except $CuBH_4$ and $U(BH_4)_4$, the borohydrides meet the 2010 target of 6.00 wt%. However, none meets the 2015 target of 9.00 wt%. The best candidate would be $Be(BH_4)_2$ with a gravimetric hydrogen storage of 8.82 wt% (with $x = 4$).

Some authors envisaged an alternative to the aqueous solution of the borohydrides. In a fuel cell system, the water produced by the fuel cell can be redirected to the borohydride and then the gravimetric hydrogen storage capacity can increase by a factor 2 in relation to the theoretical value. Kojima et al. [61] and Şahin et al. [62] have reported values of 37.02 and 14.94 wt% for $LiBH_4$ and KBH_4, respectively. Kong et al. [34] have reported the maximum deliverable hydrogen for several ionic hydrides and especially for KBH_4, $NaBH_4$, $Ca(BH_4)_2$ and $LiBH_4$ (14.94, 21.32, 23.12 and 37.02 wt%). Figure 7 shows these values, i.e. the gravimetric hydrogen storage capacities of the borohydrides if water produced in a fuel cell is recycled. This way to view the hydrogen storage shows great improvements in the gravimetric hydrogen storage capacities. The theoretical capacities vary from 10.28 wt% for $CaBH_4$ to 41.68 wt% for $Be(BH_4)_2$.

The values given in Figure 7 do not take into account the restriction due to the mass of the reaction by-products. For example, the molecular weight of $NaBH_4$ is of 37.83 g mol^{-1} while those of $NaBO_2$ and $NaBO_2.2H_2O$ are of 65.80 and 101.83 g mol^{-1} respectively (molecular weight of BH_4, BO_2 and $BO_2.2H_2O$ of 14.84, 42.81, 78.84 g mol^{-1}). In a first approximation, one may consider the density of $NaBH_4$ as being similar to that of either $NaBO_2$ or $NaBO_2.2H_2O$. For a given volume of tank, the maximal mass of the filled tank should be driven by the mass of the by-products. The tank should then be either 58%- or 37%-filled with $NaBH_4$, what decreases accordingly the gravimetric hydrogen storage capacity of $NaBH_4$ given in Figure 7 to either 12.15 or 7.89 wt%. In accordance with this argument, Figure 8 shows the gravimetric hydrogen storage capacities of the borohydrides when considering $NaBO_2$ as the by-product and taking into account its mass in the storage system. Except $CuBH_4$, $U(BH_4)_4$, $Hf(BH_4)_4$ and KBH_4, the borohydrides meet the 2015 target of 9.00 wt%. Figure 9 shows the gravimetric hydrogen storage capacities of the borohydrides when considering $NaBO_2.2H_2O$ as the by-product and taking into account its mass in the storage system. Only seven of the borohydrides meet the 2015 target of 9.00 wt% while, except $CuBH_4$,$U(BH_4)_4$, $Hf(BH_4)_4$ and KBH_4, all meet the 2010 target of 6.00 wt%. This analysis is interesting on two accounts. First, it puts the announced capacities (Figure 7) into perspective. Second, it shows that this alternative method is rather promising. Evidently, this method is worth developing.

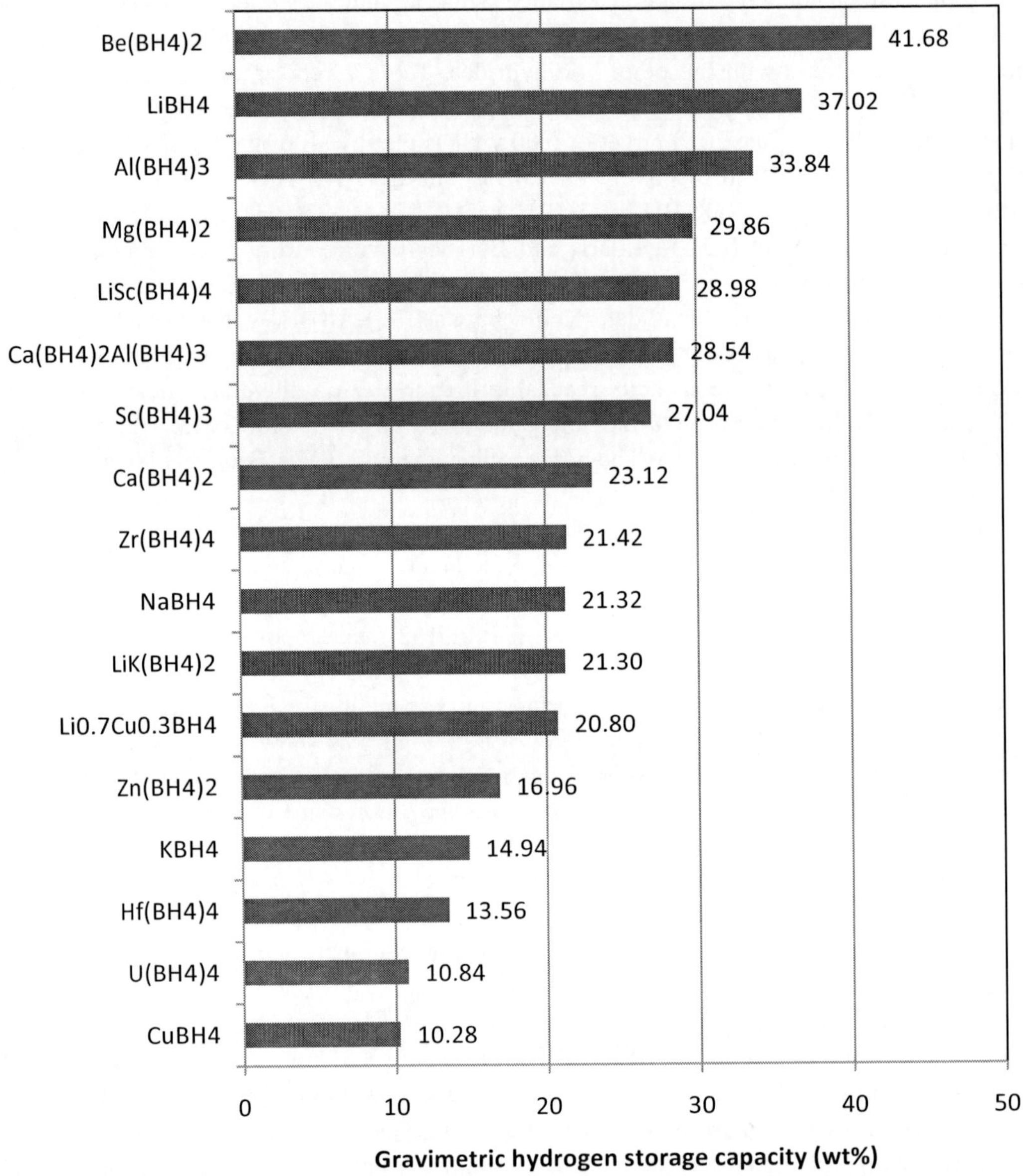

Figure 7. Gravimetric hydrogen storage capacities of borohydrides with water coming from a fuel cell.

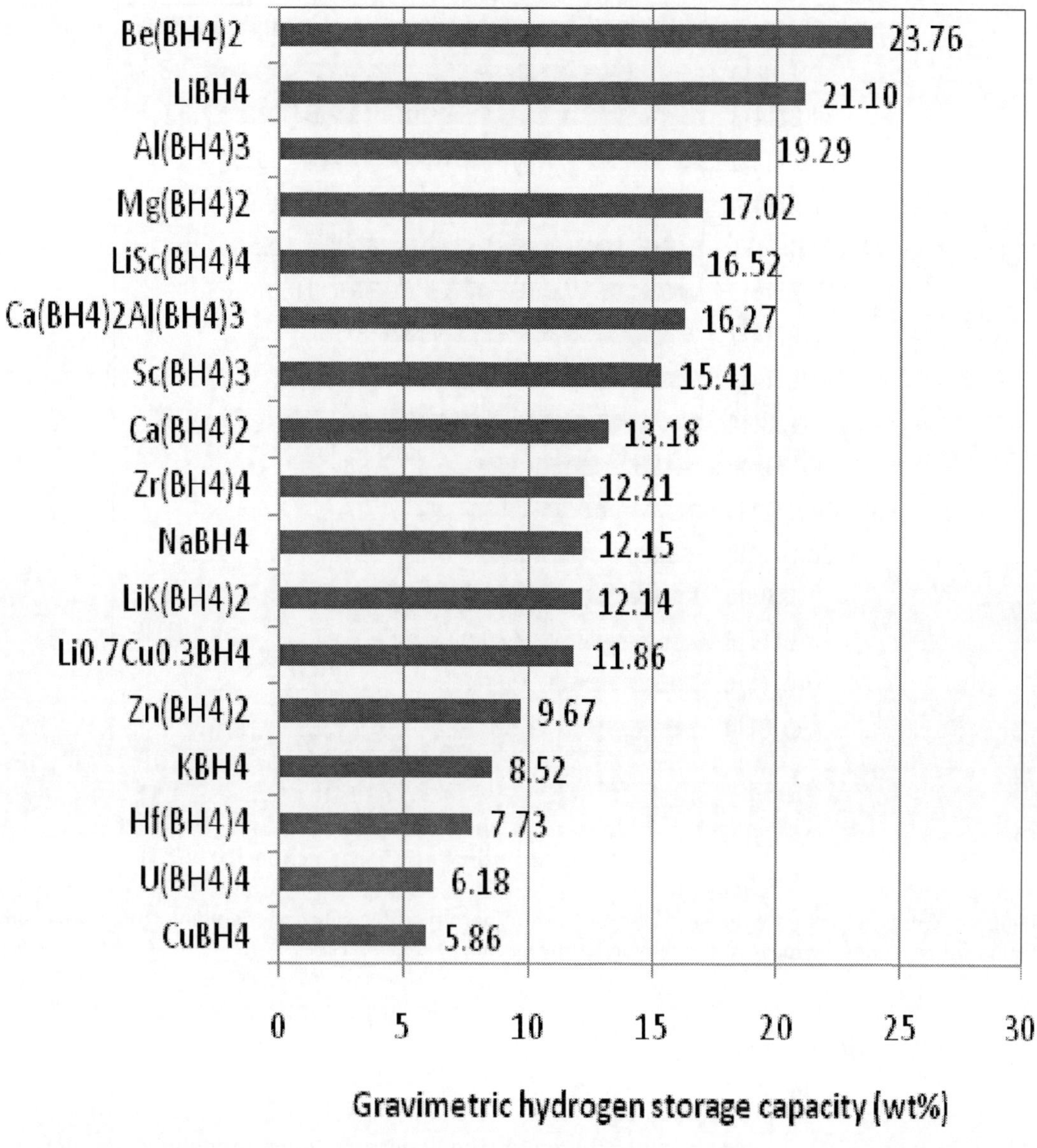

Figure 8. Gravimetric hydrogen storage capacities of borohydrides with water coming from a fuel cell, taking into account the mass of the anhydrous by-product NaBO2.

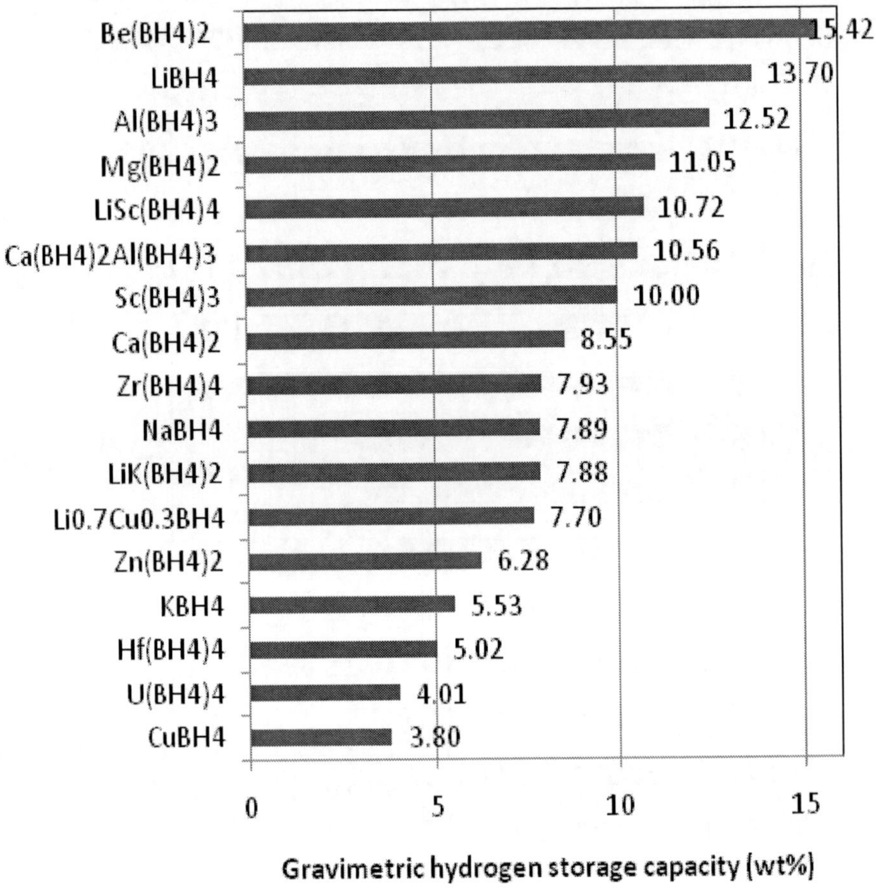

Figure 9. Gravimetric hydrogen storage capacities of borohydrides with water coming from a fuel cell, taking into account the mass of the dihydrated by-product NaBO2.2H2O.

3.4 Safety Data

The subject of the present sub-section is to discuss the safety of the borohydrides (when this information is available) because there are some contradictions throughout the literature, i.e. the open scientific literature and the material safety data sheets from the chemicals suppliers.

Let us begin with sodium borohydride. There is unanimity about its safety as e.g. in the references [11, 28]:

- The sodium borohydride solutions would be non-flammable, stable in air, easy-to-handle and non-toxic.
- Its reaction products, H_2 and $NaBO_2$, would be environmentally safe.

However, the material safety data sheets point out that $NaBH_4$ is not so safe, listing the following risks: toxic, causing severe burns, polluting, highly flammable, corrosive, and liberating toxic and highly flammable gas. A risk scale indicates that $NaBH_4$ has 3 for health, 4 for fire and 2 for reactivity (given the following risk scale: 0 for no risk and 4 for extremely dangerous). From that moment, one may wonder about the alleged safety of $NaBH_4$. For Kong et al. [34], $NaBH_4$ is potentially hazardous.

Let us see the other borohydrides. For lithium borohydride, the hazards identification gives 3 for health, 3 for fire and 2 for reactivity. Lithium borohydride is very irritant, toxic, corrosive and flammable. Its dissolution in water is very exothermic, what favors the hydrolysis and so the production of hydrogen. According to Schlesinger and co-workers [45], $LiBH_4$ is not flammable and is not affected by dry air. In the case of KBH_4, which is irritant and flammable, the hazards identification gives 2 for health, 3 for fire and 2 for reactivity. To the best of our knowledge, no material safety data sheet is available for the other borohydrides given in Table 1. Through the open literature, some safety information can be found. For example, Schlesinger and co-workers, who worked a lot on different borohydrides, have reported the following information:

- $Be(BH_4)_2$ inflames violently in air and reacts vigorously with water [43].
- $Al(BH_4)_3$ reacts readily with air, inflames spontaneously giving a blue-white flash of unusual brilliance and is rapidly hydrolyzed at ordinary temperatures to give hydrogen [44].

The two examples above show that the borohydrides are not benign compounds and that their safety has to be carefully considered.

It is quite difficult to conclude this section because of the discrepancy of safety information. One may recall that Kong et al. [34] have discussed the suitability of hydrides as candidate hydrolysis hydrides and, using, among other criteria, the safety criterion, concluded that the borohydrides are unsuitable because of potential problems in e.g. reaction controllability. Maybe a less harsh conclusion should be given:

- First, it is obvious that on the basis of the sole safety criterion all of the borohydrides listed here should not be viewed as potential hydrides for storing hydrogen.
- Second, the new borohydrides like $Ca(BH_4)_2$ should be investigated in order to identify any hazards.
- Third, the borohydrides that are NH_x-releasing like $Li_4BH_4(NH_2)_3$, highly reactive like $Be(BH_4)_2$ and "exotic"[5] like $U(BH_4)_4$ should perhaps be avoided.
- Fourth, in the current situation, $NaBH_4$ and $LiBH_4$, despite identified hazards, are the borohydrides that should be reasonably developed because their features are rather well known.

3.5 Conclusion

The borohydrides, 18 surveyed in the present section, are recognized as being promising hydrogen storage materials. This is particularly true for the most studied compound sodium borohydride which theoretical gravimetric hydrogen storage capacity of 10.66 wt% is well above the 2015 target of 9.00 wt% set by the U.S. Department of Energy. However, a

[5] By 'exotic', the authors would like to emphasize the irrelevancy of atoms such as U in the synthesis of new borohydride compounds.

theoretical gravimetric hydrogen storage capacity, because it is theoretical, is an optimistic view and overestimates the effective capacity of the hydride.

4. Ammoniaborane and its Derivatives

4.1 Ammoniaborane

Ammoniaborane (NH_3BH_3) contains 19.6 wt% hydrogen, which is higher than the capacity of lithium borohydride, and much higher than that of sodium borohydride [8, 24, 30, 58, 64]. Synthesized in the mid-1950s [29], ammoniaborane is under investigations since several years but it has attracted a great attention in 2007 when the U.S. Department of Energy recommended a no-go for sodium borohydride as hydrogen storage material for automotive applications [13]. This recommendation has been besides accompanied by a focus on ammoniaborane.

Ammoniaborane is a colorless tetragonal crystal at room temperature [65]. Its solid state structure exhibits short BH···HN intermolecular contacts due to an interaction constituting a dihydrogen bond [29, 66]. Nevertheless, ammoniaborane can be destabilized and then part of stored of hydrogen can be recovered. As for sodium borohydride, the hydrogen release from ammoniaborane can be carried out via either thermolysis or hydrolysis [24, 58]. However, ammoniaborane [67] does not reversibly store hydrogen.

Through the thermal decomposition (thermolysis) of ammoniaborane, 2.2 equivalent of hydrogen can be released up to 200 °C [29, 64]. Actually the thermal decomposition involves several reactions. Net decomposition initiates at about 70 °C and reaches a maximum at about 110 °C [68]. About 1 mol(H_2) per mol(AB) is thus released, what implies a capacity of 6.5 wt%. The second mol(H_2) per mol(AB) is released up to 200 °C [64] (capacity of 13.0 wt%). The third hydrogen can be recovered at temperatures up to 500 °C [67]. At such temperatures, boron nitride forms but this is an unpromising material for the regeneration step needed for any storage/release cycle [8]. Solutions have been proposed for improving both effective storage capacities and hydrogen releasing rates [8, 64, 69]: for example, thermolysis in ionic liquid or organic solvent, insertion into a micro-/meso-porous material, and metal-based materials catalysis.

Ammoniaborane is water-soluble (33.6 g per 100 g(H_2O) [69]) and its hydrolysis can then be considered [29, 64]:

$$NH_3BH_3 + 2\ H_2O \rightarrow BO_2^- + NH_4^+ + 3\ H_2 \quad (7)$$

The low solubility of ammoniaborane implies a gravimetric hydrogen storage capacity of only 9.8 wt% [69]. This value is further decreased because the maximum H_2/ammoniaborane ratio is 3 (although a maximum of 6 would theoretically possible) because of the formation of NH_4^+. This means therefore a capacity of 4.9 wt%. This last value is even below that of highly concentrated aqueous solution of sodium borohydride.

4.2 Amidoboranes

Both high-temperature hydrogen release and irreversibility are significant drawbacks for both borohydrides and ammoniaborane. Hamilton et al. [24] have then remarked that research is underway to discover new B–N hybrid compounds that are capable of on-board reversible hydrogen storage. A promising example is metal amidoborane (MNH_2BH_3). The metal M is used to potentially decrease the enthalpy of hydrogen loss, to eliminate formation of volatile co-products (hydrogen purity), and to help control of hydrogen release. Several metal amidoboranes have been synthesized and assessed so far:

- Lithium amidoborane ($LiNH_2BH_3$), with a capacity of 13.6 wt % [70, 71]; 10.9 wt% evolved at about 90 °C [71].
- Sodium amidoborane ($NaNH_2BH_3$), with a capacity of 9.5 wt%; 7.5 wt% evolved at about 90 °C [71, 72].
- Calcium amidoborane ($Ca(NH_2BH_3)_2$), with a capacity of 10.0 wt% [70].

According to Hamilton et al. [24], the range of metals used in metal amidoborane complexes can be expanded. Xiong et al. [71] have confirmed and have added that the anion $[NH_2BH_3]^-$ is likely to bond with other metal or metalloid elements, which will not only open up significant opportunities for hydrogen storage materials development, but will also bring considerable insights to B–N chemistry.

Amidoboranes are actually new compounds in research devoted to hydrogen storage via complex hydrides. Hence, research is at the early beginning.

4.3 Safety Data

An important aspect for chemicals intended to large-scale utilizations is safety. The few existing safety reports (e.g. material safety data sheets) about ammoniaborane are incomplete (see *e.g.* reference [73]). Lane [74] has discussed safety aspects relating to ammoniaborane. Ammoniaborane is a white solid. It is recognized as being safe and stable at room temperature if stored in closed vessel and anhydrous medium. It is moisture sensitive, hydrolyzing and generating hydrogen. Aqueous solutions of ammoniaborane are highly stable under inert atmosphere [29], but undergo slow hydrolysis reactions in air, which is presumably associated with the catalytic effect of carbon dioxide [75]. Oxidizing agents and acids must be avoided when handling it. For being stable and safe, ammoniaborane has to be of high purity. Ammoniaborane decomposes at very low temperatures. This feature has a significant attractive consequence when hydrogen releasing is conducted thermally.

Regarding the newly synthesized alkali-metal amidoboranes, very few safety information is available throughout the open literature. One may however cite Xiong et al. [71]. These authors have reported that the alkali-metal amidoboranes materials are environmentally harmless, non-flammable, non-explosive and stable at room temperature and pressure. However that may be, we will be more cautious about the safety of such new materials. Evidence is essential and no relative action has been initiated yet.

Conclusion

Among the great variety of boron compounds, boron hydrides are attractive hydrogen-rich materials that have a significant potential as hydrogen storage materials. This is especially true for some of the borohydrides (for example, sodium borohydride) and the B,N-containing boranes (namely, ammoniaborane and amidoboranes). Note that all of the boron hydrides do not reversibly store hydrogen.

The borohydrides are often cited as materials offering a safe alternative for the storage of hydrogen. Single- or mixed-cation borohydrides (e.g. $NaBH_4$ and $LiK(BH_4)_2$, respectively) have been synthesized. They all have high hydrogen contents, from about 8 to about 20 wt%, most of them surpassing the targets set by the U.S. Department of Energy (namely, 6.00 wt% for 2010 and 9.0 wt% for 2015). However, the gravimetric hydrogen storage capacity that is often given is the theoretical one, and such value is an optimistic view and overestimates the effective capacity of the hydride. Stored hydrogen can be generated by either thermolysis or hydrolysis, but the thermolysis is a quite complex process because of the stability of the borohydrides.

Ammoniaborane (NH_3BH_3) contains 19.6 wt% of hydrogen. This capacity is higher than the capacity of lithium borohydride (18.5 wt%), and is much higher than that of sodium borohydride (10.6 wt%). Stored hydrogen can be released by either thermolysis or hydrolysis but the former reaction is more appropriate to ammoniaborane. Indeed, in the case of the latter reaction, the low solubility and the incomplete dehydrogenation of ammoniaborane imply a gravimetric hydrogen storage capacity of only 4.9 wt%.

Amidoboranes are actually new compounds in the research devoted to hydrogen storage via complex hydrides. They are derivatives of ammoniaborane. For example, lithium can replace one hydrogen atom and then a new amidoborane is obtained: namely, lithium amidoborane ($LiNH_2BH_3$), with a capacity of 13.6 wt %. The range of metals used in metal amidoborane complexes can be expanded. Research devoted to such new compounds is at the early beginning.

The safety of the boron hydrides has been briefly tackled. The first observation has been that there is a discrepancy in safety information, or even no available information for some of them. However that may be, most of the boron hydrides are not as safe as it is often noticed (for example, for sodium borohydride). It is therefore important to assess the risks due to the utilization of such compounds as well as to assess their effective gravimetric hydrogen storage capacities.

References

[1] Adams, W. M. *The future of sustainability: re-thinking environment and development in the twenty-first century*, Publication of IUCN The World Conservation Union. http://cmsdata.iucn.org/downloads/iucn_future_of_sustainability.pdf.

[2] Scharlemann, J. P. W.; Laurance, W. F. *Science* 2008, *319*, 43-44.

[3] Sartbaeva, A.; Kuznetsov, V. L.; Wells, S. A.; Edwards, P. P. *Energy Environ. Sci.* 2008, *1*, 79-85.

[4] van den Berg, A. W. C.; Areán, C. O. *Chem. Commun.* 2008, 668-681.

[5] Schlapbach, L.; Züttel, A. *Nature* 2001, *414*, 353-358.

[6] Principi, G.; Agresti, F.; Magdalena, A.; Russo, S. L. *Energy* 2008, doi:10.1016/j.energy.2008.08.027.
[7] Guo, Z. X.; Shang, C.; Aguey-Zinsou, K. F. *J. Eur. Ceram. Soc.* 2008, *28*, 1467-1473.
[8] Crabtree, R. H. *Energy Environ. Sci.* 2008, *1*, 134-138.
[9] Fakioğlu, E.; Yürüm, Y.; Veriroğlu, T. N. *Int. J. Hydrogen Energy* 2004, *29*, 1371-1376.
[10] Demirci U. B.; Miele, P. *C. R. Chim.* 2008, doi:10.1016/j.crci.2008.08.002.
[11] Amendola, S. C.; Sharp-Goldman, S. L.; Janjua, M. S.; Spencer, N. C.; Kelly, M. T.; Petillo, P. J.; Binder, M. *Int. J. Hydrogen Energy* 2000, *25*, 969-975.
[12] Çakanyildirim, Ç.; Gürü, M. *Int. J. Hydrogen Energy* 2008, *33*, 4634-4639.
[13] U.S. Department of Energy Hydrogen Program. (2007). *Independent review. November 2007. Go/no-go recommendation for sodium borohydride for on-board vehicular hydrogen storage*. NREL/MP-150-42220. http://www.hydrogen.energy.gov/.
[14] Demirci U. B.; Akdim, O.; Miele, P. *Int J Hydrogen Energy* 2009, doi:10.1016/j.ijhydene.2009.01.038.
[15] Grimes, R.N. In *Advanced Inorganic Chemistry, Sixth Edition*; Cotton, F. A.; Wilkinson, G.; Murillo, C. A.; Bochmann, M.; John Wiley and Sons, Inc.; Wiley-Interscience: 1999; pp. 131-174.
[16] Greenwood, N. N. In *Comprehensive Inorganic Chemistry*; Bailar, J. C.; Emeléus, H. J.; Nyholm, Sir R.; Trotman-Dickenson, A. F.; Ed.; Pergamon Press, Oxford: 1973; Vol. 1, pp. 665-991.
[17] Lipscomb, W. N. *Chemistry* 1976, 224-245.
[18] Lipscomb, W. N. *J. Inorg. Nucl. Chem.* 1959, *11*, 1-8.
[19] Schaeffer, R.; Moody, D. C.; Dolan, P. J. *Pure Appl. Chem.* 1974, *39*, 423-430.
[20] Beall, H.; Gaines, D.F. *Inorg. Chim. Acta* 1999, *289*, 1-10.
[21] Johnson, A. L.; Kresiński, R. A.; López, C. A. *Annu. Rep. Prog. Chem. Sect. A* 2007, *103*, 54-89.
[22] Johnson, A. L. *Annu. Rep. Prog. Chem. Sect. A* 2008, *104*, 81-96.
[23] Brown, H. C. (1979). *From little acorns to tall aoks-from boranes through organoboranes*. http://www.chem.purdue.edu/hcbrown.
[24] Hamilton, C. W.; Baker, R. T.; Staubitz, A.; Manners, I. *Chem. Soc. Rev.* 2009, 38, 279-293.
[25] Züttel, A. *Mitig. Adapt. Strat. Glob. Change* 2007, *12*, 343-365.
[26] Züttel, A.; Borgschulte, A.; Orimo, S.I. *Scripta Mater.* 2007, *56*, 823-828.
[27] Filinchuk, Y.; Chernyshov, D.; Dmitriev, V. *Z. Kristallogr.* 2008, *223*, 649-659.
[28] Wee, J. H.; Lee, K. Y.; Kim, S. H. *Fuel Process. Technol.* 2006, *87*, 811-819.
[29] Wang, P.; Kang, X. D. *Dalton Trans.* 2008, 5400-5413.
[30] Demirci U. B.; Miele, P. *Environ. Int.* 2009, *35*, 626-631.
[31] Lo, C. T. F.; Karan, K.; Davis, B. R. *Ind. Eng. Chem. Res.* 2007, *46*, 5478-5484.
[32] Nakamori, Y.; Miwa, K.; Ninomiya, A.; Li, H.; Ohba, N.; Towata, S.I.; Züttel, A.; Orimo, S.I. *Phys. Rev. B* 2006, *74*, 045126-045129.
[33] Pauling, L. *The nature of the chemical bond*; 3rd Ed.; Cornell UP: Ithaca, 1960.
[34] Kong, V. C. Y.; Foulkes, F. R.; Kirk, D. W.; Hinatsu, J. T. *Int. J. Hydrogen Energy* 1999, *24*, 665-675.
[35] Au, M.; Spencer, W.; Jurgensen, A.; Zeigler, C. *J. Alloys Compd.* 2008, *462*, 303-309.
[36] Li, H. W.; Kikuchi, K.; Nakamori, Y.; Miwa, K.; Towata, S.; Orimo, S. *Scripta Mater.* 2007, *57*, 679-682.
[37] Matsunaga, T.; Buchter, F.; Miwa, K.; Towata, S.; Orimo, S.; Züttel, A. *Renew. Energy* 2008, *33*, 193-196.

[38] Chłopek, K.; Frommen, C.; Léon, A.; Zabara, O.; Fichtner, M. *J. Mater. Chem.* 2007, *17*, 3496-3503.
[39] Miwa, K.; Ohba, N.; Towata, S.; Nakamori, Y.; Orimo, S. *J. Alloys Compd.* 2005, *404-406*, 140-143.
[40] Kim, J. H.; Jin, S. A.; Shim, J. H.; Cho, Y. W. *J. Alloys Compd.* 2008, *461*, L20-L22.
[41] Rönnebro, E.; Majzoub, E. H. *J. Phys. Chem. B* 2007, *111*, 12045-12047.
[42] Jeon, E.; Cho, Y. W. *J. Alloys Compd.* 2006, 422, 273-275.
[43] Burg, A. B.; Schlesinger, H. I. *J. Am. Chem. Soc.* 1940, 62, 3425-3429.
[44] Schlesinger, H. I.; Sanderson, R. T.; Burg, A. B. *J. Am. Chem. Soc.* 1940, *62*, 3421-3425.
[45] Schlesinger, H. I.; Brown, H. C. *J. Am. Chem. Soc.* 1940, *62*, 3430-3435.
[46] Schlesinger, H. I.; Brown, H. C.; Schaeffer, G. W. *J. Am. Chem. Soc.* 1943, *65*, 1786-1787.
[47] Schlesinger, H. I.; Brown, H. C. *J. Am. Chem. Soc.* 1953, *75*, 219-221.
[48] Official website of the Nobel foundation. http://www.nobelprize.org.
[49] Schlesinger, H. I.; Brown, H. C.; Finkholt, A. E.; Gilbreath, J. R.; Hoekstra, H. R.; Hyde, E.K. *J. Am. Chem. Soc.* 1953, *75*, 215-219.
[50] Borisov, A. P.; Makhaev, V. D. *Russ. J. Inorg. Chem.* 1988, *33*, 3030-3035.
[51] Nickels, E. A.; Jones, M. O.; David, W. I. F.; Johnson, S. R.; Lowtron, R. L.; Sommariva, M.; Edwards, P. P. *Angew. Chem. Int. Ed.* 2008, *47*, 2817-2819.
[52] Meisner, G. P.; Scullin, M. L.; Balogh, M. P.; Pinkerton, F. E.; Meyer, M. S. *J. Phys. Chem. B* 2006, *110*, 4186-4192.
[53] Chater, P. A.; Anderson, P. A.; Prendergast, J. W.; Walton, A.; Mann, V. S. J.; Book, D.; David, W. I. F.; Johnson, S. R.; Edwards, P. P. *J. Alloys Compd.* 2007, *446-447*, 350-354.
[54] Titov, L. V.; Eremin, E. R. *Russ. Chem. Bull.* 1975, *24*, 1095-1096.
[55] Smith, Ch. I.; Jones, M. O.; Johnson, S. R.; David, W. I. F.; Edwards, P. P. In *ISHE, International symposium on materials issues in a hydrogen economy*; Richmond, VG, 2007; W32.
[56] Zhao, J. C.; Cui, J.; Gao, Y.; Kniajansky, S.; Soloiveichik, G. *DOE Annual Progress Report* 2007, 370-374.
[57] Züttel, A.; Wenger, P.; Rentsch, S.; Sudan, P.; Mauron, P.; Emmenegger, C. *J. Power Sources* 2003, *118*, 1-7.
[58] Umegaki, T.; Yan, J. M.; Zhang, X. B.; Shioyama, H.; Kuriyama, N.; Xu, Q. *Int. J. Hydrogen Energy* 2009, *34*, 2303-2311.
[59] Marrero-Alfonso, E. Y.; Gray, J. R.; Davis, T. A.; Matthews, M. A. *Int. J. Hydrogen Energy* 2007, *32*, 4723-4730.
[60] Marrero-Alfonso, E. Y.; Gray, J. R.; Davis, T. A.; Matthews, M. A. *Int. J. Hydrogen Energy* 2007, *32*, 4717-4722.
[61] Kojima, Y.; Kawai, Y.; Kimbara, M.; Nakanishi, H.; Matsumoto, S. *Int. J. Hydrogen Energy* 2004, *29*, 1213-1217.
[62] Şahin, Ö.; Dolaş, H.; Özdemir, M. *Int. J. Hydrogen Energy* 2007, *32*, 2330-2336.
[63] Kojima, Y.; Suzuli, K. I.; Fukumoto, K.; Sasaki, M.; Yamamoto, T.; Kawai, Y.; Hayashi, H. *Int. J. Hydrogen Energy* 2002, *27*, 1029-1034.
[64] Peng, B.; Chen, J. *Energy Environ. Sci.* 2008, *1*, 479-483.
[65] Klooster, W. T.; Koetzle, T. F.; Siegbahn, P. E. M.; Richardson, T. B. Crabtree, R. H. *J. Am. Chem. Soc.* 1999, *121*, 6337-6343.
[66] Richardson, T. B.; de Gala, S.; Crabtree, R. H. *J. Am. Chem. Soc.* 1995, *117*, 12875-12876.
[67] Langmi, H. W.; McGrady, G. S. *Coord. Chem. Rev.* 2007, *251*, 925-935.

[68] Sit, V.; Geanangel, R. A.; Wendlandt, W. W. *Thermochim. Acta* 1987, *113*, 379-382.
[69] Stephens, F. H.; Pons, V.; Baker, R. T. *Dalton Trans*. 2007, 2613-2626.
[70] Wu, H.; Zhou, W.; Yildirim, T. *J. Am. Chem. Soc*. 2008, *130*, 14834-14839.
[71] Xiong, Z.; Wu, G.; Chua, Y. S.; Hu, J.; He, T.; Xu, W.; Chen, P. *Energy Environ. Sci*. 2008, *1*, 360-363.
[72] Xiong, Z.; Yong, C. K.; Wu, G.; Chen, P.; Shaw, W.; Karkamkar, A.; Autrey, T.; Jones, M. O.; Johnson, S. R.; Edwards, P. P.; David, W. I. F. *Nature Mater*. 2008, *7*, 138-141.
[73] *Sigma Aldrich. Material safety data sheet of AB. Product number 287717.*
[74] Lane, C. F. *Ammonia-borane and related N-B-H compounds and materials: safety aspects, properties and applications, U.S. Department of Energy Chemical Hydrogen Storage Center.* http://www1.eere.energy.gov/hydrogenandfuelcells/pdfs/nbh_h2_storage_survey.pdf.
[75] Chandra, M.; Xu, Q. *J. Power Sources* 2006, *159*, 855-860.

In: Boron Hydrides High Potential Hydrogen Storage Material ISBN:978-1-61668-361-0
Editors: Umit B. Demirci and Philippe Miele

Chapter 3

LITHIUM BOROHYDRIDE: SYNTHESIS, PROPERTIES AND THERMAL DECOMPOSITION

Raphaël Janot

Laboratoire de Réactivité et Chimie des Solides, UMR 6007 CNRS, Université de Picardie Jules Verne, 33, rue St Leu, 80039 AMIENS Cedex, France, Tel: +33.3.22.82.79.01, fax: +33.3.22.82.75.90, Email : raphael.janot@u-picardie.fr

ABSTRACT

Hydrogen is currently regarded as one of the most suitable energy carriers for the future sustainable economy, but its full implementation requires paradigm shifts in its production, distribution and storage. It is beyond the scope of this chapter to cover these three sectors; we only want to focus on the safe storage, which stands as a great challenge for various applications. While attractive, the implementation of fuel cell technology in EV transportation is, for instance, beset by our inability to come out with efficient means to safely store large amounts of hydrogen. Such applications require hydrogen absorbing materials having a gravimetric capacity of at least 6 wt.% and a volumetric capacity higher than 70 g/l. Metal-hydrides have long been considered as potential candidates, but their performances have been deceiving, so that a lot of attention has been recently focused on lightweight materials like alanates or borohydrides, which are frequently termed as complex hydrides. Among them, $LiBH_4$ has been the subject of intense studies, which have led to attractive and numerous findings; so this material deserves in itself an entire chapter reuniting the main methods of synthesizing $LiBH_4$ and discussions about its physico-chemical properties, e.g. crystal structure, ionic conduction, thermodynamic stability. $LiBH_4$ undergoes an orthorhombic to hexagonal structural transition at 116°C leading to a compound with a high ionic conductivity. This compound is thermally decomposed via a complex mechanism of hydrogen release forming a new $Li_2B_{12}H_{12}$ salt. Moving towards applications, different approaches to enhance the hydrogen release kinetic are reviewed; the most promising one being the inclusion of $LiBH_4$ into mesoporous carbons, which leads to materials capable of releasing 12 wt.% of hydrogen at high rates and at temperatures as low as 200°C, making this material somewhat unique for hydrogen desorption. Nevertheless, the practical future of this material is in jeopardy if we remain unable to improve the feasibility of its re-hydrogenation process, which is presently very poor. Owing to late developments, optimism must, however, prevail.

Besides hydrogen storage applications, $LiBH_4$ could have a great impact within the field of Li-batteries, due to its high ionic conduction and its ability to lead to a Li-based electrolyte.

1. INTRODUCTION

Hydrogen is currently regarded as one of the most suitable energy carriers for the future sustainable economy, but the safe storage of hydrogen remains a great challenge, especially for transportation applications (e.g. the future fuel-cell powered cars). Due to safety problems encountered with conventional storage techniques (cryogenic liquid or compressed gas), the use of solid-state materials, which reversibly absorb (or adsorb) hydrogen is promising. Today, the research for new compounds having high capacities (both by volume and by weight) and showing fast hydrogen absorption/desorption at low temperature is tremendous. Indeed, many metals or alloys readily absorb hydrogen, even at room temperature for some of them, but their gravimetric hydrogen capacities are limited to about 1.5 wt.%, a value far from that desired for on-board hydrogen storage. Magnesium exhibits a much higher capacity (7.6 wt.% based on the formation of MgH_2), but the absorption/desorption kinetics are slow and, in addition, a temperature near 350°C is required to obtain a one bar equilibrium hydrogen pressure.

Alkali metal alanates (formed with AlH_4^- or AlH_6^{3-} complex anions) gained a lot of attention in the last years, due to their high capacities at moderate temperatures (below 200°C). The most investigated alanate is $NaAlH_4$ with a theoretical capacity of 5.6 wt.% and fast kinetics when Ti-based compounds are used as catalysts. However, due to a limited reversibility (3.6 wt.% at 150°C) as the decomposition of $NaAlH_4$ into Na_3AlH_6 + Al mixture is the only reversible reaction under practical pressure-temperature conditions, $NaAlH_4$ does not satisfy all the requirements for mobile applications.

Following a completely different process, materials with very high specific surface areas (activated carbons, MOFs, zeolithes) are able to store large amounts of hydrogen (up to 7 wt.% at 77 K). This storage phenomenon is related to the physisorption of H_2 molecules onto the surface of the materials. As this adsorption process occurs through the formation of weak Van der Waals interaction, it only happens at low temperatures, as that of liquid nitrogen (77 K). Many efforts are currently focused on the increase of these interactions in order to obtain near room temperature adsorption processes, but without significant success at the present. In addition, these materials with high surface areas have generally very low densities and the resulting volumetric storage capacities are disappointing.

Therefore, although many materials are able to reversibly store hydrogen, none of them can reach high capacities (by volume and by weight) at temperatures required for mobile applications. This observation has led to considerable attentions on metal borohydrides, due to their very high hydrogen contents. These compounds called complex hydrides consist in a BH_4^- covalent group, which is ionically bonded with alkali or alkaline-earth cations. Among them, lithium borohydride ($LiBH_4$, also named lithium tetrahydroborate) appears promising for hydrogen storage, due to both its high gravimetric (18.4 wt.%) and volumetric (121 $kg.m^{-3}$ H_2) capacities. In this chapter, we will review the physico-chemical properties of lithium borohydride and we will discuss about the possibility of using $LiBH_4$ as a future hydrogen storage material.

2. Synthesis Methods

At the industrial scale, $LiBH_4$ is today exclusively produced from $NaBH_4$, so it is of importance to briefly present in a first paragraph how is synthesized $NaBH_4$. Usually, boron is provided as a borate and the hydrogenation is performed through a transfer of hydrogen from an alkali metal hydride. The most common method for the commercial production of $NaBH_4$ is based on the use of a mineral oil suspension of NaH as hydriding agent. Sodium dispersion may be hydrogenated directly in the mineral oil and this oil provides also heat sink, which facilitates the temperature control of the exothermic reaction with borates [1]. The reaction is usually carried out at 250-270°C using trimethyl borate as initial reactant and the synthesis can be resumed according to the following equation:

$$4\ NaH + B(OCH_3)_3 \rightarrow NaBH_4 + 3\ NaOCH_3 \quad (1)$$

Concerning the synthesis of $LiBH_4$ itself, a wet chemical reaction between $NaBH_4$ and LiCl is made in ethers [2-3], usually in the presence of isopropyl amine, according to the following reaction:

$$NaBH_4 + LiCl \rightarrow LiBH_4 + NaCl \quad (2)$$

This metathesis reaction leads to the formation of sodium chloride salt and this by-product must be separated from $LiBH_4$. This requires a specific solvent for either $LiBH_4$ or the by-product but, most solvents have a finite solubility in the $LiBH_4$ product and are persistent. The solvent-free synthesis of $LiBH_4$ is therefore of a high interest and several recent studies have been devoted to the development of such dry synthesis methods.

Although the direct synthesis of $LiBH_4$ from the elements is thermodynamically favoured with an enthalpy of formation of -194.2 $kJ.mol^{-1}$, the synthesis of $LiBH_4$ requires elevated temperature and pressure. In a patent from 1958 [4], the synthesis from lithium and boron is reported under 150 bars of hydrogen pressure and a temperature of 650°C. However, these results have been reproduced with many difficulties. Obviously, a significant activation barrier must be overcome to initiate the formation of $LiBH_4$, due to the chemical inertness of boron. This inertness is revealed by the very high melting and boiling temperatures of boron (exceeding 2000°C and 3900°C, respectively) and, in addition, boron is generally passivated by an oxide layer. The difficulty in synthesizing $LiBH_4$ from the Li and B elements is also due to the formation of a LiH passivating layer at the surface of the Li particles upon hydrogenation. With a simple Li/B mixture, lithium is first hydrided forming a passivating LiH layer, which hinders a further hydrogenation and the reaction with boron. Actually, in this case, the absorption rate only continues after the LiH melting at 670°C.

To eschew this phenomenon, the use of lithium-boron alloys has been reported [5]. LiB_3 and Li_7B_6 intermetallic compounds have been prepared by heating a Li/B mixture at 330°C and 450°C, respectively. As the initial Li/B mixture had an atomic ratio of 8:5 in this study, the intermetallic compounds were obtained with lithium in excess. These LiB_3/Li and Li_7B_6/Li mixtures were then hydrogenated at 700°C under 150 bars of hydrogen. It has been shown that the hydrogen absorption rates for these mixtures are much faster than that of a simple Li/B mixture, especially in the case of the LiB_3/Li mixture which can be fully hydrogenated in 35 hours [5]. The exact mechanism for the faster hydrogen absorption of the LiB_3/Li mixture is not yet fully understood, but it seems that it is purely a kinetic effect and not due to thermodynamic reasons. The LiB_3 alloy can be viewed as a boron lattice where lithium is dissolved into the interstitial sites. This leads to a stressed and expanded boron structure and causes the decrepitation of boron particles. The boron inertness is therefore partially overcome and this allows a faster $LiBH_4$ formation.

Another dry process for the synthesis of $LiBH_4$ has been recently proposed [6]: the high energy ball-milling of LiH and boron in a hydrogen atmosphere. LiH powder and B pieces were milled under 10 bars of hydrogen for milling durations ranging up to 138 h. Boron pieces were used instead of powder to reduce the natural passivating oxide layer and, therefore, to increase the reactivity. The synthesis reaction, which is exactly the reverse of the thermal decomposition of $LiBH_4$, is the following:

$$LiH + B + 3/2\ H_2 \rightarrow LiBH_4 \qquad (3)$$

The maximum yield of $LiBH_4$ formation reported in this study is about 27 % [6]. $LiBH_4$ has been extracted from the as-milled sample by dissolution in ether and the X-ray diagram of the resulting compound is in full agreement with that of pure $LiBH_4$. Additional experiments with increased hydrogen pressures, longer milling times and/or harder milling tools should enhance the yield of $LiBH_4$ formation.

3. Crystal Structures

The crystal structures of the light metal borohydrides have been recently reviewed [7]. Four different $LiBH_4$ phases are known: two at ambient pressure and two at high pressures. The first report in 1947 [8] on the crystal structure of $LiBH_4$, at room temperature and ambient pressure, describes $LiBH_4$ in terms of tetrahedral BH_4^- anions and Li^+ cations, but the structure was not refined and the atomic positions were not given. The first detailed structural study was made on polycrystalline $LiBH_4$ using synchrotron X-ray diffraction [9]. At room temperature, $LiBH_4$ crystallizes in an orthorhombic unit cell with the Pnma space group (a=7.17858 Å, b= 4.43686 Å, c=6.80321 Å, Z=4). The calculated density of $LiBH_4$ is 0.667 $g.cm^{-3}$ according to these structural parameters. As the temperature increases, the structure undergoes a first-order transition around 116°C and becomes hexagonal with the $P6_3mc$ space group (a=4.27631 Å, c=6.94844 Å, Z=2). This transition is easily visualized by DSC with the presence of an endothermic peak at 116°C (cf. Figure 1). Upon cooling, a small hysteresis of about 10°C is noticed for the recrystallization of $LiBH_4$ in the orthorhombic unit cell.

This structural transition has been also clearly evidenced by Raman spectroscopy [10]. In the 1200-1400 cm^{-1} range, three B-H modes appear in the room temperature modification of $LiBH_4$, due to the ν_2 splitting, while only one band subsists at 1302 cm^{-1} above the transition temperature. In the same way, two strong bands at 2275 and 2300 cm^{-1} (ν_3 and ν_1, respectively) are visible below the transition temperature, whereas only one band at 2299 cm^{-1} is present for the high temperature phase. Clearly, this evolution is consistent with the increased point group symmetry of the BH_4^- anion in the hexagonal structure. At room temperature, the point group symmetry of the BH_4^- anion is C_s (space group Pnma). Seven internal BH_4- vibrations out of the nine Raman active normal vibrations are experimentally observed [10]. Above the structural transition temperature, the point group symmetry of the BH_4^- anion is C_{3v} (space group $P6_3mc$): only five Raman bands are observed. In addition to the temperature dependence of the Raman bands positions, the line widths of the peaks are largely increased above the transition temperature. This observation reveals that large amplitude vibrational motions of the quasi-rigid BH_4 tetrahedra occur around their trigonal axis in the high-temperature hexagonal phase [10].

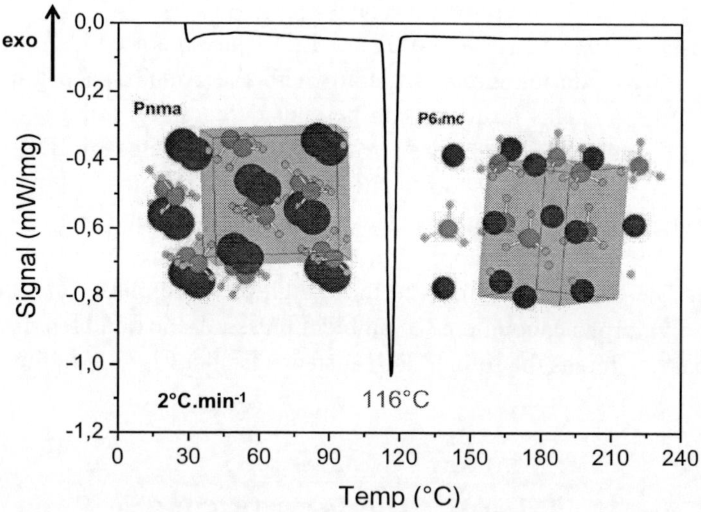

Figure 1. DSC profile of $LiBH_4$ showing the structural transition at 116°C between the LT orthorhombic phase (Pnma) and the HT hexagonal phase ($P6_3mc$).

The crystal structures of $LiBH_4$ have been recently further investigated, either by synchrotron X-ray diffraction on single crystals, or by neutron diffraction on deuterated compounds ($LiBD_4$). Single crystal investigation at 225 K of the low-temperature phase shows that the BH_4 group has an ideal tetrahedral geometry [11]. This contrasts with the strongly distorted tetrahedra reported in the earlier study on a polycrystalline material [9]. Actually, it was somewhat amazing that the Li^+ cation can induce deformations to the covalently bonded and presumably rigid tetrahedral BH_4^- anion. It is important to note that, although not observed in $LiBH_4$, an interaction of a strongly polarizing cation can lead to deformations of the BH_4 group, as encountered with $Al(BH_4)_3$ borohydride [12]. Whatever the origin of such deformations in $LiBH_4$, they could be reliably detected by diffraction on single crystals or simulated by DFT calculations.

In the low-temperature orthorhombic phase of $LiBH_4$, the observed H-B-H angles range from 108.8° to 109.9° and are in very good agreement with the ideal tetrahedral angle of 109.5°. The B-H bond lengths show a narrow spread from 1.104 to 1.131 Å [11]. Additionally, neutron diffraction experiments also show that the BH_4 group is undistorted with B-H distances in the 1.208-1.225 Å range and H-B-H angles between 107.2 and 111.7° [12]. The discrepancy between the B-H bond lengths determined by X-ray and neutron diffractions experiments comes from the well-known displacement of the electron cloud (observed by X-ray diffraction) relative to an average nuclear position (seen by neutron diffraction) [7].

The $P6_3mc$ symmetry of the high-temperature phase has been also recently revisited by synchrotron X-ray diffraction on single crystals [12] and by neutron diffraction on the isotopically substituted $LiBD_4$ compound [13]. These studies confirm the $P6_3mc$ space group, but they also reveal large and anisotropic displacements of the hydrogen atoms. The atomic displacement indicates, in addition to the thermal vibrations, an orientational disorder of the BH_4 group, as previously noticed by Raman spectrocopy [10]. The large vibration amplitudes agree with an apparent shortening of the B-H bonds to 1.07 Å, as determined by synchrotron X-ray diffraction [12], to be compared to the 1.10-1.13 Å bonds lengths measured for the rigid BH_4 tetrahedra of the low-temperature phase.

At high pressures, two other polymorphs of $LiBH_4$ are reported [14]. $LiBH_4$ crystallizes under 2.4 GPa in a pseudo-tetragonal structure with the Ama2 space group (a=6.4494 Å, b=5.307 Å, c=5.2919 Å). This structure can be considered as on orthorhombically distorted anti-structure of PtS, where Li atoms are tetrahedrally coordinated by BH_4 groups. The BH_4 groups have here a square-planar coordination by four Li atoms. Above 10 GPa, $LiBH_4$ undergoes a structural transition and becomes isostructural to cubic $NaBH_4$ with the Fm-3m space group (a=5.109 Å). The Li atoms and the BH_4 groups are octahedrally coordinated in this cubic phase. The B-H bond lengths in these high pressure phases are similar to those of the two $LiBH_4$ polymorphs encountered at ambient pressure: the bond length is around 1.17 Å for the Ama2 phase, whereas the refined B-H distance is about 1.08 Å in the cubic phase at 18 GPa [14].

4. Ionic Superconduction

The orthorhombic (low temperature phase) to hexagonal (high temperature phase) transition in $LiBH_4$, taking place at about 116°C, was recently demonstrated to be accompanied by a threefold jump in electrical conductivity, reaching a value of the order of 10^{-3} $S.cm^{-1}$ just above the transition temperature as presented in Figure 2 [15]. This value of electrical conductivity has been obtained by AC complex impedance spectroscopy. Both for the LT and HT phases, only single arcs were observed on the impedance plots, indicating that neither the response of the electrode interface nor the grain boundary was detected. Below and above the transition, the temperature dependence of the electrical conductivity follows Arrhenius behaviours. The activation energies for the conduction were found at 0.69 and 0.53 eV for the LT and HT phases, respectively [15]. Obviously, the jump in electrical conductivity is fully reversible, with a small hysteresis of about 10°C, which is in good agreement with the hysteresis observed by DSC experiments tracing the structural transition.

In the same study [15], it was shown by ^{7}Li NMR that the high conductivity of the HT hexagonal phase is mainly due to lithium ionic conduction. Actually, the spin-lattice relaxation time increases with temperature, but presents a steep decrease upon the structural transition at 116°C. Above this temperature, the temperature dependence of the relaxation time allows the calculation of the activation energy of the lithium ionic motion: a value of 0.56 eV has been found, which agrees well with the activation energy calculated from the complex impedance results. In addition, using the Nernst-Einstein equation, electrical conductivities calculated from the ^{7}Li NMR correlation times were found very close to those measured by the AC impedance method: the calculated conductivities are 2.10^{-3} $S.cm^{-1}$ and

10^{-2} S.cm^{-1} at 120°C and 180°C, respectively. From these results, it was concluded that the high electrical conductivity of the HT phase originates from the lithium superionic conduction.

Further studies by Nuclear Magnetic Resonance (NMR) have been recently carried out in order to better clarify the atomic motions in $LiBH_4$ [16-17]. The analysis of the temperature dependence of the ^1H and ^{11}B spin-lattice relaxation times shows that there are two types of rotational motions of the BH_4 tetrahedra in the LT orthorhombic phase [16], most likely the 2-fold and 3-fold reorientations. Actually, the reorientation motions of the BH_4 groups are fast: the jump rates of the reorientations reach 10^{11} s^{-1} near the upper limit of stability of the LT phase (110°C). Above the transition temperature, the measured ^1H and ^{11}B spin-lattice relaxation rates are governed by an additional low frequency process e.g. the translational diffusion of Li cations. This phenomenon is clearly evidenced by looking at the ^7Li NMR spectra, which are greatly sharpened above the structural transition temperature [17]. In the HT phase, the contributions due to the rotational motion of BH_4 tetrahedra are small and the motions of the Li cations are much faster than the hydrogen diffusion. These NMR results confirm that the HT phase exhibits very rapid lithium diffusion and that $LiBH_4$ is a good lithium ionic conductor.

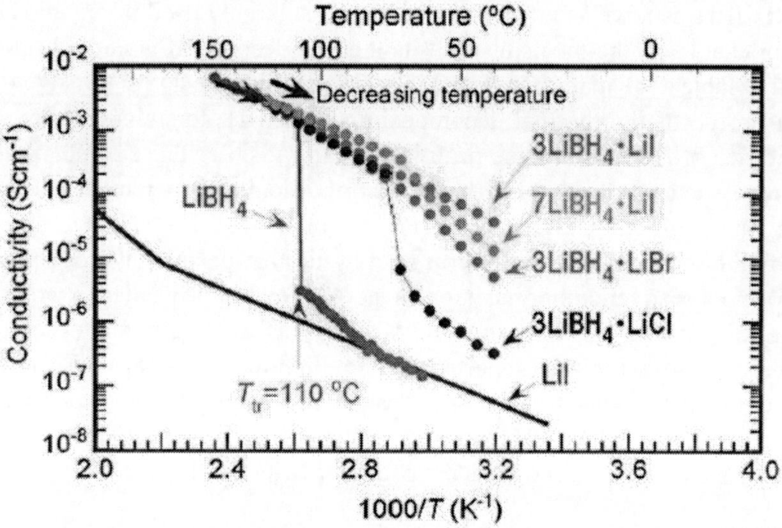

Figure 2. Electrical conductivities obtained by AC impedance measurements versus temperature for $LiBH_4$, LiI and $LiBH_4$-LiX mixtures (with X=Cl, Br, I), [Reprinted with permission from Ref. 19. Copyright 2009 American Chemical Society].

This lithium superionic conduction could potentially lead to the development of $LiBH_4$ as an electrolyte for all solid-state lithium batteries. However, only the HT polymorph of $LiBH_4$ presents a high ionic conductivity meaning that a temperature higher than 115°C is needed. Recent papers from Orimo's group [18-19] have shown that the HT phase can be stabilized at lower temperatures (possibly at room temperature) by mixing $LiBH_4$ with lithium halides (LiX with X=Cl, Br, I). Especially, a $LiBH_4$-LiI mixture with a 3:1 molar ratio mechanically milled under argon presents some very interesting features. The 3 $LiBH_4$ + LiI sample does not present a structural transition from room temperature to 150°C. It seems by X-ray diffraction that a solid-solution between $LiBH_4$ and LiI is formed: all the reflections of the X-ray diagram recorded at room temperature can be indexed in the hexagonal lattice of the HT phase of $LiBH_4$. The LiI reflections are not visible and the unit cell volume is slightly larger than that of pure $LiBH_4$, which accounts for a substitution of BH_4^- (ionic radius of 2.05 Å) by I^- anion (2.11 Å).

As a result, the electrical conductivity measured by AC impedance spectroscopy of the 3 $LiBH_4$ + LiI sample is around 10^{-4} $S.cm^{-1}$ at room temperature [19], whereas a value below 10^{-7} $S.cm^{-1}$ is obtained for LiI and the low-temperature phase of $LiBH_4$ (cf. Figure 2). These results are confirmed by ^7Li NMR, where narrow peaks are observed on the spectra of 3 $LiBH_4$ + LiI, due to the fast lithium ionic motion. The ^7Li NMR line shape remains the same from room temperature to 150°C, giving evidence of the room temperature stabilization of the superionic conductor phase with the hexagonal unit cell. Further characterizations are still needed to fully explain this stabilization phenomenon and to evaluate the real opportunity of using such $LiBH_4$-based materials as solid electrolytes for lithium batteries.

5. Thermal Decomposition

Although $LiBH_4$ is a well-known compound and is largely used in organic chemistry due to its reducing character, the mechanism of its thermal desorption is not yet fully understood and is still the subject of many studies. Figure 3 exhibits the DSC (Differential Scanning Calorimetry) profile of the thermal decomposition of $LiBH_4$ for temperatures up to 600°C: two sharp and two broad endothermic peaks are clearly visible. The first sharp peak at 116°C is related to the structural transition between the orthorhombic low temperature phase (Pnma) and the hexagonal high temperature phase ($P6_3mc$) of $LiBH_4$. The HT phase melts at 286°C, as revealed by the second sharp peak, forming a liquid that desorbs hydrogen in two steps as seen by the two broad endothermic signals at 425 and 550°C. This two-step hydrogen desorption is known and the formation of "$LiBH_2$" as an intermediate compound of the desorption process was initially reported in the literature [20-21]. It is now quite well admitted, according to both Raman spectroscopy and ^{11}B NMR experiments, that the intermediate compound is rather a mixture of LiH and lithium dodecaborane ($Li_2B_{12}H_{12}$) [22] and that the two-step desorption follows the equation written below:

$$LiBH_4 \rightarrow 1/12\ Li_2B_{12}H_{12} + 5/6\ LiH + 13/12\ H_2 \rightarrow LiH + B + 3/2\ H_2 \qquad (4)$$

The last broad peak around 550°C observed on the DSC curve corresponds to the formation of the final solid desorbed material (e.g. the second step of the above equation), which consists in a mixture of LiH and amorphous boron. Due to the high stability of LiH (ΔH_f = -90.7 $kJ.mol^{-1}$), its desorption proceeds at temperatures above 700°C. As a consequence, the complete recovery of the whole hydrogen content of $LiBH_4$ (18.4 wt.%) is impossible and the practical hydrogen release is limited to 3 hydrogen atoms per $LiBH_4$ formula unit e.g. a hydrogen capacity of 13.8 wt.%.

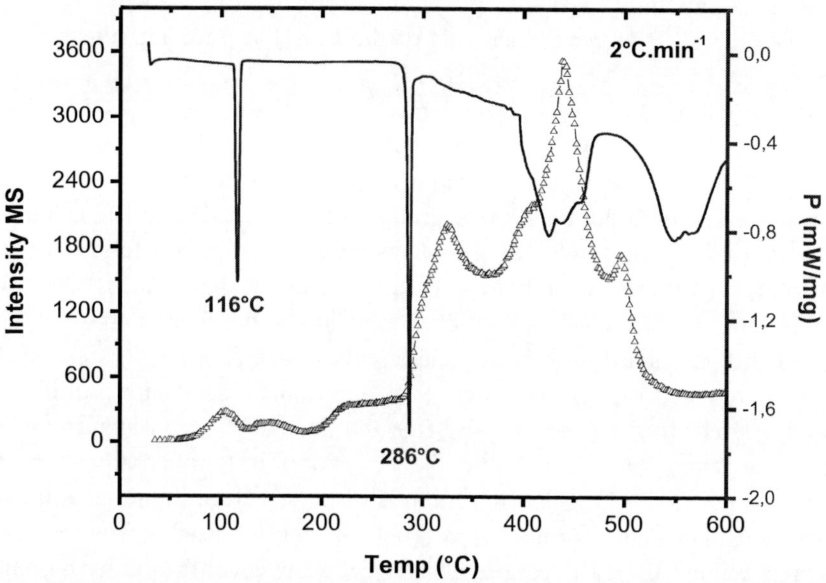

Figure 3. DSC profile (in blue) and MS curve of hydrogen emission (in black), for bulk $LiBH_4$ heated at 2°C.min^{-1}.

The enthalpies determined by DSC for the structural transition and melting are 4.2 and 7.6 kJ.mol^{-1}, while values in the 55-65 kJ.mol^{-1} range can be deduced for the hydrogen desorption enthalpy from the two broad peaks at 425 and 550°C together. This is lower than the enthalpy of 91.7 kJ.mol^{-1} [23] expected for the transformation of liquid $LiBH_4$ into LiH + B. The origin of this discrepancy is not yet understood, but cautions should be exercised in discussing enthalpies of solid-gas reactions measured by DSC. On the other hand, the entropy of the desorption process of $LiBH_4$ has been calculated at 238.7 J.K^{-1}.mol^{-1} H_2 [23], which is a surprisingly high value, since the standard entropy change for hydrogen is about 130 J.K^{-1}.mol^{-1} H_2 (change between H_2 gaseous molecules and H atoms in a metallic lattice). The reasons for this very high entropy are still unknown today.

The two-step hydrogen desorption of $LiBH_4$ is also clearly evidenced by tracing the hydrogen emission at a constant heating rate with a mass spectrometer (cf. Figure 3). After the melting of $LiBH_4$ at 280°C, a significant hydrogen desorption occurs giving a peak with a maximum at 320°C, which can be attributed to the formation of an intermediate Li-B-H compound, most likely $Li_2B_{12}H_{12}$. The second peak at 445°C corresponds to the final desorption leading to a mixture of LiH and amorphous boron. The hydrogen desorption is fully completed at 550°C.

The exact nature of the intermediate compound formed upon the $LiBH_4$ thermal decomposition has been further identified both theoretically and experimentally. By first-principle calculations, the stability for some possible intermediate compounds such as LiB_3H_8 and $Li_2B_nH_n$ with n=5-12 has been reported [24]. The computational results indicate that the monoclinic $Li_2B_{12}H_{12}$ phase is the most stable among the candidates and consists in $(B_{12}H_{12})^{2-}$ anions and Li^+ cations. In another study, the local atomic structure of the $Li_2B_{12}H_{12}$ phase has been experimentally investigated by Raman spectroscopy. It has been shown that new B-H bands at 600-1000 cm^{-1} and 2500 cm^{-1}, which do not correspond to $LiBH_4$, appear above 470°C. These bands are fairly consistent with the phonon density of states theoretically calculated for the monoclinic $Li_2B_{12}H_{12}$ phase and correspond to the bending modes and the stretching modes of the B-H covalent bonds of the $(B_{12}H_{12})^{2-}$ anionic clusters, respectively [23].

The crystal structure of $Li_2B_{12}H_{12}$, which might present some interests as Li-based salt for Li-ion batteries, has been recently resolved by powder X-ray diffraction on a polycrystalline material [25]. Actually, all the reflections can be indexed in a cubic cell, and not a monoclinic as previously assumed, with the Pa-3 space group (a=9.5771 Å, Z=4). The calculated density of $Li_2B_{12}H_{12}$ is 1.178 g.cm^{-3} according to these structural parameters. The structure of $Li_2B_{12}H_{12}$ is related to those of the heavy alkali-metal dodecaboranes $M_2B_{12}H_{12}$ with M=K, Rb, Cs [26]: it can be understood as a rational variation of the face-centred cubic potassium compound. Because Li^+ is a significantly smaller cation than K^+ (0.60 Å vs. 1.33 Å), the unit cell parameter is reduced by approximately 9 % and, more importantly, a different internal symmetry is favoured. In the case of $K_2B_{12}H_{12}$, the $(B_{12}H_{12})^{2-}$ anion is surrounded by 8 K^+ cation within the cubic geometry, while the K^+ cation is tetrahedrally surrounded by 4 $(B_{12}H_{12})^{2-}$ anions. For $Li_2B_{12}H_{12}$, the Li^+ cation is displaced from the tetrahedral site and lies on a near-trigonal planar site formed by 3 $(B_{12}H_{12})^{2-}$ anions, each of which resides in the octahedral cage defined by 6 Li^+ cation. This structure was confirmed by a combination of neutron vibrational spectroscopy and first-principle calculations [25].

From these studies, it is now well established that $Li_2B_{12}H_{12}$ is formed as an intermediate compound upon the thermal desorption of $LiBH_4$. Further decomposition leads to the formation of LiH and amorphous boron. The overall enthalpy of desorption (from the low-temperature phase of $LiBH_4$ to the LiH + B mixture) is highly endothermic [23]: its value of +103 kJ.mol^{-1} (+69 kJ.mol^{-1} H_2) makes the rehydrogenation of LiH + B very difficult. Actually, this rehydrogenation process has been only experimentally completed at 350 bars H_2 and 600°C [27]. Therefore, $LiBH_4$ cannot be used for hydrogen storage applications when used as a pristine bulk material.

6. Hydrogen Release Enhancement by Catalysts Addition

In order to reduce the dehydrogenation temperature of $LiBH_4$, many experiments have been focused on the effect of catalysts addition. Züttel et al. have especially reported, in a pioneering work about the possibility of using $LiBH_4$ as a hydrogen storage material [20], that the hydrogen desorption is strongly enhanced when $LiBH_4$ is mixed with large amounts of SiO_2. More specifically, a mixture of 25 wt.% $LiBH_4$ and 75 wt.% SiO_2 has been prepared in this study: its thermal desorption spectrum exhibits the same features as that of pristine $LiBH_4$ with two main peaks of hydrogen release, but the peaks are shifted to lower temperatures. The first hydrogen desorption peak starts already at 200°C (so below the melting point of $LiBH_4$) and the second peak starts at about 450°C and is very sharp. It was concluded in this paper that SiO_2 catalyzes the decomposition reaction of $LiBH_4$ [20], although the mechanism of catalysis is not mentioned.

More recently, the modification of the hydrogen desorption behaviour of $LiBH_4$ by additions of various metal oxides or metal chlorides has been further undertaken [28-30]. $LiBH_4$ was thoroughly ball-milled with V_2O_3, TiO_2, ZrO_2, SnO_2, $TiCl_3$, and $MgCl_2$ powders and the thermal desorption spectra of the resulting materials were confronted to that of

$LiBH_4$. About the oxide-modified $LiBH_4$ materials, the best results have been obtained with the 75 wt.% $LiBH_4$ + 25 wt.% TiO_2 sample, which presents a total desorption of 9 wt.% of hydrogen in good agreement with the initial $LiBH_4$ content and the formation of LiH + B as final desorption products. The material desorbs through two dehydriding stages: (1) from 100°C to 300°C, about 3 wt.% of hydrogen is released, (2) and an additional hydrogen release of about 6 wt.% occurs from 300°C to 550°C.

Interestingly, it is also reported in this study [30] that the TiO_2-modified $LiBH_4$ can be rehydrogenated at 600°C and 100 bar H_2 in only one hour, whereas the rehydrogenation of pristine $LiBH_4$ is negligible under the same conditions. However, it is noted that the rehydrogenation produces not only $LiBH_4$, but also Li_3BO_3 that does not participate in the following absorption-desorption processes, due to its high thermodynamic stability. As a matter of fact, upon the first rehydrogenation, only 0.82 mole of $LiBH_4$ is produced for each mole of dehydrided $LiBH_4$. Upon the following absorption-desorption cycles, the amount of Li_3BO_3 increases leading to a fast loss of storage capacity.

About the chloride-modified $LiBH_4$, it is shown that the hydrogen release starts at only 60°C. For the $LiBH_4$ + 0.2 $MgCl_2$ + 0.1 $TiCl_3$ sample, a total of 5 wt.% of hydrogen is released through at least four different dehydrogenation steps [30]. Actually, conversely to the oxide-modified $LiBH_4$ for which the oxide does not react with $LiBH_4$ but simply acts as a catalyst, here, a chemical reaction occurs upon milling between $LiBH_4$ and the $TiCl_3$ and $MgCl_2$ chlorides according to the following reactions:

$$3\ LiBH_4 + TiCl_3 \rightarrow Ti(BH_4)_3 + 3\ LiCl \quad (5)$$

$$2\ LiBH_4 + MgCl_2 \rightarrow Mg(BH_4)_2 + 2\ LiCl \quad (6)$$

The low-temperatures hydrogen desorption steps encountered for chloride-modified $LiBH_4$ are due to the decomposition of $Ti(BH_4)_3$ and $Mg(BH_4)_2$ at about 60°C [31] and 260-280°C [32], respectively. These decomposition processes are irreversible and the rehydrogenation of the chloride-modified $LiBH_4$ materials is therefore impossible. As a conclusion, if the hydrogen release from $LiBH_4$ can be strongly improved in term of decrease of the decomposition temperature, both by milling with oxides or chlorides, $LiBH_4$ cannot be used for storage applications, due the very poor feasibility of the rehydrogenation process.

7. MODIFICATION OF HYDROGEN RELEASE BY CONFINEMENT INTO NANOPOROUS CARBONS

The other strategy for kinetic improvement of the $LiBH_4$ hydrogen desorption is to decrease the particles size down to the nanoscale range. Owing to the shortening of the hydrogen diffusion paths, nanoparticles can promote faster hydrogen absorption/desorption kinetics. Such approach was initially put forward about hydrogen storage materials in the study of the hydrogen release of ammonia borane (NH_3BH_3) incorporated into a mesoporous silica (SBA-15, pores diameter = 8 nm) [33]. It has been shown that the kinetic of hydrogen release is largely improved compared to neat ammonia borane and, in addition, is was suggested that the mechanism of hydrogen desorption is modified, in spite of the relatively large particles size, as the amount of borazine $N_3B_3H_6$ contaminating the hydrogen desorption flow (borazine is a cyclic molecule, which is iso-electronic with benzene) is largely reduced. These results have been further improved by using a mesoporous carbon (pores diameter in the 2-20 nm range) as the host instead of silica [34].

The preparations of $NaAlH_4$ and $LiBH_4$ nanoparticles supported on carbons have been also independently reported [35-36]. For the former, the particles were dispersed in carbon fibres through an impregnation and drying technique [35], whereas, for the latter, the hydride was directly ball-milled with mesoporous carbons [36]. In both cases, although the hydride was not really incorporated into the pores of the carbonaceous host, but rather simply deposited, a strong decrease in the onset temperature of hydrogen desorption was also found together with a much faster hydrogen release.

The real inclusion of $LiBH_4$ nanoparticles into the mesopores of a carbonaceous host, with the hope of avoiding sintering of particles upon heating, has been recently reported [37-39]. In this study, mesoporous carbons called aerogels (pores diameter in the 10-40 nm range), prepared from formaldehyde and resorcinol condensation, have been impregnated with molten $LiBH_4$ at 280-300°C. The $LiBH_4$ loadings of the resulting composites reach 28 wt. % and it has been shown that the onset temperature for hydrogen desorption of the carbon-encapsulated $LiBH_4$ nanoparticles is about 200°C, whereas the onset temperature for bulk $LiBH_4$ coincides with its melting point (280°C). This coincidence in bulk $LiBH_4$ results from the higher diffusion constants in liquid state, which is supposed to increase the hydrogen diffusion rate and the growth of the dehydrogenated compounds (LiH + B). According to preliminary DSC results, a similar effect probably occurs for carbon-encapsulated $LiBH_4$, since the melting temperature of the nano-sized $LiBH_4$ particles seems to be decreased [39].

In addition, the equilibrium hydrogen pressure for $LiBH_4$ confined in the carbon aerogels has been determined 10 times higher at 300°C than that of bulk $LiBH_4$ (3 bar and 0.3 bar, respectively), which suggests that a significant modification of the enthalpy of desorption has been induced by the confinement [39]. These results are somewhat surprising, since the pores size of the carbons aerogels is relatively high (above 10 nm) and it was not expected that the thermodynamic stability for 10 nm-sized $LiBH_4$ particles would have been modified. In the same study, the rehydrogenation of the $LiBH_4$-carbon composites has been successfully carried out under 100 bar H_2 at only 400°C, which represents a significant progress compared to the conditions needed for the rehydrogenation of bulk $LiBH_4$ (350 bars-600°C) [27]. These remarkable results show that both the kinetic and thermodynamic have been modified through this approach of confinement into nanoporous carbons.

Pursuing along that line, porous carbons with smaller pores size have been synthesized and a new room temperature method for the preparation of composites with high $LiBH_4$ contents has been developed [40]. This new process enables an easier handling and easier scale-up, as compared to the impregnation with molten $LiBH_4$ at 300°C, while preventing the loss of hydrogen upon impregnation, hence maintaining the full hydrogen desorption capacity of the so-prepared $LiBH_4$-carbon composites.

Porous carbons have been prepared by the template method using SBA-15 silica as matrix and sucrose as carbon precursor [41-42]. These poorly graphitized carbons exhibit a well-defined hexagonal arrangement (a=10 nm) of mesopores, which can be easily visualized by TEM. The pores of these mesoporous channels (diameter around 4 nm) have the shape of the walls of the SBA-15 silica and are interconnected with micropores (diameter < 2 nm). Therefore, a 3D porosity is present and enables the cohesion of the carbon replica. The BET specific surface area of these porous carbons is around 1000-1100 $m^2.g^{-1}$ as measured by nitrogen adsorption at 77 K. The corresponding pore size distribution is centred on 4 nm (mesopores), but smaller pores (micropores) are also present. The total porous volume (micro + meso) reaches about 1.1 $cm^3.g^{-1}$. These porous carbons show a good purity with an oxygen content below 2 atomic %.

The synthesis of the $LiBH_4$-carbon composites has been achieved thanks to a process of liquid infiltration of the mesoporous carbons by $LiBH_4$ dissolved in ethers. If a large amount of $LiBH_4$ can be solvated by THF (e.g. 2.0 M solution is commercially available), the removal of the solvent remains difficult by evacuation under primary vacuum at room temperature, due to the formation of stable etherates. Unfortunately, an additional heating to decompose these etherates can lead simultaneously to the beginning of the hydrogen desorption from $LiBH_4$ and, therefore, the recovery of pure $LiBH_4$ from THF solutions is almost impossible. To avoid the formation of such stable etherates, $LiBH_4$ has been dissolved in heavier ethers, especially methyl tert-butyl ether (MTBE). This solvent enables the solubilization of small amounts of $LiBH_4$ (e.g. 0.1 M solution of $LiBH_4$ in MTBE has been successfully prepared) with an easy recovery of pure $LiBH_4$ crystals by simple drying under vacuum at room temperature. The steric effect induced by the tert-butyl group reduces the interaction between the $LiBH_4$ species and the O atoms of the ether and, thus, hinders the formation of too stable etherates.

Several steps of impregnation/drying of the porous carbons have been performed in order to increase the $LiBH_4$ contents and to obtain heavily $LiBH_4$-loaded carbons. Upon impregnation, the BET specific surface area is drastically reduced to 16 m^2/g regardless of the $LiBH_4$ loading content and the mesoporous porosity is not yet observable, which seems to show that the carbon porosity is filled with $LiBH_4$ crystals. Figure 4 presents the X-ray diagram of a 33:67 $LiBH_4$/carbon composite (ratio expressed in weight), which interestingly does not present any reflection except those attributed to the sample holder, due to the confinement of $LiBH_4$ to the nanoscale into the mesoporous host.

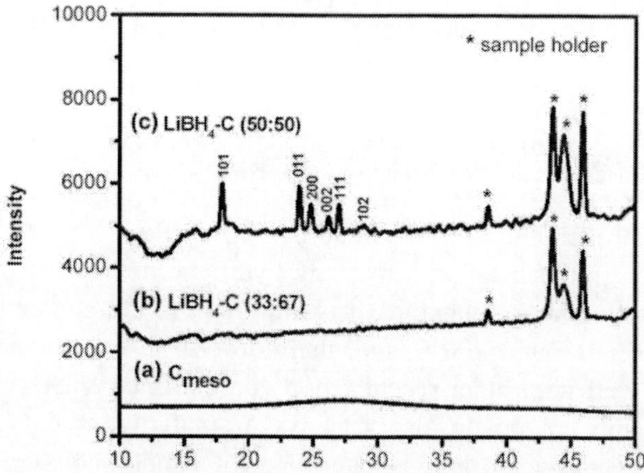

Figure 4. X-ray diffractograms of (a) mesoporous carbon obtained by the template method, (b) 33:67 $LiBH_4$/carbon composite and (c) 50:50 $LiBH_4$/carbon composite. The reflections due to the sample holder (Be) are shown as bold lines.

The DSC profile of the 33:67 composite is shown as a bold curve in Figure 5. The first two endothermic peaks at 118 and 284°C correspond to the structural transition and melting of $LiBH_4$, respectively. The temperatures are almost identical to those of neat $LiBH_4$, which suggests that the thermodynamic parameters governing these two transitions are not significantly modified. One single broad endothermic peak related to hydrogen desorption is observed at 335°C, whereas it must be recalled that the hydrogen release for neat $LiBH_4$ occurs in two steps above 400°C (see the dotted curve of Figure 5). Interestingly, a quantitative analysis shows that the ratio between the areas of the hydrogen desorption peak and the melting peak is exactly the same in both cases (by cumulating the areas of the two hydrogen desorption signals for neat $LiBH_4$). These results suggest that the overall desorption enthalpy of $LiBH_4$ is not reduced by the confinement, although an important modification occurs with the absence of any formation of intermediate compounds (as $Li_2B_{12}H_{12}$) upon desorption of the composite. Finally, it is worth mentioning that no exothermic peak, which could be related to oxidation or carbide formation, is detected by DSC.

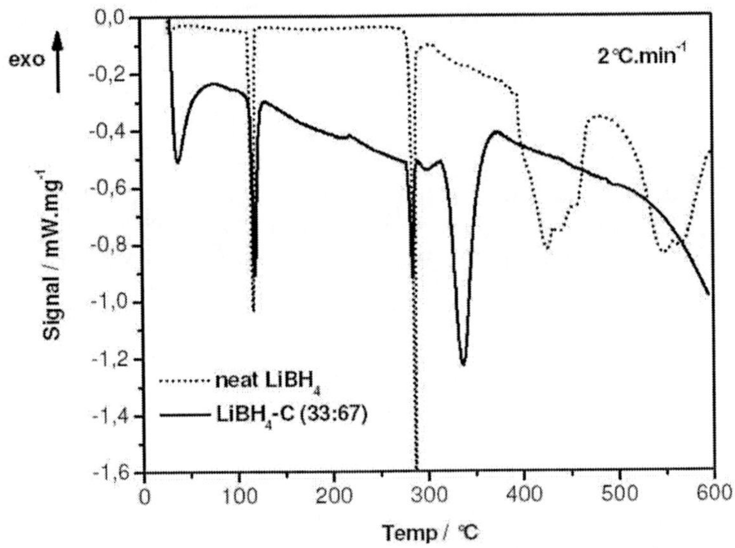

Figure 5. DSC profiles (2°C/min) for neat $LiBH_4$ and 33:67 $LiBH_4$/carbon composite. Note the strong modification of the hydrogen desorption behaviour above 300°C.

The curve of hydrogen emission recorded by TPD-MS (cf. Fig. 6) is in good agreement with the DSC results given ahead with a strong hydrogen desorption peak centred at 320°C for the 33:67 composite. At this juncture, it is important to notice that the onset temperature for hydrogen release is around 200°C and, therefore, 80°C lower than that of neat $LiBH_4$. These results agree well with other recent reports on confined $LiBH_4$ nanoparticles [37-39]. The TPD curve of the composite also displays a small hydrogen release at a very low temperature (65°C). Its exact origin is unclear as it cannot be ascribed to the structural transition of $LiBH_4$, which still occurs around 116-118°C after impregnation, as shown by DSC. In fact, a recent paper reports a similar feature for air-exposed $LiBH_4$ samples [43] and the authors have suggested that such a low temperature hydrogen release could be nested in the decomposition of a complex between water and $LiBH_4$. Here, all the composites were kept in an argon filled glove box and were not exposed to air for their characterizations. Such a complex formation with water is therefore unlikely, but cannot be totally excluded. It is our belief, that the small hydrogen release at 65°C is rather related to the decomposition of small amounts of etherates (e.g. complex between ether and $LiBH_4$), which were not completely removed upon the drying process (primary vacuum at RT for 12 hours). Although such etherates formation was not observed by X-ray diffraction for neat $LiBH_4$, the presence of mesoporous carbons could favour their stability. Due to the very low particles size of $LiBH_4$ when incorporated into the porous carbon, X-ray diffraction is unfortunately unable to evidence such etherates. Additional characterizations using local probe techniques like IR spectroscopy and NMR, which are the most exquisite ways to monitor traces of organic species, are needed.

The releases of species other than hydrogen (boranes, hydrocarbons...) are not detected by TPD-MS experiments. We have to emphasize the absence of diborane (B_2H_6) release for these $LiBH_4$/carbon composites, as it is reported in the literature that B_2H_6 strongly contaminates the hydrogen desorption flow of $LiBH_4$/silica gel mixtures [44]. This discrepancy could be explained by the fact that $LiBH_4$ is simply hand-mixed with silica-gel in this study and, therefore, that $LiBH_4$ is not really incorporated into the mesopores. In addition, it is worth mentioning that the interactions between $LiBH_4$ and silica and the interactions between $LiBH_4$ and carbon are probably different, but a detailed characterization of the surface of the porous hosts (especially the nature of the surface groups) is necessary to draw a definitive conclusion.

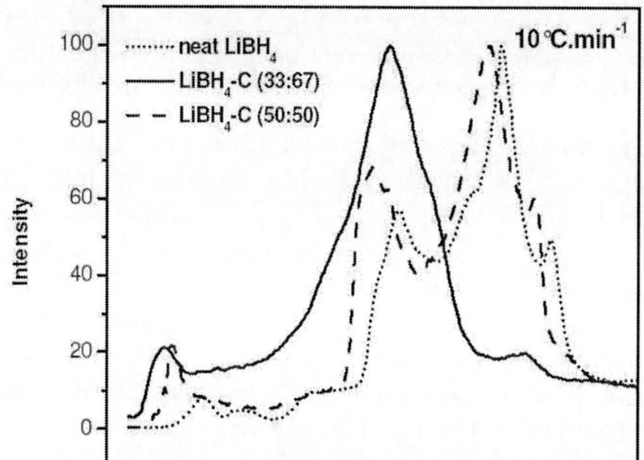

Figure 6. MS curves of hydrogen emission for neat $LiBH_4$, 33:67 $LiBH_4$/carbon composite and 50:50 $LiBH_4$/carbon composite (10°C.min^{-1}).

The quantification of the hydrogen desorption for the 33:67 $LiBH_4$/carbon composite is given in Figure 7 (squares) together with that of neat $LiBH_4$ (triangles). The final hydrogen desorption capacity at 500°C for neat $LiBH_4$ reaches 11.3 wt.%, which corresponds to 82 % of the theoretical capacity (13.8 wt.% based on the formation of LiH + B). The full achievement of the desorption process for neat $LiBH_4$ requires higher temperatures and/or longer durations. The very pleasant effect provided by the $LiBH_4$ confinement to the nanoscale is the capability to release hydrogen at temperatures as low as 200°C (cf. Fig. 7); a temperature for which neat $LiBH_4$ does not show any sign of hydrogen desorption. The total hydrogen desorption capacity at 500°C is 4.0 % per weight of composite or near 12.0 % per weight of $LiBH_4$. Note that already 3.4 wt. % of hydrogen is desorbed at 300°C for the composite, whereas only 0.2 wt. % is released for bulk $LiBH_4$ at the same temperature.

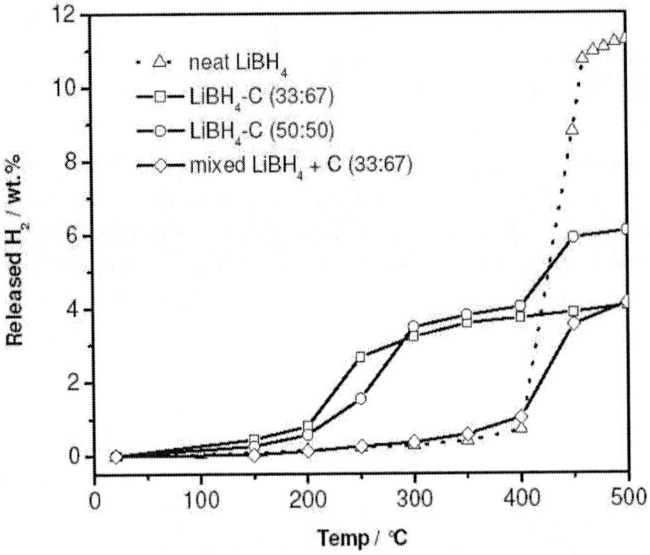

Figure 7. Amounts of hydrogen released at different temperatures, for neat $LiBH_4$, 33:67 $LiBH_4$/carbon composite, 50:50 $LiBH_4$/carbon composite, and simple mixing of 33 wt.% $LiBH_4$ and 67 wt.% carbon.

After desorption at 500°C, the X-ray diagram of the 33:67 $LiBH_4$/carbon composite still does not display any reflection, which reveals that the desorbed material (LiH + B) is also embedded into the carbon mesoporosity. The carbonaceous host is therefore able to maintain the confinement of the material during the desorption process, although this one occurs mainly from the liquid state. Combining the results obtained with different techniques (DSC, TPD, volumetry), the positive attributes of confining $LiBH_4$ into a nanometric porosity (here 4 nm in diameter) unambiguously appear, with the most important one being the ability to start releasing hydrogen at temperatures even below the melting temperature of $LiBH_4$ (284°C as measured by DSC). Then, hydrogen strongly evolves from the liquid state and directly leads in a single step to the final desorbed material (LiH + B), without any formation of intermediate compound, as encountered with pure $LiBH_4$. This is indicative of the feasibility to modify the hydrogen release mechanism path for a hydrogen storage material when introducing size (i.e. nanoscale) as an additional variable besides structure and composition.

To check the effect of confinement on the improvement of the hydrogen desorption performances, a sample was simply made by hand-mixing 33 wt.% of $LiBH_4$ with 67 wt.% of the same mesoporous carbon obtained by the template method. The curve of the resulting hydrogen desorption (cf. diamonds, Fig. 7) traces the one obtained for neat $LiBH_4$: the hydrogen release of the hand-made mixture does not occur below 400°C, which confirms the strong benefit of confining $LiBH_4$ to the nanoscale range. Besides the amount of desorbed hydrogen, another key parameter for applications is the hydrogen desorption kinetic. Here the kinetics have been measured by thermogravimetry under a constant backup hydrogen pressure of one bar. The staggering enhancement of the hydrogen desorption for the 33:67 $LiBH_4$/carbon composite is nicely visualized when looking at the kinetics at 300°C (cf. Figure 8). For the composite, 3.4 wt.% of hydrogen is released in 90 min versus 0.3 wt.% for the hand-mixed sample. After 180 min, solely 0.7 wt.% of hydrogen is desorbed for the mixture showing the insignificant effect of carbon addition on the improvement of the $LiBH_4$ hydrogen release.

Our first attempts of rehydrogenation under 100 bars H_2 at 300°C of the 33:67 $LiBH_4$/carbon composite were disappointing: the hydrogen absorption was barely detectable even after several hours of hydrogen exposure. Hydrogenation at higher temperatures (like 600°C as required for bulk $LiBH_4$ rehydrogenation [27]) is required. This poor absorption behaviour reveals that, even if the kinetic and maybe the thermodynamic of the $LiBH_4$ hydrogen desorption is modified by nano-confinement, the chemical inertness of boron cannot be easily overcome. The pore's size (4 nm) of the mesoporous carbon is probably too large to induce a large change in surface energies and does not allow the decrease in the large endothermic character of the desorption reaction. As a result, it appears that the synthesis of carbons with a smaller pores size is still needed to the extent that liquid infiltration can proceed. These results are somewhat in contradiction with the thermodynamic modification noted for $LiBH_4$ particles grown in carbon cryogels with pores size in the 10-40 nm range [39]. Further experiments are needed to understand this discrepancy. Actually, the synthesis of a large variety of carbons with narrow pores size distributions, ranging from 0.5 nm to 5 nm, in under way to try to correlate the $LiBH_4$ particles size and the thermal desorption temperatures.

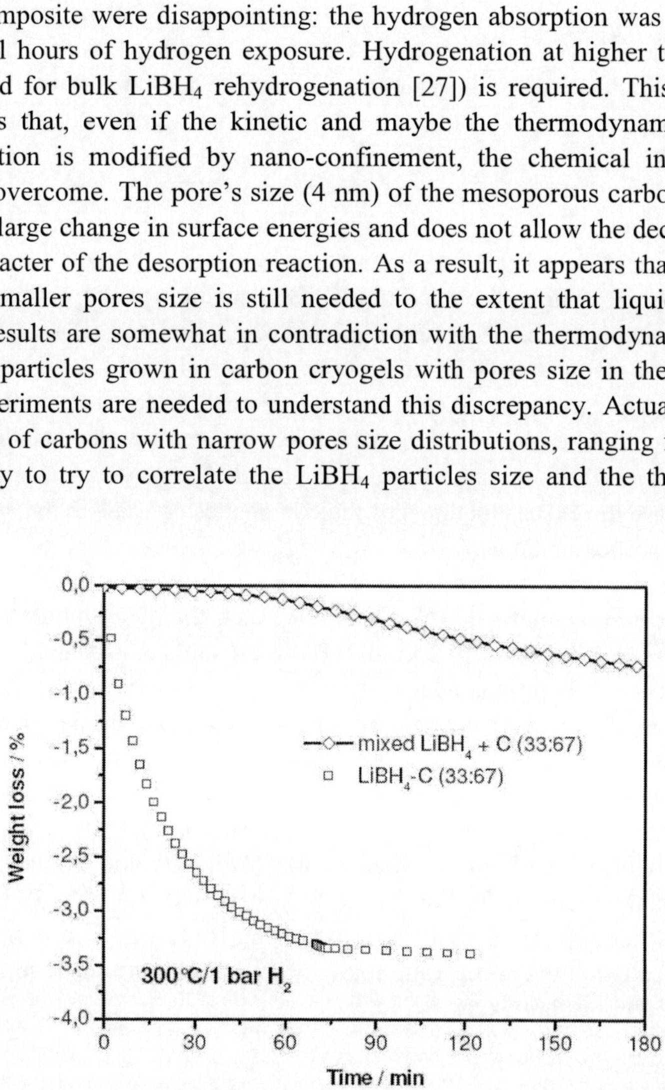

Figure 8. Kinetics of hydrogen release at 300°C under 1 bar H_2 for 33:67 $LiBH_4$/carbon composite and a simple mixing of 33 wt.% $LiBH_4$ and 67 wt.% carbon.

In order to increase the total hydrogen desorption capacity, composites with higher $LiBH_4$ loadings were also prepared [40]. Conversely to the X-ray diagram of the 33:67 $LiBH_4$/carbon composite, the one corresponding to the 50:50 ratio clearly exhibits sharp reflections assigned to the low temperature phase of $LiBH_4$ (cf. Figure 4c), suggesting that one part of $LiBH_4$ is not incorporated into the mesopores of the carbonaceous host, but remains as a bulk material onto the carbon surface. Such hypothesis was confirmed by looking at the TPD trace for the 50:50 $LiBH_4$/carbon composite (cf. Fig. 6), which looks like that of neat $LiBH_4$ with at least four hydrogen releases at about 75, 300, 430 and 480°C. Besides the small hydrogen desorption occurring at 75°C (which origin might be related to etherates decomposition as explained before), it should be noted that the main hydrogen release occurs through several steps above 300°C. From the amounts of hydrogen released at different temperatures (cf. circles, Fig. 7), two domains can be described: one below 400°C assigned to the desorption of $LiBH_4$ incorporated into the mesopores and the other one above 400°C corresponding to $LiBH_4$ simply deposited onto the carbon surface. The total hydrogen desorption capacity at 500°C reaches 6.0 wt.%, which is in good agreement with the 50:50 weight ratio.

Among the 6.0 wt. % desorption capacity of the 50:50 composite, 4.1 wt.% of hydrogen are desorbed at 400°C, which is a much higher value than that of bulk $LiBH_4$ (0.6 wt.%). If this desorption capacity of 4.1 wt. % is entirely attributed to nanosized $LiBH_4$ included in the porosity of the carbon matrix, the maximum $LiBH_4$ loading of the porous carbon is 36 %. Interestingly, this value comes close to the calculated amount of $LiBH_4$ (42 %) that could be hosted into a carbonaceous matrix having an opened porosity of 1.1 $cm^3.g^{-1}$ fully filled with $LiBH_4$ (crystallographic density = 0.66 $g.cm^{-3}$). The small discrepancy with the experimental loading (36 %) could be due to the difficulties in wetting the smaller pore (e.g. the micropores, Ø < 2 nm) with the etheric solutions of $LiBH_4$. Anyway, it remains that the best approach to increase the $LiBH_4$ loadings of the composites consists in the synthesis of carbon hosts with higher porous volumes.

Last, although not yet evaluated, the effect of the improvement of heat transfer, due to the carbon matrix, must be commented. As a matter of fact, the hydrogen desorption reaction of $LiBH_4$ is highly endothermic (+69 $kJ.mol^{-1}$ H_2) [23] and, consequently, involves a large quantity of heat. As the thermal conductivity of $LiBH_4$ is low (around 1 $W.m^{-1}.K^{-1}$ [45]), the heat transfer is poor and appears as a rate limiting step for the hydrogen release. The presence of carbon with high thermal conductivities should largely improve the heat transfer. If the thermal conductivity of mesoporous carbons is low (even below that of $LiBH_4$ with values in the 0.1-0.5 $W.m^{-1}.K^{-1}$ range for commercial activated carbons), it is well known that the thermal conductivity of carbons increases with their graphitization degree. Thus, the development of new processes for the synthesis of carbons with both a high porous volume and a significant thermal conductivity is a promising route that is presently explored. This finding could lead to the preparation of new $LiBH_4$/carbon composites with high absorption/desorption performances.

Conclusions

Since many decades, lithium borohydride ($LiBH_4$ also called lithium tetrahydroborate) is well known in organic chemistry as a reducing agent, especially for the reduction of esters. In addition to this old application, $LiBH_4$ gains today a lot of interest in the scientific community, due to its very high hydrogen content, which could give to this compound a possible use as hydrogen storage material. In this chapter, we discussed about the last advances made on this compound concerning many aspects: synthesis methods, ionic conductivity, hydrogen release kinetic, and rehydrogenation conditions.

The development of new solvent-free synthesis methods of $LiBH_4$ has been recently reported. The synthesis by direct hydrogenation of the elements is very difficult due to the boron inertness, but it has been shown that this issue must be partially overcome by starting from lithium-boron alloys. Another promising route for the $LiBH_4$ synthesis consists in the ball-milling of LiH with boron under hydrogen atmosphere. Although the yield of $LiBH_4$ formation reported in the first study is not high (only 27 %), the optimizations of the milling conditions should largely enhance this value in a near future.

$LiBH_4$ crystallizes in an orthorhombic unit cell (Pnma) at room temperature and undergoes a structural transition at about 116°C leading to a hexagonal phase ($P6_3mc$). This structural transition is accompanied by a jump of the electrical conductivity by three orders of magnitude: the conductivity reaches 10^{-3} $S.cm^{-1}$ immediately above the transition temperature. It has been shown, especially by ^{7}Li NMR, that this conductivity is purely ionic, which makes pristine $LiBH_4$ a good lithium ionic conductor above 116°C. Interestingly, the HT phase of $LiBH_4$ has been recently stabilized at room temperature by synthesizing $LiBH_4$-LiI solid solutions. These materials have ionic conductivities of the order of 10^{-4} $S.cm^{-1}$ at 25°C and could be used as solid electrolytes in lithium-ion batteries.

Although $LiBH_4$ is a well-known compound, its thermal decomposition follows a complex process, which is still today the subject of many investigations. Briefly, after the $LiBH_4$ melting at 280°C, the dehydriding reaction proceeds slowly from the liquid state and leads in two steps to a mixture of amorphous boron and lithium hydride. The intermediate compound formed upon the decomposition process has been recently identified as being lithium dodecaborane ($Li_2B_{12}H_{12}$) both by computational approaches and by experimental results like Raman spectroscopy and ^{11}B NMR. The overall desorption process is highly endothermic (~70 $kJ.mol^{-1}$ H_2) and the rehydrogenation of the (LiH + B) mixture has been only reported at 350 bars H_2 and 600°C. According to these values, it is evident that bulk $LiBH_4$ cannot be used for hydrogen storage applications.

The hydrogen release kinetic of $LiBH_4$ can be strongly enhanced by ball-milling with metal oxides (SiO_2) or chlorides ($TiCl_3$). In the first case, the oxides act really as catalysts (no reaction with $LiBH_4$ upon milling) and can lower the onset dehydrogenation temperature of $LiBH_4$ to less than 200°C. The exact mechanism of the catalytic effect with oxides is however not yet identified. About the chlorides-modified $LiBH_4$, the process is very different since a chemical reaction occurs between $LiBH_4$ and the chlorides ($TiCl_3$) upon milling. The resulting borohydride, like $Ti(BH_4)_3$ for instance, can be decomposed at remarkably low temperatures (60°C), but these reactions are fully irreversible.

Another strategy for the enhancement of the desorption kinetic is the synthesis of $LiBH_4$ nanoparticles into porous hosts. This approach has been achieved using various meso(-micro)porous carbons (aerogels, silica replica) and various methods for the carbon impregnation (molten $LiBH_4$ at 300°C, $LiBH_4$ dissolved in ethers). In all cases, the rate of hydrogen release of $LiBH_4$ has been strongly enhanced together with a decrease of the onset dehydrogenation temperature to about 200°C (so below the melting point of $LiBH_4$). In addition to the kinetic effect, a modification of the desorption mechanism is noticed: it seems that the formation of intermediate compounds like $Li_2B_{12}H_{12}$ does not occur upon the hydrogen release for the carbon-encapsulated $LiBH_4$ nanoparticles.

The possible change in thermodynamic of the desorption reaction, for such nano-sized $LiBH_4$ particles, must be confirmed by further studies. The synthesis of a large variety of carbons with narrow pores size distributions ranging from 0.5 nm to 5 nm is an exciting challenge in order to try to correlate the $LiBH_4$ particles size and the thermal desorption temperatures. Nevertheless, the strong benefit of the approach consisting in the inclusion of $LiBH_4$ nanoparticles into porous carbons is already demonstrated, as the feasibility to modify the hydrogen release mechanism path for a hydrogen storage material, when introducing size (i.e. nanoscale) as an additional variable besides structure and composition, has been shown.

ACKNOWLEDGMENTS

Dr. S. Cahen and J.B. Eymery are gratefully acknowledged for their works on the synthesis of $LiBH_4$ nanoparticles by impregnation of mesoporous carbons. Pr. J.M. Tarascon is thanked for the critical reading and the suggestions made to this chapter.

REFERENCES

[1] Schlesinger, H.I.; Brown, H.C., US Patent 2.461.661 (15/02/1949)
[2] Schlesinger, H.I.; Brown, H.C.; Abraham, B.; Bond, A.C.; Davidson, N.; Finholt, A.E. et al., *J. Am. Chem. Soc.* 1953, 75(1), 186
[3] Brown, H.C.; Choi, Y.M.; Narasimhan, S., *Inorg. Chem.* 1981, 20, 4456
[4] Goerrig, D., Deutsches Patent DE000001077644A (1960)
[5] Friedrichs, O.; Buchter, F.; Borgschulte, A.; Remhof, A.; Zwicky, C.N.; Mauron, P.; Bielmann, M.; Züttel, A., *Acta Mater.* 2008, 56, 949-954
[6] Agresti, F.; Khandelwal, A., *Scripta Mater.* 2009, 60, 753-755
[7] Filinchuk, Y.; Chernyshov, D.; Dmitriev, V., *Z. Krystallogr.* 2008, 223, 649-659
[8] Harris, P.M.; Meibohm, E.P., *J. Am. Chem. Soc.* 1947, 63, 1231
[9] Soulié, J.P.; Renaudin, G.; Cerny, R.; Yvon, K., *J. All. Comp.* 2002, 346, 200-205
[10] Gomes, S.; Hagemann, H.; Yvon, K., *J. All. Comp.* 2002, 346, 206-210
[11] Filinchuk, Y.; Chernyshov, D.; Cerny, R., *J. Phys. Chem. C* 2008, 112, 10579-10584
[12] Aldridge, S.; Blake, A.; Downs, A.; Gould, R.O.; Parsons, S.; Pulham, C.R., *J. Chem. Soc. Dalton Trans.* 1997, 1007-1012
[13] Hartman, M.R.; Rush, J.J.; Udovic, T.J.; Bowman, R.C.; Hwang, S.J., *J. Solid State Chem.* 2007, 180, 1298-1305
[14] Filinchuk, Y.; Chernyshov, D.; Nevidomskyy, A.; Dmitriev, V., *Angew. Chemie Int. Ed.* 2008, 47, 529-532
[15] Matsuo, M.; Nakamori, Y.; Orimo, S.; Maekawa, H.; Takamura, H.; *Applied Phys. Let.* 2007, 91, 224103
[16] Skripov, A.V.; Soloninin, A.V.; Filinchuk, Y.; Chernyshov, D., *J. Phys. Chem. C* 2008, 112, 18701-18705
[17] Corey, R.L.; Shane, D.T.; Bowman, R.C.; Conradi, M.S., *J. Phys. Chem. C* 2008, 112, 18706-18710
[18] Matsuo, M.; Takamura, H.; Maekawa, H.; Li, H.W.; Orimo, S., *Applied Phys. Let.* 2009, 94, 084103
[19] Maekawa, H.; Matsuo, M.; Takamura, H.; Ando, M.; Noda, Y.; Karahashi, T.; Orimo, S., *J. Am. Chem. Soc.* 2009, 131, 894-895

[20] Züttel, A.; Wenger, P.; Rentsch, S.; Sudan, P.; Mauron, P., Emmenegger, C., *J. Power Sources* 2003, 118, 1-7
[21] Züttel, A.; Rentsch, S.; Fischer, P.; Wenger, P.; Sudan, P.; Mauron, P., Emmenegger, *J. All. Comp.* 2003, 356-357, 515-520
[22] Orimo, S.; Nakamori, Y.; Ohba, N.; Miwa, K.; Aoki, M.; Towata, S.; Züttel, A., *Applied Phys. Let.* 2006, 89, 219201-219203
[23] Züttel, A.; Borgschulte, A.; Orimo, S., *Scripta Mater.* 2007, 56, 823-828
[24] Ohba, N.; Miwa, K.; Aoki, M.; Noritake, T.; Towata, S.; Nakamori, Y., Orimo, S.; Züttel, A., *Phys. Rev. B* 2006, 74, 75110
[25] Her, J-H.; Yousufuddin, M.; Zhou, W.; Jalisatgi, S.; Kulleck, J.; Zan, J.; Hwang, S-J.; Bowman, R.; Udovic, T., *Inorg. Chem.* 2008, 47, 9757-9759
[26] Tiritiris, I.; Schleid, T., *Z. Anorg. Allg. Chem.* 2003, 629, 1390
[27] Orimo, S.; Nakamori, Y.; Kitahara, G.; Miwa, K.; Ohba, N.; Towata, S.; Züttel, A., *J. All. Comp.* 2005, 404-406, 427-430
[28] Au, M.; Jurgensen, A., *J. Phys. Chem. B* 2006, 110, 7062-7067
[29] Au, M.; Jurgensen, A.; Zeigler, K., *J. Phys. Chem. B* 2006, 110, 26482-26487
[30] Au, M.; Spencer, W.; Jurgensen, A.; Zeigler, K., *J. All. Comp.* 2008, 462, 303-309
[31] Hoekstra, H.R.; Katz, J.J., *J. Am. Chem. Soc.* 1949, 71, 2488-2492
[32] Wiberg, E.; Bauer, R., *Z. Naturforsch.* 1950, 5b, 397
[33] Gutowska, A.; Li, L.; Shin, Y.; Wang, C.; Li, X; Linehan, J.; Scott Smith, R.; Kay, B.; Gutowski, M.; Autrey, T., *Angew. Chem. Int. Ed.* 2005, 44, 3578-3582.
[34] Feaver, A.; Sepehri, S.; Shamberger, P.; Stowe, A.; Autrey, T.; Cao, G., *J. Phys. Chem. B* 2007, 111, 7469-7472.
[35] Baldé, C.P.; Hereijgers, B.; Bitter, J.; De Jong, P., *Angew. Chem. Int. Ed.* 2006, 45, 3501-3503.
[36] Zhang, Y.; Wang, A.; Sun, L.; Fan, M.; Chu, H.; Sun, J.; Zhang, T., *Int. J. Hydrogen Energy* 2007, 32, 3976-3980.
[37] Vajo, J.; Olson, G., *Scripta Mater.* 2007, 56, 829-834.
[38] Vajo, J.; Salguero, T.; Gross, A.; Skeith, S.; Olson, G., *J. All. Comp.* 2007, 446-447, 409-414.
[39] Gross, A.; Vajo, J.; Van Atta, S.; Olson, G., *J. Phys. Chem. C* 2008, 112, 5651-5657.
[40] Cahen, S.; Eymery, J.B.; Janot, R.; Tarascon, J.M., *J. Power Sources* 2009, 189, 902-908.
[41] Ryoo, R.; Joo, S.H.; Jun, S., *J. Phys. Chem. B* 1999, 103, 7743-46.
[42] Parmentier, J.; Saadhallah, S.; Reda, M.; Gibot, P.; Roux, M.; Vidal, L.; Vix-Guterl, C.; Patarin, J., *J. Phys. Chem. Solids* 2004, 65, 139-146.
[43] Mosegaard, L.; Moller, B.; Jorgensen, E.; Bösenberg, U.; Dornheim, M.; Hanson, J.; Cerenius, Y.; Jakobsen, H.; Besenbacher, F.; Jensen, T., *J. All. Comp.* 2007, 446-447, 301-305.
[44] Kostka, J.; Lohstroh, W.; Fichtner, M.; Hahn, H., *J. Phys. Chem. C* 2007, 111, 14026-14029.
[45] Talyzin, A.V.; Andersson, O.; Sundqvist, B.; Kurnosov, A.; Dubrovinsky, L., *J. Solid State Chem.* 2007, 180, 510-517.

In: Boron Hydrides High Potential Hydrogen Storage Material ISBN:978-1-61668-361-0
Editors: Umit B. Demirci and Philippe Miele

Chapter 4

Hydrogen Cycle with Sodium Borohydride

***Çetin Çakanyildiriıldırım*[1] *and Metin Guruürü*[2]**
Gazi University, Eng. Fac., Chemical Eng. Dep., 06570 Maltepe-Ankara/Turkey

Abstract

In this chapter, production of $NaBH_4$ as hydrogen storage material is discussed. Linking this with other processes, which constitutes a system to recycle hydrogen is shown. Difficulties and advantages of $NaBH_4$ synthesis by natural available boron source kolemanite and borax minerals are discussed in details. It is presented that basic chemicals suffice for $NaBH_4$ production in a procedure designed to work below 275°C. That proves the feasibility of $NaBH_4$ production. Offered procedure may be used to compensate for the loss of materials during the recycle. In this procedure, hydrogen including substances has been used to remove oxygen and bond remaining hydrogen to the by-product $NaBO_2$ of dehydrogenation reaction. The wet and dry applications of producing $NaBH_4$ have been also discussed. Moreover a shorter procedure can be proposed to directly obtain $NaBH_4$ from borax mineral. Among these processes, dehydrogenation is critical but it requires plenty of water. Hence, new types of catalysts are important to reduce the amount of water required during dehydrogenation. Catalyst types and their production techniques are also given in this chapter. Performance of the catalysts during the hydrolysis of $NaBH_4$ is tried to explain according to the dehydrogenation conditions such as temperature, concentration of the reactants and by-product effect.

Keywords: $NaBH_4$; Hydrogen storage; Dehydrogenation; Renewable energy.

1. Importance of Sodium Borohydride in Hydrogen Cycle

The decreasing reserves of natural gas, coal, and petroleum and the increasing consumption rate of these resources highlight the importance of renewable energy. Although

1Tel.: +90 312 582 3544; E-mail: cetinc@gazi.edu.tr
2Tel.: +90 312 582 3555; E-mail: mguru@gazi.edu.tr

the consumption of fossil fuels is restricted because of their high carbon/sulfur contents, these restrictions are not always effective. Thus, environmental problems, such as acid rain, ozone depletion, and climate change, arise from the presence of CO_2, SO_x, and NO_x in the emission gases. Using clean and renewable energy sources can alleviate most of these problems. For this purpose, hydrogen seems to be the most appropriate energy carrier. Hydrogen fuel cells have been successful in effectively producing energy. Yet a proton source to feed the cells and more investigations are needed for an uninterrupted energy (i.e. electricity) production. In this context, scientists are driven to discover new ways to produce renewable energy carrying systems [1].

Hydrogen occupies the center stage in these efforts as an environment friendly clean fuel, producing no pollution, giving high amounts of energy that is easy to convert to different forms. Practicable use of hydrogen energy is hostage to the easy and feasible hydrogen producing processes and hydrogen storage. Main storage and transportation methods are:

- Storage of hydrogen within a pressurized steel tank,
- Storage as a liquid under the cryogenic conditions,
- Storage of hydrogen on activated carbon, in carbon nanotube, in graphite nanofiber,and
- Storage in alloys suitable to adsorption and desorption of hydrogen in their structures [2, 3].

The mass ratio of stored hydrogen to its container is an important parameter and it should ideally be at least 10%.

Metal borohydrides can adsorb hydrogen and store it in solid form. These compounds release hydrogen by thermal decomposition or catalytic desorption. Theoretically, thermal decomposition yields only 10 wt% of the hydrogen in the storage medium, unfortunately this ratio is practically limited at 1 %. Catalytic approach is advantageous since the hydrogen of the water is also used [4, 5]. Some of the metal borohydrides and their properties are listed in Table 1.

Table 1. Properties of some metal borohydrides.

Property	$LiBH_4$	$NaBH_4$	KBH_4	$RbBH_4$	$CsBH_4$
Melting point, K	541	778	858	-	-
Decomposition temp., K	548	673	773-857	873	873
Density, g cm^{-3}	0.66	1.07	1.18	1.71	2.40
H_2, wt%	18.51	10.66	7.47	4.02	2.73
Molecular weight	21.78	37.83	53.94	100.31	147.75
Reactivity index	-	1.55	1.49	1.49	1.50
Crystal energy, kJ mol^{-1}	792.0	697.5	657.0	648.0	630.1
ΔH°, kJ mol^{-1}	-184	-183	-243	-246	-264
ΔS°_{298}, kJ mol^{-1}	-128.7	-126.3	-161.0	-179.0	-192.0

Metal borohydrides are the best storage materials in view of their high storage capacity and security. Therefore, these chemicals could be indispensable to supply hydrogen to the fuel cells in near future. Because of this reason, economic and effective production technologies have to be developed. Two (dry and wet) methods are common to synthesize metal borohydrides. According to the desired product and reactants, a suitable method can be used. Especially, the hydrogen recycles system in Figure 1 or variants thereof must be the concern of scientists. Figure 1 shows an environmental friendly recycling system for metal borohydride production. The process starts with mined raw boron mineral and continues with the addition of basic chemicals such as H_2SO_4 and methanol to synthesize the commercially valuable product $NaBH_4$. This method can be used for both conversion of metal metaborates to their borohydrides [6, 7] and production of metal borohydrides from their elements [8]. However some elements such Na behaves sticky in ball milling application and resists reacting. Therefore wet method may be the suitable procedure in $NaBH_4$ production. Nevertheless studies performed in the presence of B_2O_3 show that $NaBH_4$ production by means of ball milling is possible even elemental Na are used as reactant.

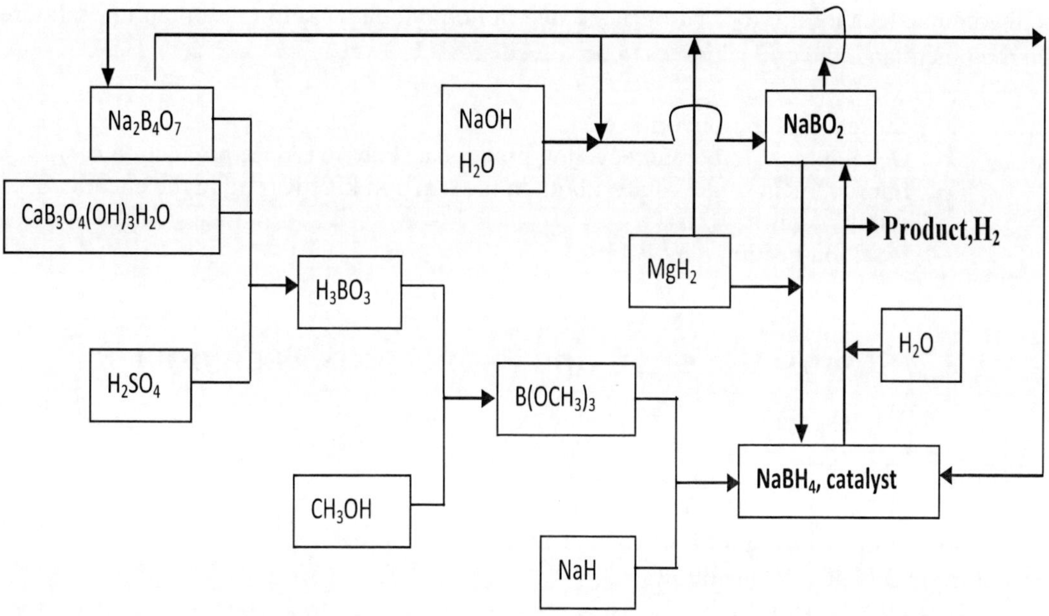

Figure 1. Hydrogen cycle system.

In beyond applications, hydrogen has been identified as an attractive candidate for energy carrier by operating it as an energy source in proton exchange membrane fuel cell (PEMFC). Hydrogen generated from steam reforming of natural gas, the common industrial process to produce hydrogen gas, is not suitable for feed to proton exchange membrane fuel cell devices due to inevitable poisoning of proton exchange membrane owing to the by-products of this process, such as carbon dioxide and carbon monoxide. The carbon contamination reduces the performance of proton exchange membrane fuel cell by deactivation of catalyst. Therefore, complex hydrides are attractive hydrogen supplying candidates to fuel cells at room temperature. Among the metal hydrides, aqueous sodium borohydride ($NaBH_4$) seems to be an ideal hydrogen source because it is stable, non-flammable, nontoxic in nature, and with hydrogen storing capability of 10.8 wt%.

Hydrogen generation through sodium borohydride ($NaBH_4$) hydrolysis has attracted much attention. The total hydrolysis of 1 g of borohydride releases at about 2.5 l of hydrogen in standard conditions. This operation must occur in a solution which pH value is arranged to 12-14 by addition of NaOH or KOH. However usage of KOH for stabilization results KBH_4 synthesis, decreasing the hydrogen generation rate and so performance of dehydrogenation. Stable aqueous solutions for hydrogen generation are in the concentration range of 1–30 wt% of sodium borohydride.

Sodium borohydride is easy to handle with respect to the other chemical hydrides. At room temperature only a small percentage of the theoretical amount of hydrogen is released by the $NaBH_4$ hydrolysis reaction, catalytically activated. Hydrogen produced by this reaction, could be directly sent to the anode of a fuel cell, leading to a simple electrical energy generator. A sodium borohydride solution (35 wt% of $NaBH_4$ solution equivalent to 7.7 wt% of hydrogen) is chemically stable in non-flammable, non-explosive and basic environment, not producing pollutants or by-products responsible for greenhouse effect. Finally, it seems to be the unique technology for hydrogen storage in liquid (aqueous) form at room temperature. On the other hand, the critical issues to be solved are:

- Current high borohydride cost,
- Development of borate recycling process in order to reduce production cost,
- Improvement of the catalyst both in the reaction kinetics and in the chemical stability, and
- Reaction heating removal [9].

2. Sodium Borohydride Production Techniques

2.1. Borax Reactions

Borax is mined in the form of its hydrides. Mined borax is grinded before it is reacted with sulfuric acid at 90°C as follows:

$$2CaO.3B_2O_3.5H_2O + 2H_2SO_4 + 6\ H_2O \rightarrow 2\ (CaSO_4.\ 2H_2O) + 6H_3BO_3 \quad (1)$$

Filtration and crystallization are required to purify H_3BO_3. Since the water in borax hydrides cause some problems in further steps of alkaline or alkaline earth borohydride production, it is sometimes removed by a series of reactions:

$$1/4\ Na_2B_4O_7.10H_2O + 1/2\ NaOH + 5/4\ H_2O \rightarrow NaBO_2.4H_2O` \quad (2)$$

$$NaBO_2.4H_2O \rightarrow NaBO_2 + 4H_2O \quad [270\ ^\circ C] \quad (3)$$

2.2. Trimetyl Borate Reactions

Trimethyl borate $B(OCH_3)_3$ is the main reactant in most of the alkaline or alkaline earth borohydride production processes named as wet method. It is a weak Lewis acid, with various applications in organic synthesis. Trimethyl borate is stored and sold as an azeotrope of methyl alcohol. Generally, the commercial product contains 70% trimethyl borate and the rest is alcohol but 99% pure form is also available. Density of pure trimethyl borate is 0.915 g/ml and density measurements can determine the purity of this chemical. Also, titrimetric method [10] or instrumental measurement methods are feasible.

Trimethyl borate compound is necessary to form sodium borohydride and can be synthesized by an esterification reaction of boric acid with methyl alcohol in the presence of dehydrants at about 65°C:

$$B(OH)_3 + 3\ CH_3OH \rightarrow \leftarrow B(OCH_3)_3 + 3\ H_2O \quad (4)$$

$$Na_2B_4O_7.10H_2O + 2\ H_2SO_4 + 16\ CH_3OH \rightarrow 4\ [B(OCH_3)3CH_3OH] + 2\ NaHSO_4 + 17\ H_2O \quad (5)$$

The last reaction is an alternative way of producing trimethyl borate with borax [11]. Moreover, trimethyl borate can be produced with 1,1-dimethyloxicyclohexane or 2,5-dimethyloxi-2,5- dimethyl-1,4-dioxan instead of methanol [12].

During the trimethyl borate production, water is formed and it reverses the reaction, so it has to be taken away from the system. For this purpose, reaction vessel must be filled with enough amount of dehydrant. The most important parameters of selecting dehydrant is its stability while the reaction happens. There are many types of dehydrants: calcium chloride ($CaCl_2$), lithium chloride (LiCl), zinc chloride ($ZnCl_2$) and some of the artificial products. From a cost point of view, dehydrant must be chosen economically. Calcium chloride seems to be a good option. It is well known that 50-60 g of calcium chloride is adequate to dehydrate approximately 255 g of reactant. On the other hand, lithium chloride is twice more expensive but just 25 g of lithium chloride is sufficient to do the same amount of dehydration. Besides, the first azeotropic product trimethyl borate is purer and more if compared with synthesized lithium chloride. Zinc chloride is advantageous because of its rapid affinity for moisture but the low intake capacity of zinc chloride limits its own application. Artificial dehydrants may be financially beneficial and the recycle of them may be achieved simply by thermal desorption. Furthermore, at the reaction temperature trimethyl borate-methanol azeotrope can be removed from the reactor in gas form and condensed by using a basic water cooled condenser. This removal increases the reaction yield.

2.3. $NaBH_4$ Reactions

Sodium borohydride ($NaBH_4$) is a versatile reducing agent in various organic and inorganic processes [13]. Sodium borohydride is also expected to be the hydrogen carrier and supplier for fuel cells and energy production systems. It can be produced with trimethyl borate and sodium hydride reactants; this reaction is named Rohm-Hass process:

$$4\ NaH + B(OCH_3)_3 \rightarrow NaBH_4 + 3\ NaOCH_3 \quad (6)$$

This reaction is generally performed in an autoclave. Reaction chamber must be swept by an inert gas such as nitrogen, helium or argon before the reaction. Sodium borohydride can also be produced by sodium metaborate ($NaBO_2$) and hydrogen reaction in the presence of oxide adsorbent elements.

If the produced amount of sodium borohydride is more than laboratory scale, slowly adding trimethyl borate and rapidly transferring the created heat provide more secure circumstances. The reactants must be kept at about 225-275°C and under high pressure during the reaction [14]. The pProduct of the reaction must be purified. Generally, a solvent is used to solve the desired or undesired product and then crystallization or filtration process takes place. Each solvent should be applied at its optimum temperature. Therefore, such solvents improve the purification process. Ammonia and ethylene diamine can be used as a solvent. In addition, Sodium borohydride solubility in diether glycol changes radically with the temperature. Therefore, diethylene glycol dimethyl ether has been offered to be a good solvent nevertheless sodium atom of Sodium borohydride is very attractive for the three oxygen atoms of the ether.

Another way of sodium borohydride production is reacting dehydrated borax and silica:

$$Na_2B_4O_7 + 16Na + 8H_2 + 7\ SiO_2 \rightarrow 4\ NaBH_4 + 7\ Na_2SiO_3 \quad (7)$$

The stoichiometric lack of sodium in reaction 7 must be compensated during the reaction. Sodium carbonate (Na_2CO_3) is a suitable chemical to tolerate sodium inadequately since it causes no oxide. The main step of this process occurred at < 3 bar hydrogen atmosphere and temperatures of 450-500 °C. Product was taken by leaching with ammonia [15]. Also, elemental sodium and diboran can yield sodium borohydride in the diethyl ether if reacted for 24 hours.

$$2\ Na + 2\ B_2H_6 \rightarrow NaBH_4 + NaB_3H_8 \quad (8)$$

Sodium hydroxide and diborane reaction is an alternative way of production [16].

$$4\ NaOH + 2\ B_2H_6 \rightarrow 3\ NaBH_4 + NaBO_2 + 2\ H_2O \quad (9)$$

3. Dehydrogenation of $NaBH_4$ and its Usage as Hydrogen Carrier

Hydrogen gas can be generated by the thermal and catalytic hydrolysis of sodium borohydride. Nevertheless dehydrogenation of sodium borohydride is a chemical change resulting in exchange between the hydrogen atoms of fuel and oxygen of water. Thus, most of the investigations are concentrated in catalytic hydrolysis of sodium borohydride but thermal hydrolysis and its limited performance must be described to understand the importance of the catalytic one.

3.1. Thermal Dehydrogenation

Hydrogen atom can be evaluated from MBH_4 structures by means of high temperature. In this procedure thermal desorption temperature must be kept below the melting point of the MBH_4. However, under hydrogen atmosphere MBH_4 melting temperatures can be regarded as indexes of hydrogen desorption (decomposition) temperatures. The differential scanning calorimetry (DSC) profiles of MBH_4 (M=Li, Na and K) during the heating rate of 10 °C min^{-1} under 0.1 MPa hydrogen are given in Figure 2. Sodium and potassium borohydrides has one melting reaction at 497 °C and 607 °C respectively whereas lithium borohydride has two reactions. One belongs to structural changes at 107 °C, and the other is melting at 277 °C, respectively [17]. By contrast, thermal hydrogen evolution temperatures of lithium borohydride are given at about 340 and 500 °C [18]. Thus thermal desorption temperature have to be decreased below melting temperature level by additives or cation substitutions using smaller sized and/or higher valenced-cations with larger electro negativities. For example silica, magnesium, lithium and copper can be used for this purpose. However, high-temperature working conditions and low yields decrease the attractiveness of thermal desorption.

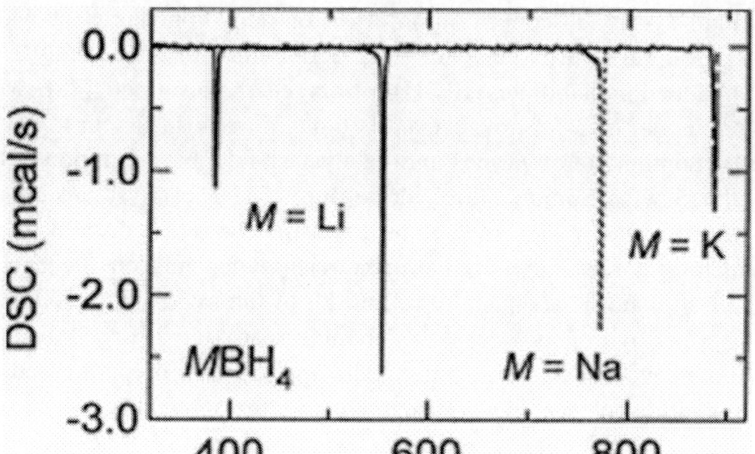

Figure 2. Differential scanning calorimetry (DSC) profiles of MBH_4 (M=Li, Na, and K) under 0.1 MPa hydrogen. The heating rate is 10 °C min^{-1} (Permitted by Orimo et al. [17]).

3.2. Catalytic Dehydrogenation

Hydrogenation of sodium borohydride is a chemical process for which the catalytic approach is much more favored method [19, 20]:

$$NaBH_4 + (2 + x)\,H_2O \rightarrow 4\,H_2 + NaBO_2.xH_2O \qquad (10)$$

Schlesinger et al. proved that catalyst could significantly catalyses the sodium borohydride hydrolysis reaction. In order to achieve that addition of acids may be applied [21]. However, limited number of investigations on homogeneous catalysts has been carried out due to the difficulty of the controlling failures. Keçeli and Özkar [22] offered using ruthenium (III) acetylacetonate as the homogeneous catalyst for sodium borohydride hydrolysis reaction. It is considered that use of homogeneous catalyst for sodium borohydride hydrolysis is good for continuous hydrogen production but not suitable for the hydrogen supply on demand because sodium borohydride hydrolysis reaction usually become uncontrollable due to the difficulty to halt reaction process [23].

3.2.1. Alkaline Hydrolysis Solution

Sodium hydroxide must be added to hydrolysis process to suppress the self-hydrolysis of sodium borohydride. Researchers suggest different mass and mole percentage for sodium hydroxide addition. According to the reactor type and amount of reactants, NaOH can be added in various amounts but generally in the range of 5-15 wt% of reactants. The excess amount of NaOH decreases the hydrogen yield [24]. This effect particularly appears on Ru based catalyst. Researchers explain this effect on the basis of reduction in water activity [25, 26] and occupation of active sites with OH^- instead of BH_4^- [27]. However, investigations proved that sodium hydroxide concentration effect on dehydrogenation obeys zero reaction order. Moreover, sodium hydroxide renews the active sites of used catalyst and enhances desorption of hydrated metaborates $B(OH)_4^-$ from catalyst surface [28]. Mostly sodium hydroxide in the amount of 3-5 % of reactant is thought to be sufficient to control hydrogen release [29, 30]. Sodium borohydride, and catalyst should be continuously mixed in the reactor. Generally 5 wt% Sodium borohydride and 1 wt% catalyst are enough to obtain considerable amount of hydrogen. As soon as the distilled water is introduced to the system, hydrogen production starts. Time of hydrogen release depends on the catalyst type and amount. Just 2 ml of water is adequate to release all of the hydrogen in sodium borohydride and the some of water hydrogen. Because of sodium hydroxide presence, during the process pH value remains between 12-14. In some of the applications, preparing sodium hydroxide and sodium borohydride solution in advance might easy the applications.

3.2.2. Water Requirement for Hydrolysis

Water amount is an important parameter because approximately 95 wt% of the reactant is occupied by water. Therefore, sodium borohydride has potential to support hydrogen storage for a portable system only if the water amount is decreased. Fuel cell applications will be possible if low-weight hydrogen engines can be produced. For an automotive hydrogen storage system, the value x in reaction (10) was given by the U.S. Department of Energy (DOE) as 0.84 for applications beyond year 2015. In order to achieve this x value, the water amount must be reduced to one-fifteenth of what is described above. Reducing water will support manufacturers working hard to create engines lighter than 132 kg, which can cover 300 miles with 8 kg of hydrogen. In order to achieve hydrogen release with less water, processes that use steam instead of water have been developed. Steam hydrolysis system has disadvantages because of by-products and $NaBO_2.H_2O$ which covers the surface of fuel and prevents the steam from reaching deeper [31]. Therefore, on board application of hydrogen generation systems by using sodium borohydride seems to be difficult to achieve, unfortunately [32]. However researchers are still spending effort to find economic solutions for this renewable and environmentally friendly energy carrier.

Hydrolysis of sodium borohydride directly means the hydration of by-product sodium metaborate. In reaction 3, thermal method to dehydrate the hydrated sodium metaborate $NaBO_2.4H_2O$ is explained. In order to decrease the water amount bounded to sodium metaborate, optimum stoichiometric ratio must be applied. The solubility of sodium borohydride in 100 g of water at 25°C is 55 g, but that of sodium metaborate is 33 g. The concentration of sodium borohydride at this temperature should be below 15 g/100 g water in order to keep sodium metaborate and sodium borohydride without precipitation. Thus minimum amount of required water to hydrolyze sodium borohydride and create a homogeneous solution can be computed. Using less water than the minimum value certainly causes precipitation of reactants over active sites of the catalyst. On the contrary, excess amount of water using increases the required heat load to remove the water of hydrolyzed by-product sodium metaborate. Thus, the quality of the reactants and gravimetric hydrogen storage capacity of the system decreases.

It is known that when 1 mole of borohydride ion BH_4^- is hydrolyzed, 1 mole of hydarated metaborate ion $B(OH)_4^-$ will be formed. It means that 2 moles of free water are held by metaborate ions BO_2^-, resulting in decreasing the water reactivity. Therefore, the water diffusion process would dominate borohydride ion hydrolysis. Consequently, hydrogen generation rate are dependent upon the sodium borohydride concentration and water concentration if high concentrated sodium borohydride solutions are applied for hydrogen generation [23]. The concentration of sodium borohydride should be as high as possible in order to improve the fuel energy density. However, when the sodium borohydride concentration was high enough, hydrated borates, as the reaction product would precipitate from the solution, which would block the active sites of the catalysts and bring about the complexity of solution transportation. On the other hand, water reactivity would influence the hydrolysis reaction mechanism, as water was another hydrogen source for the borohydride ion hydrolysis. Theoretically speaking, with decreasing the water content in the reaction system, following hydrolysis reactions would occur [23]:

$$NaBH_4 + 6\ H_2O \rightarrow 4\ H_2 + NaBO_2.4H_2O\ (\text{or } NaB(OH)_4.2H_2O) \downarrow,\ \Delta H^\circ = -272.4\ \text{kJ mol}^{-1}(NaBH_4) \quad (11)$$

$$NaBH_4 + 4\ H_2O \rightarrow 4\ H_2 + NaBO_2.2H_2O\ (\text{or } NaB(OH)_4) \downarrow,\ \Delta H^\circ = -\ 250.1\ \text{kJ mol}^{-1}(NaBH_4) \quad (12)$$

3.2.3. Catalyst Preparation and its Effect on Hydrolysis

The 3d transition elements are advantageous in catalyst synthesis because of their low cost. These elements may be reduced to metal boride form (particularly, Co-B or Ni-B) before their usage. The procedure is simply the reaction of these metal chlorides with sodium borohydride. An extra compound may be introduced to the reactor if a complex catalyst structure is needed. Reduced metal must be well filtered and washed with distilled water and ethanol. In the drying procedures inert atmosphere must be established and extreme temperature values must be avoided since high temperatures cause agglomeration and reduce the surface area and catalytic activity. Besides, reduction of the catalyst simultaneously and spontaneously occurs during the early moments of dehydrogenation. Thus cobalt chloride or nickel chloride can be introduced directly or on a support to the alkaline solution and cobalt boride Co-B or nickel boride Ni-B production can be regarded as a first step of the reaction. Nevertheless these catalysts are sensible for oxidation. Even they are stable against oxidation before they usage, they must be protected from atmospheric effect after their processing. Oxidation of these metal borides reduces hydrogen revolution rates or completely poisons the catalyst depending on the production futures. In order to avoid the catalyst can be stored under inert atmosphere or in solution after their usage.

Catalysts can accelerate hydrogen generation from $NaBH_4$. In catalytic applications, surface area of the catalyst is important to provide enough active sites for the molecules to react. Surface area of the catalyst can be arranged with synthesis circumstances. For instance Patel et al. [28] showed that rapid chemical reduction application causes limited time for particles to grow above a few nanometers. Thus Co-P-B catalyst can be synthesized with average particles size in the range between 30-40 nm and high hydrogen generation rates can be obtained up to approximately 4000 ml min^{-1} g^{-1}. Furthermore, Chen et al. [33] proved that 10^{-2} wt% sodium borohydride is the best reduction solution concentration value to operate for polymer supported ruthenium catalyst. Decreasing the reduction solution concentration limits catalyst activity because of the self-hydrolysis under such a diluted solution, and therefore no significant Ru reduction obtained. Also, titrating rate was reported to keep between 1.5-2.0 ml min^{-1} to obtain uniform and smooth ruthenium nanoclusters [33]. Active sites must be homogeneously distributed over the support. Also, support material must not be broken or dissolved in sodium borohydride alkaline solution. Activated carbon, gamma alumina (γ-Al_2O_3) and silica are known as the support materials showing best resistance in alkaline solution. Salts of 3d transition metals have extreme desorption behaviors. PtRu-$LiCoO_2$ catalyst efficiency was found to be twice as high as that of either Ru-$LiCoO_2$ or Pt-$LiCoO_2$ [29]. Zhang et al. [30] studied in detail the kinetics of sodium borohydride hydrolysis on Ru-on-carbon catalyst. At low temperatures, zero-order and, at high temperatures, first-order kinetic behaviors were obtained. Results demonstrated that the reaction consists of two steps:

- First, sodium borohydride adsorption on the surface, and,
- Second, reaction of the adsorbed species on the catalyst surface.

The Langmuir-Hinshelwood model led to the conclusion that the adsorption step was responsible for kinetic order variation [30]. Besides, exothermic character of the reaction may shape the reaction order. Hence, changes in reaction temperature cause order to shift from zero to first. Moreover concentration of sodium borohydride also affects the reaction order. There are many of research to compute the rate of dehydrogenation reaction. These studies can be grouped in two categories:

- First, rate changes with temperature (given above), and,
- Second, rate changes with sodium borohydride concentration.

According to the Co-P-B catalyst prepared by chemical reduction method, at high sodium borohydride concentration varying from 0.075 to 0.25 M it results zero order while at concentration less than 0.05M kinetics obeys first order [28]. At low sodium borohydride concentrations, the hydride/catalyst molar ratio is low and thus there are so many active sites available to serve dehydrogenation. Therefore, limiting step is the diffusion of borohydride ion to the catalyst surface and it constitutes the longest period of hydrogen revolution phenomenon. Nevertheless, at high concentrations, the zero order kinetics controls the reaction due to the borohydride ion induced dynamic saturation of the catalyst [28].

3.2.4. Supported Materials of Heterogeneous Catalysts

Hydrogen may be produced from the reformation of petroleum and nature gas with steam-methane, steam-oxygen or auto-thermal reforming. These applications require large equipments and have no chance to be performed within small machines. In comparison to the technologies listed above, sodium borohydride is more appropriate as an environmentally friendly and recyclable energy source because it is pure, controllable, stable and nonflammable. In order to see effective use of this high capacity hydrogen carrier, sodium hydroxide-stabilized sodium borohydride solution was investigated with platinum catalyst dispersed on active carbon, carbon black and alumina supports. The Pt/C catalyst was the best among the tested ones by virtue of its high surface area and its stable hydrogen liberation rate of 8.5 L min^{-1} g^{-1}(Pt). Besides, effects of the catalysts particular sizes were investigated to obtain the optimal conditions. It was found that 6 wt% of Pt and less than 2 nm sized catalyst powder increased the hydrogen release. To obtain such size, the calcination temperature of the catalyst was kept at 300 ˚C [34].

Proton exchange membrane fuel cells are popular power devices and continuous hydrogen feeders. A group of catalyst candidates (Co, Ni, Ru, Pt, and Pt/Ru) were placed on active carbon and γ-Al_2O_3. Both of the supports keep their physical properties in an alkali solution. However, pore diameters of γ-Al_2O_3 support are large enough to eliminate the adsorption resistance. One gram of Co/γ-Al_2O_3 catalyst had capacity to feed a system with 220 ml min^{-1} hydrogen flow for 5.5 h [35]. If supporting materials are connected to each other, catalysts can be physically removed from the reaction environment. This technique facilitates the control of hydrogen flow rate. Ingersoll et al. [24] used a Ni-Co-B catalyst that can liberate hydrogen even at 0 ˚C. This catalyst lost its efficiency if sodium hydroxide exceeds 15 wt% of reactant. Besides, increasing sodium borohydride amount resulted in more sodium metaborate synthesis. Therefore, optimum reactant amounts must be determined. Moreover, high sodium borohydride concentration systems did not pair well with a Ni-Co including catalyst [24].

3.2.5. Processing Magnetic Catalysts

Nickel is a type of non-noble metal that can be used as catalyst and magnetic separator since it intrinsically shows ferromagnetic nature. Liu et al. [27] prepared Ni-Ru nanocomposite catalyst to evaluate the hydrogen from sodium borohydride. Ru is known to be paramagnetic. Its magnetic interaction with permanent magnets is generally weak. Therefore existence of Ni makes catalyst behavior strongly magnetic. Those types of catalyst are operated in dehydrogenation processes to easily separate the catalyst from sodium metaborate solution by permanent magnets [27]. A fuel cell power generator has been designed and operated with modified cobalt commercial catalysts by Oronzio et al. [9]. The performance of the designed magnetic-containment reactor was determined in a test-rig and 25 ml min^{-1} hydrogen flow rate was observed with 90 % hydrogen efficiency. Catalyst stability and activity is described as satisfactorily enough to use in mobile hydrogen generators [9].

3.2.6. Co-Ni Couple Catalysts and Reaction Mechanism

Co-Ni-B can be synthesized by chemical reduction as Co-B and Ni-B catalysts in the presence of sodium borohydride. Co-Ni-B catalyst performance exceeds that of both Ni-B and Co-B and can achieve by itself because of the electronic synergetic effect operate with Co and Ni atoms. Therefore activation energy of the reaction decreases to 34 kJ mol^{-1}. In order to understand this phenomenon, occurred mechanism may be described as:

- Firstly, borohydride ions are chemisorbed on the metal surface of the Ni and Co atoms meanwhile the hydride ion H^- is transferred from borohydride ion BH_4^- to an unoccupied adjacent Ni or Co metal atom; and
- Finally, this hydrogen atom acquires a negative charge in form of electron from the metal and leaves the site in hydride form (H^-), which reacts with the water molecule to produce hydrogen and hydroxide ion OH^-. Hydroxide ion OH^- reacts with boron in BH_3 to produce the $BH_3(OH)^-$ ion.

The cycle of hydrogen absorption on the metal sites continues till $BH_3(OH)^-$ forms $B(OH)_4^-$ [36]. A suggestion for occurring reactions is given in Figure 3. Molecular hydrogen is released during the full cycle. Mole ratios of Co and Ni constituting the Co-Ni-B catalyst are an important parameter to study on. $Co_{0.85}$-$Ni_{0.15}$-B suggested by Fernandes et al. [36] as the best catalyst composition so far. Catalyst prepared with respect to this ratio results high hydrogen generation rates when compared with the other mole ratios or metal borides. Most of the catalyst agglomerate at high temperature circumstances. This obligates circumstances of the catalytic investigations to be at mild temperatures and limits some on board applications. At 500°C Co-Ni-B catalyst surface morphology transform from well-dispersed particles to porous crystallite morphology. However, the catalyst maintains its activity. Thus Co-Ni-B catalyst can be offered for high temperature applications [36].

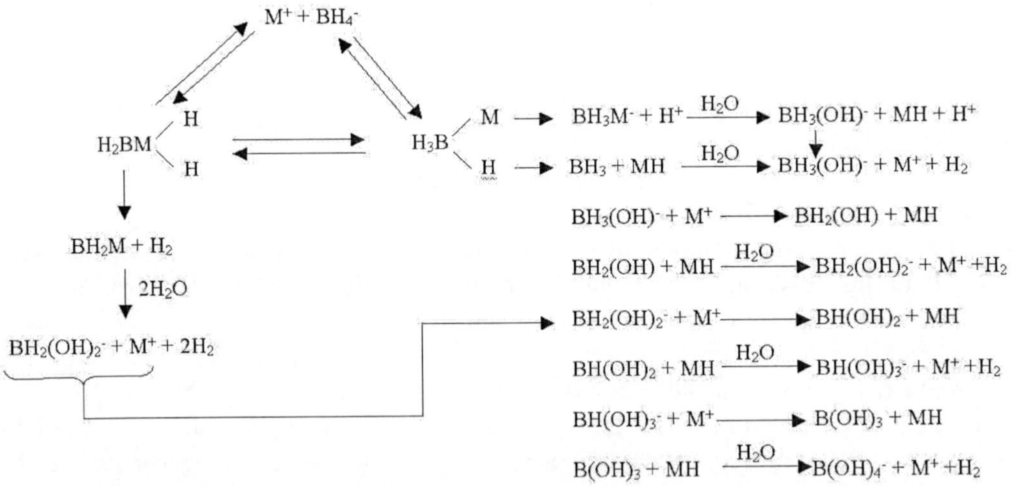

Figure 3. Interactions between borohydride, catalyst and water.

Cobalt chloride should be paid more attention among the other type catalysts because of its advantageous of activity and cost. Its hydrogen generation rate is at about of 86.3 L min^{-1} g^{-1} which is much better than either cobalt boride and other cobalt forms ($CoSO_4$,

$Co(CH_3COO)_2$, $Co(NO_3)_2$, CoF_2, $Co_3(C_3H_5O(COO)_3^{3-})_2$, $Co(CH_3COCHCOCH_3)_2$). This extreme catalytic property of $CoCl_2$ is a result of its perfect ionization to Co^{2+} and Cl^-. Chlorine is the most electronegative element thus Co^{2+} shows very electrophilic character. That property of Co^{2+} makes it attractive by borohydride anion [37].

In order to dehydrogenate sodium borohydride, another catalyst was prepared by reducing nickel (II) acetylacetonate with sodium borohydride in an aqueous solution that was stabilized by hydrogen phosphate anion. Water dispersible nickel (0) nanocluster was found to be very active even at room temperature for 1450 turnover. Hydrolysis of sodium borohydride was found first order in catalyst concentration and zero order in substrate concentration. Therefore, this catalyst seems appropriate for high concentration hydrolysis of sodium borohydride [38].

3.2.7. Electrochemical Catalysts

An alternative way of preparing catalyst for $NaBH_4$ hydrolysis is operating the electrodeposition method. Intrinsically, this technique provides some parameters to control the catalyst properties affecting the hydrogen generation rate. Kim et al. [39] showed that Co-Ni-P catalyst could be successfully deposited on copper surface by using a cathodic current density of 0.01-0.1 A cm^{-2}. Increasing the amount of cobalt in the electrodeposition solution results in greater and more three-dimensional growth of the catalyst particles on surface. Since this type of surface morphology provides more surface area for the dehydrogenation reaction, hydrogen generation rate rises. Further investigation shows that at low cathodic current densities two-dimensional surface morphology occurs. Besides, cobalt content in the Co-Ni-P electrodeposits decreased while the nickel content increased with increasing cathodic current density. Performed hydrogen evolution experiments clearly demonstrate that catalytic efficiency of the Co-Ni-P catalyst is sensitive to the cathodic current density used for the electrodeposition. The effects of the electrodeposition time on surface morphologies of Co-Ni-P catalyst synthesized on the Cu substrate were determined by grain size growth and hydrogen generation rate. Longer operation times results in more three-dimensional growths and increase the catalyst surface area. These are the effects that positively change the hydrogen generation rate. The cycle test of synthesized Co-Ni-P catalyst performed for 15 times. It was observed that after 15 tests the hydrogen generation rate decreased to 25.4% of the first one.

3.2.8. Fuel Cell Application of Catalytic Hydrolysis

In response to the demand for supporting all electronic devices perennially with renewable and clean energy, lithium batteries are replaced by fuel cells. Direct liquid fuel cells (DLFCs) have the advantage of using liquid fuel therefore thought to be the appropriate candidate for forthcoming applications. Polymer electrolyte membrane fuel cells (PEMFCs) can produce more power than direct liquid fuel cells, but hydrogen supply for polymer electrolyte membrane fuel cells is still a problem for portable applications [40]. In order to feed polymer electrolyte membrane fuel cells, a system using porous nickel supported Co-B catalyst was constructed. Catalyst preparation procedure was composed of simple doping and drying steps, so this procedure is easy to apply. Various amounts of sodium hydroxide did not affect the hydrogen generation at all in the presence of 20 wt% sodium borohydride. 12.5 wt% of sodium borohydride was adjudged to be optimal concentration for maximum hydrogen generation. The system was capable to produce more than 6 L min^{-1} hydrogen, which is sufficient to operate a 450 W polymer electrolyte membrane fuel cell stack [40]. Although the system has been very successful, purification of water vapor from hydrogen needs further investigation to prevent membrane poisoning. Dehydrogenation processes involve multi-component and multi-phase flow and multi-mode heat and mass transfer. Zhang et al. [41] have a model to predict mass and heat transfer behavior of the fuel cells. This model has two components:

- First, dealing with the non-isothermal water evaporation process, and,
- Second, dealing with pressure drop through the porous media.

Also, a one-dimensional numeric model was developed to study the homogeneous catalyst. Model succeeded to capture the water vapor generation both by chemical reaction and by mass transfer from liquid phase to gas phase. Although the model failed at extreme pressure and temperature values, it is a good starting point to theoretically investigate the water vapor separation phenomenon [41].

4. $NaBO_2$ Recycle

Hydrogen generation systems are designed to convert sodium borohydride into sodium metaborate and hydrogen, the main objective being the hydrogen release. In response to the increasing demand for processing the by-product sodium metaborate, recycling has attracted much economic attention. Sodium metaborate always contains water molecules in its structure. Water should be taken away for the success of the further processes. In order to dry sodium metaborate, thermal techniques can be applied in the presence of dehydrants. Generally, bonds of water molecules are broken at about 370 °C but for wet sodium metaborate 270 °C is enough. Studies for developing new catalysts that can evaluate hydrogen with minimum amount of water may decrease the load on the drying process.

There are different ways to convert dried sodium metaborate to sodium borohydride. Sodium metaborate can react with MgH_2 or Mg as in reaction below:

$$NaBO_2 + 2\ MgH_2 \rightarrow NaBH_4 + 2\ MgO \qquad (13)$$

This reaction was run under 0.5-7 MPa hydrogen pressure and at 350-750 °C. The optimal conditions of the reaction were determined to be 7 MPa pressure and 550 °C. It is expected that the reaction yields more product as pressure is increased, since the products are not in gas phase. On the contrary, the reaction yielded only 10% because of agglomeration [42]. Agglomeration is a common problem decreasing the yield at high temperature, solid-solid reactions. In order to overcome this, magnesium silicide ($MgSi_2$) or materials alike can be added to the reaction environment [42].

A way to recycle $NaBO_2$ is to process it with coke and methane as explained by Kojima and Haga [42]. Such techniques need catalysts or high temperatures and pressures so they are not cost-effective to prepare sodium borohydride. Thus, alternative techniques must be searched. Ball milling under hydrogen atmosphere may solve the problem. In an experiment, recycle of the potassium metaborate reaction was attempted by ball milling to obtain potassium borohydride. Sodium hydride (NaBH), magnesium hydride (MgH_2) and calcium hydride (CaH_2) were mixed with potassium metaborate ((KBO_2) under argon atmosphere. In addition, Figure 4 lists some extra chemicals that can be potentially used. Sodium hydride is a thermodynamically unfavorable reactant for the desired reaction and calcium hydride has not been found effective to yield satisfactorily. Yet, after 120 min milling magnesium hydride gave 32 % yield if stoichiometric ratio was applied. Furthermore 100 % yield was determined if 35% MgH_2 is used in excess. In order to purify the KBH_4 as shown in Figure 4, extraction with ethylenediamine and filtration steps are necessary [43].

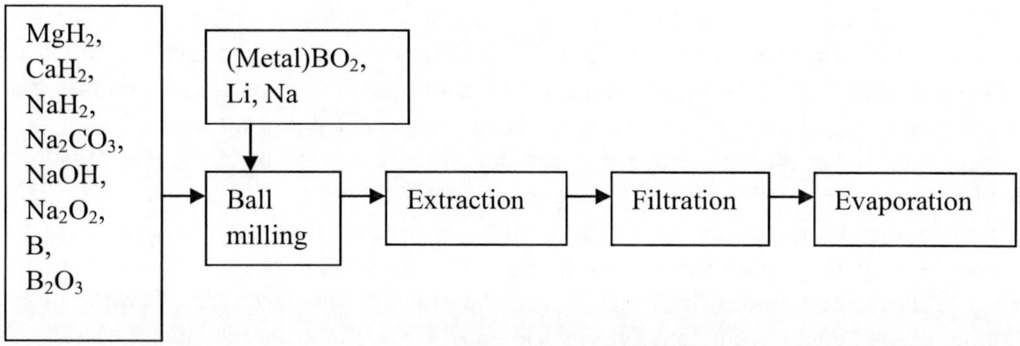

Figure 4. Procedure of metal borohydride production by means of mechano-chemical reaction.

Sodium metaborate conversion to sodium borohydride is a process of exchanging oxygen with hydrogen atoms. Different forms of hydrogen molecule can be used for this purpose. "Protide" is another state of hydrogen existing both as liquid and solid states in metal-hydrogen-complex anions. The most reactive form of hydrogen protide can play a role in dynamic hydriding and dehydriding process to obtain sodium borohydride. This process was based on the transition state of hydrogen between gaseous hydrogen and protide. The thermal difference between initial and terminal point of the reaction supported the dynamic changes of Mg-H system between hydriding to dehydriding regions. At lower temperatures, Mg-H system tended to form MgH_2. On the other hand, at higher temperatures, it tended to dissociate hydrogen under the protide state back to the gas phase. During rapid changes between hydriding and dehydriding states, any hydride formed on the extreme surface of magnesium. It means that hydrogen was only consumed by sodium borohydride generation. The extreme surface condition of magnesium during the dynamic temperature changes can lead us to consider magnesium as a catalyst converting hydrogen to hydrides (H^-). Experiments showed that no magnesium oxidehydride (MgH_2O) is produced in the reactor while reactions happen [44].

Mechano-chemical reactors are widely processed in the recycle of sodium metaborate. This type of reactors are simply closed mortar including different size of balls and operate both back and forth or rotational. High speed rotations result reactants to collide each other and yield thermodynamically favored product. Disadvantageous sides of this type of devices are mainly limited capacity of the mortar, reaction time and energy requirement. Nevertheless mechano-chemical reactors result high yield in $NaBH_4$ production operations [45]. $NaBO_2$ can be recycled back to $NaBH_4$ easily with MgH_2 by means of mechano-chemical reactors.

Alternatively, dehydrated borax or elemental sodium can be processed instead of sodium metaborate. Also Rohm-Hass (Reaction 6) process constituting a part of Figure 1 is the mainly used procedure in sodium borohydride synthesis. Except Rohm-Hass method these reactants need MgH_2 as a hydrogen provider. X-ray diffraction (XRD) and Fourier transformed infrared spectroscopy (FTIR) analysis results of these four techniques are depicted in Figure 5 and 6, respectively. As discussed before, sodium insufficiency of dehydrated borax using limited by maximum 50 % yield unless sodium including additive is added. In the XRD results given below low yield of borax is obvious. In addition, iodimetric analysis of ball-milling study performed with borax results that only 38 % of the product is sodium borohydride whereas this value reaches up to 73% for sodium metaborate. On the other hand, XRD spectra of ball-milling and Rohm-Hass studies have some differences. Firstly it is clear that the products of the ball-milling studies are amorphous. Therefore, peaks are broaden and generally hidden by others. Especially broad peak of amorphous side-product MgO hides the sodium borohydride peaks at about 40.8 two-theta. On contrast peaks belonging to the crystalline product of Rohm-Hass process are sharp and clear.

Performed investigations prove that reaction time is critical parameter in ball milling reaction. Moreover, yield continuously increases with working time. That may be explained by the extreme surface properties of magnesium element existing in the reactor during the process. Long working time creates opportunity for the reactant to reach Mg surface and react further. Thus, sodium borohydride yield can be up-grated by increasing the working time. In addition, excess amount of magnesium hydride positively affects on reaction. Generally, 10-40% excess amount of magnesium hydride using results satisfactorily yields for sodium borohydride such are over 70% [46].

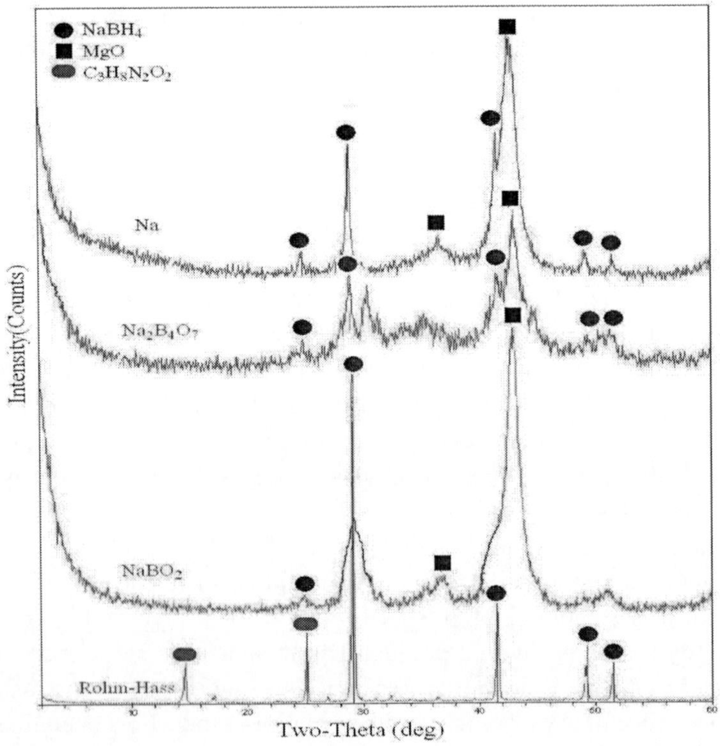

Figure 5. XRD spectra of $NaBH_4$ synthesized with Na, $Na_2B_4O_7$, $NaBO_2$ and trimethyl borate.

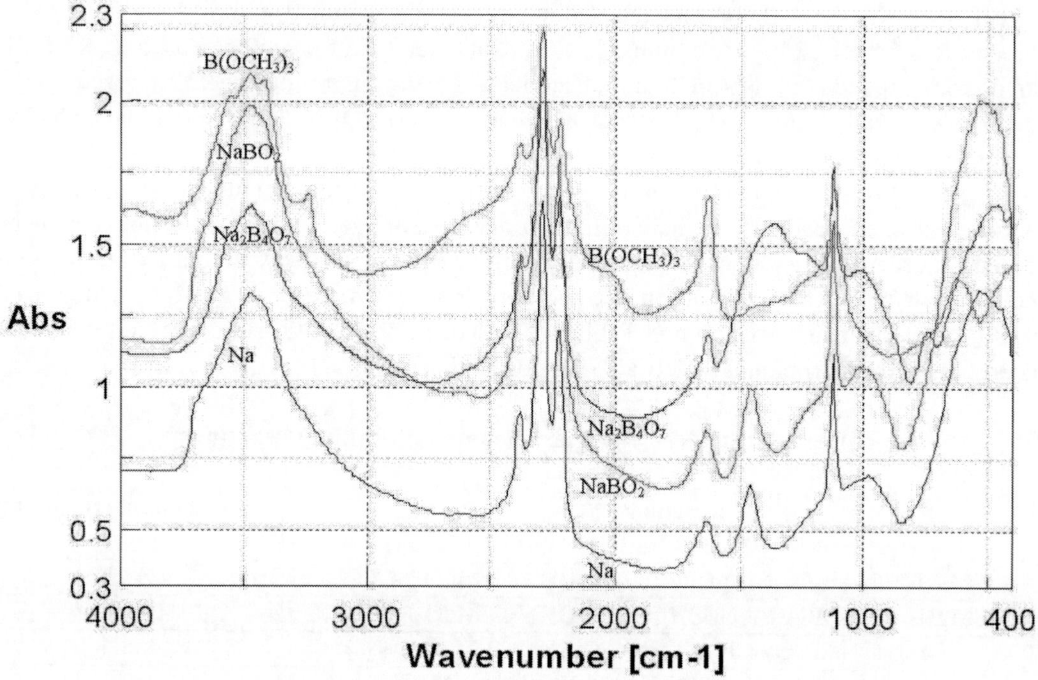

Figure 6. FTIR spectra of synthesized $NaBH_4$ with Na, $Na_2B_4O_7$, $NaBO_2$ and $B(CH_3)_3$.

5. $NaBO_2$-Borax Conversion

The only disadvantage of using sodium borohydride as a hydrogen storing material is its cost. It is not easy to react with boron minerals and conversion of by-product such as sodium metaborate to sodium borohydride needs further research. Electrochemical techniques can be used to obtain borax from hydrolysis reaction by-product sodium metaborate. Instead of metaborate, borax can be a starting material for synthesis of sodium borohydride. Thermodynamically borax tends to result in sodium borohydride easily if compared with metaborate. Gibbs free energy to obtain sodium borohydride from borax is -239.7 kJ mol^{-1} but it is -269.7 kJ mol^{-1} to obtain it from metaborate. Thus less energy consumption is possible if borax is considered as a starting reactant [47]. For this purpose Pd, boron-doped diamond (BDD), Au and Pt were used as electrodes with +1.54 V potential for 5 hours. Pd and BDD electrodes were determined to need less voltage for high yields when compared to Pt electrode. BDD electrode was offered to be the best one since it consumes less energy for water electrolysis [47].

Wet borax may be processed to obtain $NaBO_2$ and remove its water molecules by endothermic application as shown in reactions (2) and (3). Besides, the metaborate may be directly processed by ball milling with hydrides as in Figure 4.

$$(4\text{-}9)\ MgH_2 + Na_2B_4O_7 + X \rightarrow 4\ NaBH_4 + (8\text{-}9)\ MgO + Y \qquad (14)$$

In reaction (14), X represents the sodium including chemical entering the reaction to compensate for the lacking of sodium. Na_2CO_3, NaOH and Na_2O_2 can be chosen as X or it is also possible to add no Na including chemical. According to the chemical used as X, stoichiometric coefficient of MgH_2 varies between 4-9 and CO_2 or hydrogen are taken as by-product represented by Y in reaction 14 [7].

NaOH and Na_2O_2 contain oxide-hydroxide groups. These functional groups are likely to produce undesired products. In ball milling processes, oxygen and water vapor should be avoided. In order to keep them away, usually a glove-box controlled with inert atmosphere is used to wash chemicals with argon. In addition, low Gibbs free energy for reaction with Na_2CO_3 is an advantage as shown in Table 2. Consequently, Na_2CO_3 is the preferable one for this process and experimental results also supported this [7].

Table 2. Gibbs free energies of some ball-milling reactions.

Reactions	ΔG°, kJ mol^{-1}($NaBH_4$)
4 MgH_2 + $Na_2B_4O_7$ → 2 $NaBH_4$ + 4 MgO + B_2O_3	-239.7
8 MgH_2 + $Na_2B_4O_7$ + Na_2CO_3 → 4 $NaBH_4$ + 8 MgO + CO_2	-254.4
9 MgH_2 + $Na_2B_4O_7$ + 2 NaOH → 4 $NaBH_4$ + 9 MgO + 2 H_2	-360.6
9 MgH_2 + $Na_2B_4O_7$ + Na_2O_2 → 4 $NaBH_4$ + 9 MgO + H_2	-438.4
$Na_2B_4O_7$ + 7 MgH_2 + 2 NaH → 4 $NaBH_4$ + 7 MgO	-269.3
$Na_2B_4O_7$ + 8 MgH_2 + 2 Na → 4 $NaBH_4$ + 7 MgO + Mg	-301.6

According to the reaction (14), a bypass may be drawn in Figure 1 to connect borax and $NaBH_4$. Thus, the process cost and in turn the chances of implementation will improve. Alternatively, borax may be converted to metaborate and alternative procedures may be developed to reach $NaBH_4$.

6. Characterization of $NaBH_4$

Characterization techniques are necessary to understand whether the required product is synthesized or estimate the yield of the process. There are a number of ways to determine the sodium borohydride existence. However, the best testing way must be the operating the sample on the process it will be used in. Thus a basic dehydrogenation system, given in Figure 7, may be the easiest and fastest way to characterize if the sample is sodium borohydride. Actually 1.18 NL of hydrogen must be evaluated by catalytic dehydrogenation of 1 g of sodium borohydride. When the hydrogen comes from water taken in to account, this value reaches up to 2.5 NL.

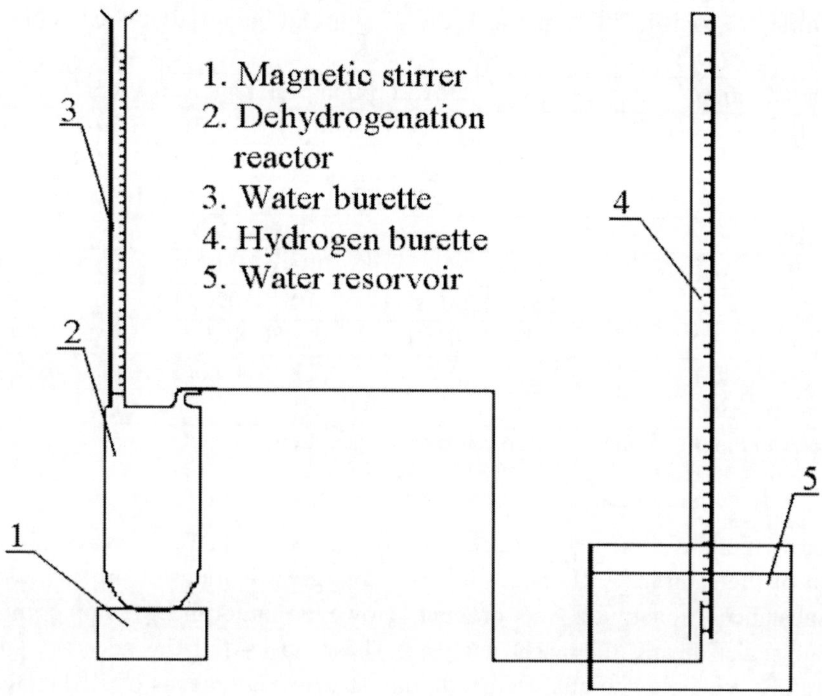

Figure 7. Dehydrogenation set-up.

FTIR analysis can be considered as another way of detecting sodium borohydride existence. This technique is advantageous because of its less sample requirement and short analysis time. Also, B-H atomistic vibrations of the borohydride anion can be detected by Raman spectroscopy at about 2320 cm^{-1} similar to FTIR, which sense the B-H bond vibration at 2280 cm^{-1}. Therefore, both instruments can be operated to accomplish the characterization. Furthermore, Raman spectroscopy provides to find crystalline structures of the sample and analyze them. Thus it may be thought more comprehensive technique nevertheless FTIR spectroscopy is mostly advantageous because it can be achieved with use of only basic KBr pellets. FTIR analysis of the samples previously depicted as XRD spectra in Figure 5 are given in Figure 6. In FTIR analysis bigger peak area and longer peak height indicates purer sample. Therefore adsorption peak areas and peak height of the samples must be computed by instrument software and noted during the analysis. These data of FTIR spectra are given in Table 3. Also, a schedule of possible metal borohydride FTIR peak areas region isare summarized in Table 4.

Table 3. Peak processing of FTIR peaks between 2400-2200 cm^{-1}.

Reactant to synthesize $NaBH_4$	FTIR peak area	FTIR peak height
Elemental Na	111.0	1.1930
$Na_2B_4O_7$	168.0	1.4905
$NaBO_2$	154	1.4236
$B(OCH_3)_3$	100.5	0.6616

Table 4. Possible FTIR peaks indicating metal borohydride existence.

Functional group	Adsorption peak region area (cm^{-1})		
BH	2630-2350 (s)	1075-1010*	
BH_2	2630-2350 (s)	1205-1140 (s)	975-945 (w)
BH_4^-	2400-2200 (s)	1130-1040 (s)	
BO	1400-1300		
BO_2^-	1360-1300 (s)		
$B_4O_7^{2-}$	1370-1330 (s)	1090-1070 (m)	1000-990 (s)
CO_3^{2-}	1450-1410 (s)	880-850	720-680*
MO* (M=Metal)	1100-900 (s)		
Crystalline water	3600-3000 (s)	1670-1600 (s)	

s: strong adsorption; m: moderate adsorption; w: weak adsorption
*: Broaden

Titrimetric method firstly introduced by Lyttle et al. [10]. It provides quantitative computation of the sample. XRD analysis also gives great amount of information about the chemical substances constitutes the product. However, amorphous sample morphologies results broad peaks during the XRD analysis. These peaks fail the accuracy of the XRD result. Therefore it is better to analyze crystalline structure by means of XRD. Besides using different spectroscopic instruments are developing some new ways to distinguish the sodium borohydride existence in the sample. At this point, UV spectroscopy can be emphasized as one of the new technique attracting attention.

CONCLUSION

In this chapter, production methods and advantageous sides of $NaBH_4$ is discussed as a candidate of clean and renewable hydrogen carrier material. Also the way to liberate the stored hydrogen form its structure is explained and characterization techniques are summarized. Moreover, idealization for a system, which can recover most of the hydrogen contained by sodium borohydride in the presence of catalyst, is studied. Recycling of by-product sodium metaborate is an attractive research area therefore thought system is also considered to achieve this indispensable part. It is likely possible to lower 50 \$ kW^{-1} energy cost of sodium borohydride by recycling and thus it can make fuel cell usage applicable in portable small devices and hydrogen cars. Environmental and commercial requirements force the scientists to regain waste material of hydrolysis. Conventional techniques have been potentially carried out furthermore new technologies are strongly needed. Among these, electrochemical conversion of sodium metaborate to borax and direct synthesis of sodium borohydride from borax mineral at extreme surface condition of magnesium seem to have the biggest potential to cut down the costs.

References

[1] Chater, P.A.; Anderson, P.A.; Prendergast, J.W.; Walton, A., Mann, V.S.J.; Book, D.; David, W.I.F.; Johnson, S.R.; Edwards, P.P. *J. Alloys Compd.* 2007, *446-447*, 350-354.
[2] Kojima, Y.; Kawai, Y.; Kimbara, M.; Nakanishi, H.; Matsumuto, S. *Int. J. Hydrogen Energy*. 2004, *29*, 1213-1217.
[3] Kojima, Y.; Suzuki, K.; Kawai, Y. *J. Power Sources*. 2006, *155*, 325-328.
[4] Kaya, S.; Gürü, M.; Ar, I. Alnıak, O.; *Proceeding 2nd International hydrogen Energy Congress and Exhibitio,* IHEC Istanbul Turkey; 13-15 July 2007, p. 211.
[5] Matsunaga, T.; Buchter, F.; Miwa, K.; Towata, S.; Orimo, S.; Züttel, A. *Renewable Energy*. 2008, *33(2)*, 193-196.
[6] Li, Z.P.; Liu, B.H.; Morigazaki, N.; Suda, S. *J. Alloys Compd.* 2003, *354*, 243-247.
[7] Li, Z.P.; Morigazaki, N.; Liu, B.H.; Suda, S. *J. Alloys Compd.* 2003, *349*, 232-236.
[8] Çakanyıldırım, Ç.; Gürü, M. *Renewable Energy*. 2008, *33*, 2388-2392.
[9] Oronzio, R.; Monteleone, G.; Pozio, A.; De Francesco, M.; Galli S. *Int. J. Hydrogen Energy*. 2009, *34*, 4555-4560.
[10] Lyttle, D.A.; Jensen, E.H.; Struck, W.A. *Anal. Chem.* 1952, *24*, 1843-1844.
[11] Schlesinger, H.I.; Brown, H.C.; Mayfield, D.L.; Gilbreath, J.R. *J. Am. Chem. Soc*. 1953, *75*, 213-215.
[12] Bowman, C.M.; Watson, E.J. Method for the preparation of trimethyl borate. US2976313, 1962.
[13] Bilici, U.; Özdağ H.; *Second International Boron Symposium*, 23-25 September 2004, Eskişehir Turkey, 119-125.
[14] Çakanyıldırım, Ç.; Gürü, M. *Renewable Energy*. 2009, *34*, 2362-2365.
[15] Cooper, H.B.H. Production of alkali metal borohydrides, US Patent No:3,473,899, 1969.
[16] Adams, R.M.; Siedel, A.R. *Boron, metallo-boron compounds and boranes*; Interscience Publisher: New York, 1964, pp. 390-392.
[17] Nakamori, O.; Orimo, S. *J. Alloys Compd.* 2004, *370*, 271-275.
[18] Züttel, A.; Wenger, P.; Rentsch, S.; Sudan, P.; Mauron, P.; Emmenegger, C. J. *Power Sources*. 2003, *118*, 1-7.
[19] Pena-Alonso, R.; Sicurelli, A.; Callone, E.; Carturan, G.; Raj, R. *J. Power Sources*. 2007, *165*, 315-323.
[20] Cho, K.W.; Kwon, H.S. *Catal. Today*. 2007, *120*, 298-304.
[21] Schlesinger, H.I.; Brown, H.C.; Finholt, A.B.; Gilbreath, J.R.; Hockstra, H.R.; Hydo, E.K. *J. Am. Chem. Soc.* 1953, *75(1)*, 215-219.
[22] Keçeli, E.; Özkar, S. *J. Mol. Catal. A*. 2008, *286*, 87-91.
[23] Liu, B.H.; Li, Z.P. *J. Power Sources*. 2009, *187*, 527-534.
[24] Ingersoll, J.C.; Mani, N.; Thenmozhiyal, T.C.; Muthaiah, A. *J. Power Sources*. 2007, *176(1)*, 450-457.
[25] Shang, Y.; Chen, R. *Energy Fuels*. 2006, *20*, 2149-2154.
[26] Hua, D.; Hanxi, Y.; Xinping, A.; Chuansin, C. *Int. J. Hydrogen Energy*. 2003, *28*, 1095-1100.
[27] Liu, C.H.; Chen, B.H.; Hsueh, C.L.; Ku, J.R.; Jeng, M.S.; Tsau, F. *Int. J. Hydrogen Energy*. 2009, *34*, 2153-2163.
[28] Patel, N.; Fernandes, A.; Miotello, A. *J. Power Sources*. 2009, *188*, 411-420.
[29] Kriskan, P.; Yang, T.H.; Lee, W.Y.; Kim, C.S. *J. Power Sources*. 2005, *143*, 17-23.
[30] Zhang, J.S.; Delgass, W.N.; Fisher, T.S.; Gore, J.P. *J. Power Sources*. 2007, *164*, 772-781.

[31] Marrero-Alfonso, E.Y.; Gray, J.R.; Davis, T.A.; Matthews, M.A. *Int. J. Hydrogen Energy*. 2007, *32(18)*, 4717-4722.
[32] Demirci, U.B.; Akdim, O.; Miele, P. *Int. J. Hydrogen Energy*. 2009, *34(6)*, 2638-2645.
[33] Chen, C.W.; Chen, C.Y.; Huang, Y.H. *Int. J. Hydrogen Energy*. 2009, *34*, 2164-2173.
[34] Xu, D.; Zhang, H.; Ye, Wei. *Catal. Comm*. 2007, *8*, 1767-1771.
[35] Ye, W.; Zhang, H.; Xu, D.; Ma, L.; Yi, B. *J. Power Sources*. 2006, *64(2)*, 544-548.
[36] Fernandes, R.; Patel, N.; Miotello, A.; Filippi, M. *J. Mol. Catal. A*. 2009, *298*, 1-6.
[37] Akdim, O.; Demirci, U.B.; Muller, D.; Miele, P. *Int. J. Hydrogen Energy*. 2009, *33(6)*, 2631-2637.
[38] Önder, M.; Özkar, S. *Int. J. Hydrogen Energy*. 2007, *32*, 1707-1715.
[39] Kim, D.R.; Cho, K.W.; Choi, Y.I.; Park C.J. *Int. J. Hydrogen Energy*. 2009, *33(6)*, 2622-2630.
[40] Kim, S.J.; Lee, J.; Kong, K.Y.; Jung, C.R.; Min, I.G.; Lee, S.Y.; Kim, H.J.; Nam, S.W.; Lim, T.H. *J. Power Sources*. 2007, *170*, 412-418.
[41] Zhang, J.; Zheng, Y.; Gore, J.G.; Mudawar, I.; Fisher, T.S. *J. Power Sources*. 2007, *170*, 150-159.
[42] Kojima, Y.; Haga, T. *Int. J. Hydrogen Energy*. 2003, *28*, 989-993.
[43] Li, Z.P.; Liu, B.H.; Nobuto, M.; Suda, S. *J. Alloys Compd*. 2003, *354*, 243-247.
[44] Suda, S.; Morigasaki, N.; Iwase, Y.; Li, Z.P. *J. Alloys Compd*. 2005, *404-406*, 643-647.
[45] Çakanyıldırım, Ç.; Gürü. M. *Int. J. Hydrogen Energy*. 2008, *33*, 4634-4639.
[46] Hsueh, C.L.; Liu, C.H.; Chen, B.H.; Chen, C.Y.; Kuo, Y.C.; Hwang, K.J.; Ku, J.R. *Int. J. Hydrogen Energy*. 2009, *34*, 1717-1725.
[47] Park, E.H.; Jeong, S.U.; Jung, U.H.; Kim, S.H.; Lee, J.; Nam, S.W.; Lim, T.H.; Park, Y.J.; Yu, Y.H. *Int. J. Hydrogen Energy*. 2007, *32*, 2982-2987.

In: Boron Hydrides High Potential Hydrogen Storage Material ISBN:978-1-61668-361-0
Editors: Umit B. Demirci and Philippe Miele

Chapter 5

POTENTIAL AND LIMITATION OF THE DIRECT BOROHYDRIDE FUEL CELL. SPECIAL EMPHASIS ON THE BOROHYDRIDE OXIDATION REACTION (BOR) MECHANISM AND KINETICS ON GOLD ELECTROCATALYSTS

Marian Chatenet[a1], Belen Molina Concha[a1], Gaëlle Parrour[a1], Jean-Paul Diard[a1], Fabio H. B. Lima[b2] and Edson A. Ticianelli[b]

[a]Laboratoire d'Electrochimie et de Physico-chimie des Matériaux et des Interfaces, LEPMI, UMR 5631 CNRS/ Université de Grenoble Université, BP 75, 38402 Saint Martin d'Hères Cedex, France

[b]Instituto de Química de São Carlos, Universidade de São Paulo, Av. Trab. São-Carlense 400, CP 780, CEP 13560-970, São Carlos, SP, Brazil

ABSTRACT

Direct borohydride fuel cells (DBFC) exhibit some potential for the powering of small nomad electronic devices, thanks to their high energy density as well as the facile and safe storage of borohydride salts. Since their development a decade ago, numerous groups have focused on the cell geometry, operating parameters and materials, yielding to appreciable performance improvements. However, from an electrocatalytic point of view, the DBFC remain hindered because (i) the borohydride oxidation reaction (BOR) is complex, (ii) its mechanism is imperfectly determined yet and (iii) no practical electrocatalyst exhibits both fast BOR kinetics and high faradic efficiency. The existence of these three major drawbacks renders mandatory more fundamental research on the BOR. In this context, the BOR mechanism and kinetics was characterized for one of the most studied model BOR electrocatalyst: polycristalline bulk gold. Particular attention was focused on the reaction pathway, using both electrochemistry and coupled physicochemical techniques: Fourier transformed infrared (FTIR) spectroscopy and on-line mass spectrometry (OLEMS).

[1]Corresponding author: Marian.Chatenet@grenoble-inp.fr +33 476 82 65 88; +33 476 82 67 77 (fax)

While one usually admits that gold promotes the complete 8-electron BOR, the present chapter, thanks to the coupling between on-line mass spectrometry and electrochemistry (OLEMS), reveals that H_2 is produced when the BOR proceeds on gold electrodes, as a result of some extent of heterogeneous hydrolysis of BH_4^-. Thus, gold cannot be considered as a perfect faradaic-efficient BOR electrocatalyst.

From this knowledge, modeling cyclic voltammetry (CV) and electrochemical impedance spectroscopy (EIS) data, a simplified reaction pathway was proposed, the theoretical behavior of which agrees with the experimental findings, including those from *in situ* FTIR spectroscopy. This pathway includes at least two irreversible electrochemical steps (EE) for BH_4^- oxidation, which compete with the electrochemical adsorption reaction (EAR) of OH^- anions at high potentials. This shows that the BOR mechanism on gold is not as simple as usually admitted, and that gold promotes two pathways as a function of the electrode potential (one *ca.* 4-electron pathway at low potentials, likely resulting from concomitant direct borohydride oxidation and heterogeneous hydrolysis reactions, whereas one quantitative *ca.* 8-electrons pathway proceeds at higher potentials).

As a result, due to its slow BOR kinetics and incomplete BH_4^- electrooxidation, one may conclude that Au is not an ideal electrocatalyst for DBFC anodes.

1. INTRODUCTION

The high production rate of portable electronic devices, coupled to their increased requirements of autonomy, renders stringent the development of high energy density generators. Unfortunately, the current state of the art portable generators (lithium-ion batteries) display insufficient energy densities, *ca.* below 200 Wh kg^{-1} [1] with limited hope of dramatic improvement. In consequence, "open generators", among which fuel cells, may consist of the required breakthrough in the field of delocalized energy production. At present, the Proton Exchange Membrane Fuel Cell (PEMFC) technology exhibits sufficient initial performances for use in stationary and portable applications. However, its commercial deployment is still hindered by [2] (i) the detrimental costs of most of its constitutive materials and assembling processes, (ii) the fact it requires very pure hydrogen fuel (of electrolytic "quality"), often originating from fossil compounds after several purification steps, (iii) the lack for relevant hydrogen mass storage/distribution solutions and (iv) its insufficient durability (see for example references [3-9]). The alkaline fuel cell technology, which has been used with success by the NASA since the 1960s, displays several advantages over the PEMFC. Firstly, the kinetics of most electrochemical reactions is faster in alkaline than in acidic medium. Secondly, such faster reaction kinetics renders possible the use of non-platinum electrocatalysts and/or alternative fuels to electrolytic hydrogen without noticeable performances loss. In that frame, several research groups have investigated sodium borohydride electrooxidation (Equation 1) since a decade [10-15], in systems often using non-platinum electrocatalysts [11, 16-20].

$$BH_4^- + 8\,OH^- \rightarrow BO_2^- + 6\,H_2O + 8\,e^- \;(E° = -1.24\text{ V } vs.\text{ NHE}) \qquad \text{Equation 1}$$

At the cathode of a Direct Borohydride Fuel Cell (DBFC), the electroreduction of oxygen (usually from air) takes place (Equation 2), unless a more exotic oxidant is used, e.g.

hydrogen peroxide (H_2O_2). Naturally, the former solution is better suited for economic reasons.

$$O_2 + 2\,H_2O + 4\,e^- \rightarrow 4\,OH^- (E^\circ = 0.401 \text{ V } vs. \text{ NHE}) \qquad \text{Equation 2}$$

The rationale for using borohydride salts to power fuel cell systems lies oin the multiple advantages of such compounds over pure hydrogen. They are (i) non toxic, (ii) easily stored as a powder, (iii) relatively stable in alkaline solution (but $NaBH_4$ spontaneously hydrolyses for pH below 12 and on a great variety of metals and oxides [21-23]), while (iv) present a rather high energy density (9.3 Wh g^{-1} at 1.64 V). The interested reader is directed towards a recent review from Demirci, who recently summarized the thermodynamic advantages of the DBFC over other direct alkaline fuel cells [24]. Now, although very interesting in principles the oxidation of sodium borohydride in DBFC presents several drawbacks. One can for example mention the sluggish anode reaction (BOR) kinetics, the possible fuel crossover down to the cathode compartment and the difficulty to evacuate the BOR byproducts (e.g., boric acid, the solubility of which is low in basic medium).

At present, the majority of the research groups involved in the DBFC focus on the building and testing of unit cells or small stacks [10-15, 25-27], while a few others investigate the electrode reactions kinetics, membrane/electrocatalysts materials or fuel/oxidant composition. In that frame, the present chapter aims at evaluating whether the DBFC is a relevant solution for chemical energy conversion into electrical energy in portable electronic devices. Based on a selected literature review, it will firstly (i) describe the "ideal" sodium borohydride electrolyte solutions to fuel a DBFC (section 30), and (ii) draw a global view about the state-of-the-art DBFC performances (section 40). Secondly, based on recent results of our group in the field of electrocatalysis applied to DBFC, section 50 will survey the BOR mechanism on one of the most studied anode material for DBFC: gold. It will notably emphasize that, (iii) although often considered as a material not promoting the heterogeneous hydrolysis of $NaBH_4$, gold actually yields consequent hydrogen production during the BOR, as detected from *in situ* on-line electrochemical mass spectrometry, OLEMS (see section 5.10), while is rather inactive towards hydrogen electrooxidation, thereby lowering the overall faradaic efficiency of the anode. The (iv) the nature of the main BOR intermediates on gold will also be investigated using *in situ* Fourier transformed infrared spectroscopy, FTIR (see section 5.20). Then (v) a relevant BOR pathway will be proposed; it is based on the experimental findings from cyclic voltammetry (CV) and electrochemical impedance spectroscopy (EIS) data as well as reaction modeling (section 5.30).

2. Experimental

2.1 Reagents and Solutions

The electrolytic solutions (NaOH + $NaBH_4$) were prepared from high purity reagents (Merck Suprapur®) and ultrapure water (18.2 MΩ cm – 3 ppb COT, Millipore Elix + Gradient). The electrolyte was purged with Ar (N45 Messer) for *ca.* 15 min. prior to each measurement, while Ar was streamed above the solution during the measurements. For all experiments, the electrochemical cell was cleaned overnight in concentrated H_2SO_4-H_2O_2 mixture and then thoroughly rinsed with ultrapure water, as detailed in reference [28]. Unless otherwise stated, the measurements have been attempted at room temperature (20-25°C).

2.2 On-Line Electrochemical Mass Spectroscopy (OLEMSlems)

All measurements were performed using a dedicated three-electrode cell designed specifically for coupled OLEMS measurements. This cell comprised (i) a base connected to a vacuum compartment, itself connected to a 2-way valve directing either to the on line mass spectrometry (MS) apparatus via a capillary or to a primary vacuum pump and (ii) an upper "electrochemical" glass compartment containing the reference (reversible hydrogen electrode, RHE) and gold counter electrode (CE). The base of the cell was separated from the mass spectrometry chamber by a porous hydrophobic membrane (polytetrafluoroethylene, PTFE) supported by a porous stainless-steel frit, onto which the "bulk" Au working electrode was sputtered (Au/PTFE, 50 nm-thick). More details about this MS-coupled electrochemical cell can be found in references [29-31]. The MS measurements were performed for $m/z = 2$ (the H_2 signature) on an Omnstar^{TM} GSD 301 apparatus controlled by the Quadstar^{TM} Analysis Software, both from Pfeiffer Vacuum. The working Au/PTFE electrode was connected to the potentiostat (PGSTAT 20, Autolab) via a gold current collector. The ring-shaped gold CE was located near the surface of the electrolyte, above (as far as possible) the working electrode, so that H_2 bubbles formed at the CE could not reach the working electrode, thus suppressing any experimental bias on the MS ($m/z = 2$) signal: the H_2 generation at the gold CE is not affecting the detection of the MS ($m/z = 2$) signal, as verified experimentally (see Figure 2). In other words, only the hydrogen produced at the gold working electrode is detected by our OLEMS system.

The electrochemical measurements consisted of *ca.* 100 s at the gold working electrode open circuit potential, ocp (*ca.* 0 to 0.1 V *vs.* RHE depending on the $NaBH_4$ concentration) and, after that, a 10 min. potential hold in the range -0.1 to 0.8 V *vs.* RHE was performed with the $m/z = 2$ signal measured on-line. The measurement of the H_2 production at each potential was performed by subtracting the absolute $m/z = 2$ signal at ocp to the absolute $m/z = 2$ signal at the end of the 10 min. potential hold. The mass spectrometer apparatus was calibrated with hydrogen evolution reaction (HER) faradaic current, obtained from 10 min. current hold at various intensities relevant to our system; the obtained K^* value served to recalculate the equivalent HER current relative to the H_2 yield during BOR on gold.

2.3 Ftir FTIR Spectroscopy

The electrochemical activity of Au electrode with respect to the BOR was studied by linear sweep voltammetry in a three-electrode spectro-electrochemical cell dedicated to the *in situ* FTIR measurements. The cell was of the "thin-film" type, with the working electrode facing a ZnSe flat window, as described in reference [32]. The infrared measurements were recorded with a Fourier transform infrared spectrometer Nicolet Nexus 670 with a MCT detector (N_2-liquid temperature). The wave number was varied from 4000 to 650 cm^{-1} with a resolution of 8 cm^{-1}. Each infrared spectrum recorded consisted of the average of 32 individual measurements. The potential was varied between -0.2 and 1.4 V *vs.* RHE at a scan rate of 2 mV s^{-1}, while simultaneously, one FTIR spectrum was acquired every 20 mV (SPAIRS technique). In order to obtain sufficient IR signal from the boron-containing species, the electrolyte solution consisted of 1 M NaOH + 1 M $NaBH_4$ in water.

2.4 CV and EIS

The electrochemical experiments were conducted using a numeric potentiostat (VMP, Bio-Logic®) in "model" operating conditions, for which the BOR could be considered stable in the timeframe of the measurement. These "model" conditions noticeably differ from those used in the coupled techniques (see sections 5.10 and 5.20). In particular, a very dilute electrolyte solution was used; it consisted of 10^{-2} M NaOH + 5 10^{-5} M $NaBH_4$ and was thermostated at 20°C. The Au electrode was cycled in the BOR range (at the exclusion of the gold oxide region) until the hydrodynamic CV becomes stable. The experimental protocol includes a first cyclic voltammogram from -0.9 to 0.3 V *vs.* NHE at v_b = 50 mV s^{-1} (Ω = 0, natural diffusion conditions), followed by hydrodynamic CV (Ω = 2500 rpm, v_b = 1 mV s^{-1}, hydrodynamic conditions) to first "stabilize" the response of the gold RDE. The CV were then acquired at 5 mV s^{-1} in hydrodynamic conditions (Ω = 400, 900, 1600 and 2500 rpm), while the EIS spectra were monitored in potentiostatic conditions using the following parameters: E_{pp} = 50 mV, 5 points per decade in frequency, frequency swept from 200 kHz to 0.1 Hz. For these EIS measurements, the "staircase potentio EIS" (SPEIS) procedure in EC-Lab® software (Bio-Logic®) was used, acquiring *ca.* 50 EIS spectra for increasing fixed electrode potential in the region of the BOR waves ($-0.55 < E < 0.1$ V *vs.* NHE).

3. The Ideal Reactant for DBFC

3.1 Fuel Composition for DBFC

Starting with the pioneering work of Amendola et *al.* [33-36], several research groups focused on DBFC research and development [12-15, 25, 37, 38]. While most of these studies deal with the unit cell or stack testing, some also focus on the ideal composition of the anolyte [21-23, 39-42], because it determines the energy density of the fuel and the BOR mechanism and kinetics. More specifically, the composition of the anolyte influences (i) the homogeneous (in solution) and (ii) heterogeneous (at the electrode) borohydride hydrolysis, as well as (iii) the faradic processes occurring at the electrode (reaction rates and pathways).

Firstly, it is well-known that borohydride electrolytic solutions (even diluted and at high pH) are prone to undergo homogeneous hydrolysis. Kreevoy and Jacobson [43] determined the rate of homogeneous hydrolysis of $NaBH_4$ solutions as a function of their pH (see Equation 3):

$$\log t_{1/2} = \text{pH} - (0.034\,T - 1.92) \qquad \text{Equation 3}$$

where $t_{1/2}$ (min.) is the half time (the time for one half of a $NaBH_4$ solution to decompose) and T (K) is the solution temperature. For example, Equation 3 Equation 3 gives $t_{1/2} = 4.3$, 43 and 269 days for 10^{-2} M (pH ≈ 12), 0.1 M (pH ≈ 13) and 1 M (pH 13.8) NaOH at 25°C, respectively. These rather large values enable to discard any consequent quantitative homogeneous hydrolysis for experiment undertaken on a daily basis. However, this formula does not take into account the influence of the borohydride anion concentration on the rate of $NaBH_4$ homogeneous or heterogeneous hydrolysis, which may bias the evaluation of the value of $t_{1/2}$. Indeed, the following sections will show that increasing the borohydride concentration favors their hydrolysis.

Secondly, not only the solution pH is important for DBFC anolyte solutions, but also the ratio of sodium hydroxide versus sodium borohydride concentrations [40]. For example, Chatenet et al. observed that the BOR kinetics on Au and Ag electrocatalysts is strongly influenced by this ratio [39]: whereas the theoretical stoechiometry forecasts that eight hydroxide ions are needed for each borohydride ion (see Equation 1), a larger excess hydroxide ion is practically necessary to efficiently favor the BOR [22]. Conversely, when the anolyte contains identical hydroxide and borohydride anions (large) concentrations (1 M NaOH + 1 M $NaBH_4$), the BOR on Au and Ag electrodes is limited by the hydroxide concentration, and their stepwise hydrolysis is no longer negligible (see Equation 4 and Equation 5).

$$BH_4^- + H_2O \rightarrow BH_3OH^- + H_2 \qquad \text{Equation 4}$$

$$BH_3OH^- + H_2O \rightarrow BO_2^- + 3\ H_2 \qquad \text{Equation 5}$$

Now, knowing high energy density for the anolyte is linked with the use of high $NaBH_4$ concentration [2], one understands that a compromise is needed between the anolyte solution stability (*i.e.* low $NaBH_4$ concentration) and energy density (*i.e.* high $NaBH_4$ concentration). Li et al. did survey potential $NaBH_4$ + NaOH anolyte solutions for an efficient storage and use in DBFC [44]. They found that using high $NaBH_4$ concentration in the borohydride solutions resulted in decreased specific gravity and melting point: -68°C is for example measured for an aqueous solution containing 11 wt% $NaBH_4$ + 20 wt% NaOH (positive effects), but consistently increased viscosity (negative effect).

Thirdly, one should remember that the composition of the anolyte has a serious impact on the unit cell operation and performances. On the one hand, it induces strong modification of the borohydride diffusion coefficient [45-48]: the higher the pH and borohydride concentration, the lower the diffusion coefficient [47, 48]. This can lower the apparent BOR kinetics (that takes into account the mass transport limitation). On the other hand, the anolyte composition somewhat affects the "true" BOR kinetics (in the absence of any mass-transport limitation) [39]. Cheng and Scott for example monitored a BOR order of 1 and 1.1 with respect to the $NaBH_4$ and NaOH concentrations respectively, thus showing large acceleration of the reaction (on gold electrodes) for concentrated anolyte solutions [49]. Similarly, Li et al. [44] found that the increase of the $NaBH_4$ concentration slightly improved the performance of

2 The $NaBH_4$ solubility depends on the solution pH and temperature [42].[42] S. Suda, in Aqueous borohydride solutions, Handbook of Fuel Cells Fundamentals, Technology and Applications, W. Vielstich, H. A. Gasteiger and A. Lamm (Ed.), Wiley, Chichester, 2003, Vol. 3, pp 115-120.

the anode (composed of Ni-ZrR alloy). However, it also detrimentally increased the polarization at the cathode (Pt/C powder in this study), which was explained by the appearance of a mixed potential, following consequent fuel crossover. These authors also found that the cell voltage was generally not related to the NaOH concentration, except when a cation exchanged membrane was used. One can, however, suspect that these conclusions highly depend on the electrocatalysts materials used at the two electrodes. Larger cell polarizations were also monitored experimentally for high concentration of NaOH, which was explained in the frame of a decreased mobility of the charge carrier (Na^+) due to the increase of the solution viscosity (see above).

3.2 Alternative Reactants

Some authors investigated the particular case of DBFC fed with hydrogen peroxide H_2O_2 as an oxidant [14, 18, 50], notably to both limit the issue of carbonatation (when using ambient air) and obtain high power density, as a result of the fast hydrogen peroxide reduction reaction [14]. One can, however, point out that such system would suffer very harsh corrosion at the cathode and is less interesting than the conventional air-fed DBFC from an energy density point of view.

In addition, one should mention the possibility to generate hydrogen from the catalytic decomposition of tetrahydroborate anion [21, 42]. Such reaction is catalyzed by numerous materials, such as: hydrogen storage alloys [51], Ni- [52, 53], Co- [53-55], Ru- [35, 36] and Rh-based [56] catalyst. These compounds may be used in the form of metal powder and supported metal nanoparticles (heterogeneous catalysis) or metal salts dissolved in the solution (homogeneous catalysis). In some cases, sodium borohydride gels (made of sodium polyacrylate) were preferred to solutions for BH_4^- storage [55]. Now, hydrogen evolution at the anode compartment of a fuel cell, although sometimes favored on purpose in order to replace the conventional hydrogen storage by that, more convenient and safer, of borohydride compounds, also naturally occurs in DBFC, as a result of the homogeneous/heterogeneous hydrolysis of the BH_4^- species (see Equation 4 and Equation 5). In that latter case, the *in situ* generation of hydrogen within the structure of the anode may become a bias, since it can induce both modification/destruction of the anode structure (due to pressure build-up) and decrease the overall fuel efficiency (due to losses of hydrogen) [26].

Finally, alternative borohydride fuels like ammonia borane (NH_3BH_3) [57-60] and dimethylamine borane (DMAB) [61-66] have also been studied both from a fundamental point of view and regarding their use in DBFC. The oxidation mechanisms for such species will not be detailed here, but they involve reaction intermediates, e.g. BH_3OH^-, which may also participate in the BOR, as frequently pointed out in the literature [39-41, 58].

3.3 Fuel Monitoring

Quantifying the tetrahydroborate anion concentration in alkaline anolyte solutions is important for the understanding of DBFC performances. The historical method to do so is based on a volumetric assay, consisting of iodimetric analysis [67]. Because such *ex situ* method cannot be applied directly in DBFC, some more recent references focused on "*in situ*" experimental means to monitor the $NaBH_4$ concentration. For example, Amendola et al. [33] and then Santos et al. [68] used a chronopotentiometric method (based on open circuit potential measurements) to monitor the sodium borohydride concentrations in alkaline solutions. Alternatively, Mirkin et al. [69] and then Colominas et al. [70] used cyclic voltammetry in natural diffusion condition to measure the tetrahydroborate anion concentration of alkaline sodium borohydride solutions: the peak current of BOR on gold is indeed proportional to $[BH_4^-]$, but can be affected by the presence of boron oxide compounds in solution. It is wise to point out that the electrode "history" may largely hinder both measurements, because the surface of the electrode can be obstructed by hydrogen bubbles or boron oxides, as lately discussed by Chatenet et al. [48] and Finkelstein et al. [71]. Moreover, the latter *in situ* mean of borohydride concentration measurement requires the knowledge of the BH_4^- diffusion coefficient in the solution (this value not being easy to determine [45, 47, 48]), which mitigates its interest. Extracting the borohydride concentration from the values of the limiting current in rotating disk electrode configuration is also possible (Levich method [39]), but would suffer the same bias. From this simple review, one understands that measuring basic parameters for borohydride containing solutions is eventually not so simple. This may account for some of the unexpected mismatch often observed between various results of the literature in the fields of BOR and DBFC testing.

As stated above, high concentration sodium hydroxide + sodium borohydride solutions are required to reach high energy density DBFC. Assuming (i) concentrated anolyte solutions display higher viscosity, as monitored by Chatenet et al. [48], for NaOH concentrations in the range 0.1 – 4 M at temperatures of 10, 25 and 40°C and (ii) the diffusion coefficient of the tetrahydroborate anion varies like the electrolyte viscosity [48] (it decreases for concentrated anolyte solutions [47]), one understands that the mass-transport of $NaBH_4$ will become more limiting when concentrated anolyte solutions are used. This can detrimentally hinder the anode apparent reaction kinetics.

4. Principle of Operation of a DBFC

4.1 Basics and Performance

The classical DBFC consists of the combination of an anode at which the tetrahydroborate anion is oxidized (Equation 1) and a cathode where the oxygen reduction takes place (Equation 2)Equation 2). The overall cell reaction (Equation 6), occurs in principle at the thermodynamic voltage of 1.64 V (in standard conditions). However, as both the oxygen reduction [72-76] and the borohydride oxidation [39, 46, 49, 77] suffer consequent kinetic limitations, the practical voltage of a DBFC is more or less on the order of 1 V.

$$BH_4^- + 2\,O_2 \rightarrow BO_2^- + 2\,H_2O \quad (E = 1.64\text{ V}) \qquad \text{Equation 6}$$

Usually, the DBFC is based on an ion-exchange membrane, the role of which is to separate the anode and cathode compartments (to suppress or at least limit the fuel and oxidant crossover), enable the necessary ion conduction to pass current, while prevent any electron flow in the internal circuit of the fuel cell, thus enforcing them to flow through the external circuit. The scheme of Figure 1 presents the basic construction and operation of a DBFC.

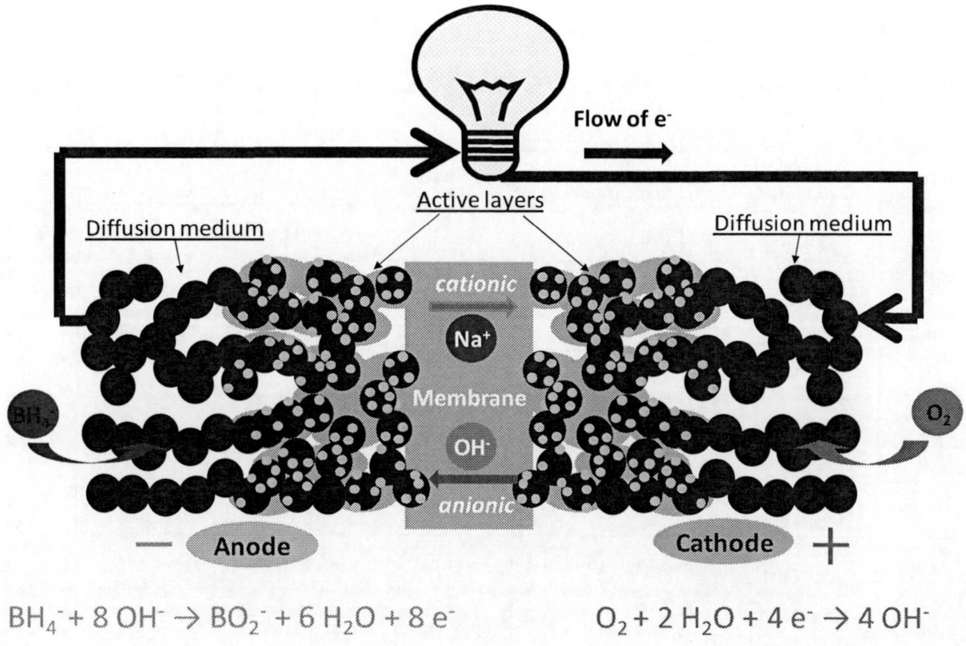

Figure 1. Schematic architecture of a DBFC baring an anionic membrane (OH^- are transported from the cathode to the anode, red arrow) or a cationic membrane (Na^+ ions are transported from the anode to the cathode, blue arrow). In all cases, the electrons flow through the external circuit (black arrow), thus powering the load.

The first DBFC system was developed by Amendola et al.; it bared an anion-exchange membrane and gold-containing anode electrocatalyst [34]. Since then, many groups have investigated the DBFC, either focusing on unit cell or electrode/membrane materials development. The interested reader is directed to useful reviews concerning the DBFC materials [78] and technology [55]. In addition, the following references are also worth to be red to learn more about anion-exchange membranes [10, 79], platinum [40, 80, 81] or non-platinum [11, 39, 80, 82] anode electrocatalysts and borohydride-tolerant cathode electrocatalysts [13, 75, 83] materials.

Table 1 presents the main performances of several DBFC systems tested in the literature. Although the membrane and electrocatalysts materials as well as the operating conditions vary, a quick stare at Table 1 shows that the output power of a DBFC is never exceeding *ca.* 0.25 W cm^{-2}, a value easily reached with PEMFC [2]. Consequently, knowing that sodium borohydride exhibits rather high energy density, the DBFC is essentially dedicated for the powering of (small) portable electronic devices [16, 18, 34, 84, 85].

Table 1. Characteristics and performances of a few DBFC systems. *P* stands for the power density and n_{e-} for the number of electrons exchanged per BH_4^- anions. "n.a." stands for "non available".

Anode	Fuel ¤	Membrane	Cathode	Oxidant	°C	*P* / W cm^{-2}	n_{e-}	Ref.
Au/C + Pt/C	25 % NaOH 5 % $NaBH_4$	Anionic (Pall-RAI)	n.a.	Air (1 bar)	0 0	0.020 0.06	6.9	[34]
PtRu/C	1 M NaOH 2 M $NaBH_4$	Anionic (ADPMorgane)	Pt/C	O_2 (1 bar)	0	0.20	n.a.	[86]
Pt/C	1 M NaOH 2 M $NaBH_4$	Anionic (ADPMorgane)	O'CAT® Evionyx	Air (1 bar)	5#	0.20	1.5*	[38]
Au/Ti mesh	2.5 M NaOH 3.3 M $NaBH_4$	Cationic (N117)	Pt/C	O_2 (1 bar)	5	0.09	n.a.	[11]
NiAlMnCo (AB5-alloy)	20 % NaOH 10 % $NaBH_4$	Cationic (N117)	Pt/C	8.9 M H_2O_2	0	0.15	6.6*	[14]
Zr-based (AB2 alloy)	20 % NaOH 10 % $NaBH_4$	Cationic (N115)	Pt/C	Air (1 bar)	0	0.22	n.a.	[37]
Ni powder	6 M NaOH 5 % $NaBH_4$	Cationic (NRE211)	Pt/C	Air (1 bar)	5#	0.04	< 4	[87]
Ni powder + Pd/C	20 % NaOH 5 % $NaBH_4$	Cationic (N112)	Pt/C	Air (1 bar)	5#	0.08+	6@	[16]
NiAlMnCo (AB5-alloy)	20 % NaOH 10 % $NaBH_4$	Cationic (N117)	FeTMPP/C	0.5 M H_2O_2	0	0.1	n.a.	[88]

¤ Concentrations given in weight%; the solvent is water.

* Values recalculated from the fuel efficiency assuming the losses due to hydrogen evolution.

@ Values recalculated from the fuel efficiency (based on coulometry for total electrolysis)

Room temperature

+ self-breathing planar DBFC stack

4.2 Influence of the Electrode Geometry and Reactant Channel Design

Section 30 showed that the composition of the anolyte had a large effect on the performances of the DBFC. In the same way, the membrane and electrode materials used also have a great impact on the power output and faradaic efficiency of a DBFC, as will be detailed in section 4.30. One should also emphasize that, like all fuel cells (open generators) the DBFC needs efficient mass-transport of both the fuel and oxidant from their storage tank (or air) down to the electrodes. The role of the each "diffusion medium" is precisely to drive the reactants to the "active layers" and to enable efficient drain of the reaction products (see Figure 1). The promotion of the mass-transport needs particular emphasis at the anode, because the BOR products, e.g. sodium borate ($NaBO_2$) or boric acid (H_3BO_3), are weakly soluble in alkaline solution (for example, the solubility of $NaBO_2$ is *ca.* 2 M in 10 wt% NaOH at 25°C [42]) and, thus, prone to precipitate within the porous structure of the anode.

In that frame, Cheng and Scott showed improved $NaBH_4$ mass-transport and faster H_2 gas bubble release using Ti mesh-supported of very open structure, compared to conventional carbon-supported anodes (both for Au and Ag electrocatalysts and at all loadings) [11]. Knowing both electrode reactions in a DBFC exhibit sluggish kinetics, one must keep in mind that "ultrathin" active layer cannot be employed, simply because reaching sufficient power output required high amount of electrocatalyst. Therefore, the usual solution consisting of using thin active layers to reduce mass-transport limitation is not relevant in DBFC, and instead open porous structures like those employed by Cheng and Scott are worth of interest. In the same way, Kim et al. monitored improved single-cell performance by adopting a corrugated anode design. Again, this is explained in the frame of a faster mass-transport within the anode structure: the evacuation of the hydrogen by-products (produced by heterogeneous hydrolysis on the anode electrocatalyst) is promoted thanks to the gap between the valleys of the corrugated anode [37].

Cheng and Scott also studied the influence of the geometry of the reactant channels pattern within the bipolar plate. In dDoing so, they demonstrated better performance for serpentine than for parallel flow field, as well as improved performances with increasing flow rates [12].

These examples show that high performances in DBFC are related not only good electrocatalytic materials, but also proper electrode and cell design. In that frame, it is also worth mentioning that the completeness of the BOR (faradaic efficiency) depends on the structure of the electrodes (and especially of the anode active layer). Indeed, Molina Concha and Chatenet recently showed that Pt/C electrocatalysts can yield complete BOR ($n_{e-} \approx 8$ per BH_4^- anion) when the active layer is sufficiently thick, *i.e.* thickness beyond 3 μm, whereas the overall faradaic efficiency drops dramatically ($n_{e-} = 2$ per BH_4^- anion) when the active layer is too thin, *i.e.* thickness below 1 μm [89]. For sure, one could argue that such findings may be related to the particular experimental conditions employed in their study. However, this nevertheless demonstrates that sufficient residence time of the molecules formed, e.g. H_2 by heterogeneous hydrolysis (see Equation 4 and Equation 5) or hydride-containing BOR intermediates, within the active layer is needed to favor the complete HOR and/or BOR. Similar transport effect within the active layer was also monitored for the ORR by Chatenet et al. [72, 73] and Schneider et al. [90]. In particular, Schneider et al. demonstrated that H_2O_2 is not only formed as an adsorbed reaction intermediate, which instantaneously further reduces to H_2O in a second step (4-electron ORR pathway), but that instead it can desorb, diffuse into the stagnant diffusion layer and then finally either re-adsorb and further react to H_2O (stepwise 2+2-electron ORR pathway) or diffuse into the bulk electrolyte and leave the reaction cell (2-electron pathway). Naturally, the latter pathway is favored for active layers of very little thickness or low electrocatalyst loading [90].

4.3 Membrane and Electrode Materials

As most fuel cell systems, the DBFC is usually based on a membrane. Table 1 shows that both cationic and anionic membranes have been used with some success for DBFC testing. It is interesting to note that the maximum power density of the DBFC systems tested so far (*ca.* 0.2 W cm^{-2}) seems not to depend on the nature of the ion-exchange groups of the membrane. However, both materials display several advantages and drawbacks, as described below.

Nafion® is by far the most popular membrane for fuel cell systems (essentially PEMFC and DMFC). This material indeed exhibits impressive electrochemical performances (e.g.

cation conductivity) and durability in the aggressive medium of a fuel cell. However, when used in a DBFC, this cation-exchange membrane favors the displacement of the anolyte towards the cathode compartment [87]: the current flow drives the Na^+ cations from the anode to the cathode, where they combine with the OH^- produced by the reduction of dioxygen (or hydrogen peroxide). As a result, NaOH is concentrated at the cathode, which can ultimately yield obstruction of the pores of the gas-diffusing electrode [3]. In opposition, cation-exchange membranes theoretically prevents the (anionic) fuel crossover [87], as a result of the Donan exclusion [91] and because the anions transported from the anode to the cathode are against the current-flow (Figure 1). These assertions remain however a theoretical construct, since the membrane is soaked with the anolyte and can transport ionic charges from bulk diffusion (in liquid electrolyte), at least to some extent. Consequently, some BH_4^- fuel crossover has been put into evidence for cation-exchange membranes [87], thereby lowering the DBFC performances through a decreased fuel efficiency and increased cathode polarization. Now, such crossover decreases both for thicker membranes and upon DBFC operation following consequent fuel consumption at the anode [27], as well documented for DMFC.

The use of an alkaline (anion-exchange) membrane renders even more critical the issue and consequences of the (BH_4^- anion) fuel crossover [44]. Moreover, one will have remarked that only a very limited number of references deals with the testing of anionic membranes in fuel cell [19, 34, 38, 86, 87], mostly because anion-exchange membrane materials are still under optimization. A direct consequence of such scarcity and incomplete development is the overall low (i) OH^- conductivities, (ii) mechanical stability and (iii) calendar stability in the DBFC environment of most anion-exchange membranes. This is naturally incompatible with any wide spreading of the alkaline membrane DBFC technology in its present state of development [2].

To end with this issue, it is worth saying that the lack of relevant ion-exchange membranes for DBFC has (indirectly) encouraged the use of alkaline battery separators to build DBFC systems. Naturally, in such architecture, the cathode compartment is wetted by the anolyte and in consequence sufficient power output is only possible provided borohydride-tolerant electrocatalysts are used at the cathode [13].

It is well beyond the point of the present contribution to survey all the potential electrocatalysts that can be used at the cathode of a DBFC. As stated above, the non-ideal separation between the anode and cathode compartments renders necessary to use a BH_4^- tolerant electrocatalyst at the cathode. In that frame, the usual (and efficient) alkaline ORR electrocatalysts like Pt, Ag and Au are not suited, since they are not at all insensitive to the presence of $NaBH_4$ [75]. On the contrary, materials like manganese oxide [13, 75] or transition-metal macrocycles (e.g. FeTMMP [88], etc.) show both sufficient activity towards ORR and tolerance to $NaBH_4$. In consequence, they behave much better than Pt and Ag-based electrocatalysts at the cathode of DBFC, regardless of the possible fuel crossover [27, 75].

As the (sluggish) borohydride electrooxidation reaction is impeding the development of DBFC, it is no surprise that numerous anode catalysts have been tested for this reaction. The early studies were based on gold- [34, 39, 40, 46, 49] or platinum- [23, 40, 81, 92] containing electrocatalysts, these materials still being very popular nowadays [19, 20, 28, 48, 58, 71, 93-96]. In addition, in order to take benefit of the interest of the alkaline medium, numerous non-Pt and non-Au electrocatalysts and alloys have been employed with some success, such as

[3] At the anode, the boron oxide can also precipitate within the porous structure of the electrode [42] Ibid..

silver [11, 39, 97-99], palladium [26, 100], ruthenium [100], nickel [82] and several AB2 or AB5 hydrogen storage materials [37, 51, 54, 101]. Ideally, these BOR electrocatalyst should exhibit fast kinetics and optimal faradaic yield. It is beyond the point of the present chapter to detail all the results obtained in the literature using these different materials, but is nevertheless worth emphasizing that the common belief is to separate the BOR electrocatalysts in two families: the first contains the materials prone to catalyze the ionization of hydrogen (e.g. Pt, Pd, Ni, hydrogen storage alloys, etc.) and the second those inactive for such reaction (e.g. Au and Ag). Whereas the former family of electrocatalysts supposedly yields faradaic efficiency strictly below 100 %, as a result of uncontrolled hydrogen gas generation by heterogeneous hydrolysis of $NaBNH_4$ (see Equation 4 and Equation 5) and further loss within the electrode structure, the latter would yield faradaic efficient materials (not promoting the heterogeneous hydrolysis of $NaBH_4$). In that frame, several studies mention the use of hydrogen evolution inhibitors (e.g., thiourea) to increase the BOR faradaic efficiency on Pt-based electrocatalysts [40, 102]. Although *a priori* interesting, such strategy displays the immense drawback of poisoning most of the electrocatalyst surface, therefore largely counterbalancing the supposed advantages in electrooxidation completeness (faradaic efficiency) by large drawbacks hindrances in terms of reaction kinetics, as well documented by Finkelstein et al. [71].

Now, comparing the literature data, one can easily observe that the number of electrons exchanged per BH_4^- anion (the most common way to quantify the BOR faradaic efficiency) vary from one study to the other, even for the simplest and most studied DBFC anode electrocatalysts. For example, it was reported that the BOR on Pt/C involves n_{e-} values per BH_4^- anion from *ca.* 2 [22, 23] to *ca.* 8 [89]. Naturally, such variability may be ascribed to the different experimental conditions employed in these studies, like anolyte composition and temperature (see section 30) or geometry of the active layer (see section 4.20). More surprisingly, recent papers provide another explanation, by contradicting the "well-admitted" findings (Pt yields low faradaic efficiency and Au high faradaic efficiency). For example, Krishnan et al. [93], Chatenet et al. [48] and Finkelstein et al. [71] all observed non-negligible H_2 production on Au during the BOR (nearly at all potentials [71]), which is only possible provided Au actually catalyses the heterogeneous hydrolysis of $NaBH_4$. Regarding Pt, while Molina Concha and Chatenet demonstrated that Pt/C electrocatalysts could yield nearly complete BOR [89] (see section 0), Finkelstein et al. showed that, in contrary to the common belief, Pt does not catalytically hydrolyzes BH_4^- to H_2 until *ca.* 1 V BOR overpotential [71]. These two examples emphasize the need for more fundamental surveys of the BOR, even on simple electrocatalyst. This is the aim of section 05.

5. Survey of the BOR on Gold

The complete BOR is complex, involving 8 electrons and 8 hydroxide anions (see Equation 1). For most catalytic materials like nickel [53, 103] or platinum [40, 62, 81], the direct BOR (Equation 1) presumably competes with the heterogeneous hydrolysis of BH_4^- (Equation 4 and Equation 5). The combination of these two processes (Equation 7, first introduced by Liu et al. [82]) lowers the practical number of electrons exchanged per BH_4^-

anion, as the H_2 generated may evolve away from the electrode, preventing its oxidation and subsequent conversion into electrical energy.

$$BH_4^- + x\,OH^- \rightarrow BO_2^- + (x - 2)\,H_2O + (4 - 0.5x)\,H_2 + x\,e^- \qquad \text{Equation 7}$$

Up to now, platinum has been one of the most studied anode material in a DBFC. However, the BOR mechanism on platinum is still uncertain (see section 4.30), even if there is consensus on one point: the overall reaction kinetics is rather fast on Pt [40, 58, 81, 89].

Another well-employed DBFC anode material, gold, is supposedly rather inactive towards both hydrogen evolution and oxidation reactions [41, 46, 80]. This property has rendered gold electrocatalysts very popular, since the common belief is that BH_4^- oxidation on gold will possibly be complete, strictly involving 8 electrons per BH_4^- species [39, 46, 49, 104]. However, although this should simplify the electrode processes, there is no consensus in the literature regarding the BOR mechanism on gold. It was nevertheless proposed that BOR on gold involves a sequence of electrochemical (E) or chemical (C) elementary steps [40, 46], the chemical steps being fast [45]. The aim of this study is ultimately to isolate the most probable reaction pathway for the BOR on gold, which supposes that both the possible intermediates can be detected using coupled physical techniques (OLEMS for H_2 detection, section 5.10 and FTIR for boron compounds detection, section 5.20) and the reaction can be modeled in agreement with the physical and electrochemical findings (section 5.30).

5.1 Is Gold Inactive Regarding the Heterogeneous Hydrolysis of BH_4^- Aanion?

It is widely admitted that gold does not promote the BH_4^- heterogeneous hydrolysis [40, 46] but instead would be a faradaic-efficient BOR catalyst [46, 47, 49, 77], thus justifying its use as DBFC anode material. However, lately some reports mentioned that Au may not yield to the complete direct BOR [48, 71], as concluded confirmed from several studies mentioning less than 7.5 electrons produced per BH_4^- anion [34, 39, 40, 93] (4.5 electrons per BH_4^- anion was even reported [82]). Such lower-than-expected faradaic efficiency could be related to the release of hydrogen, probably linked to BH_3OH^- formation [93] as a result of BH_4^- partial hydrolysis on Au (such hydrolysis was also detected for NH_3BH_3 [105]). To overcome the uncertainties regarding the catalytic activity of gold regarding the heterogeneous hydrolysis of BH_4^-, the effective H_2 generation on a gold electrode has been investigated as a function of its polarization in the BOR potential range using on-line electrochemical mass spectrometry (OLEMS).

Figure 2-A shows the linear relationship existing between the faradaic HER current and the corresponding MS ionic current ($m/z = 2$). The corresponding average calibration constant of the setup is: $K^* = 1.2\ 10^{-7}$ for $m/z = 2$. It was used to extract the actual hydrogen yield from the *in situ* OLEMS measurements during BOR. As stated in the experimental section 20, Figure 2-B clearly exhibits the absence of any quantitative MS signal ($m/z = 2$) monitored when H_2 is generated in the bulk of the solution, while keeping the Au/PTFE electrode at ocp. These conditions mimics the HER occurring on the counter electrode during BOR experiments held at the Au/PTFE electrode. The slope of MS signal on the insert of Figure 2-

B matches that monitored when the Au/PTFE is held at ocp in the absence of any HER in solution. This highlight that (i) Au actually generates some H_2 at ocp and (ii) our OLEMS setup is only sensitive to the H_2 generated at the Au/PTFE electrode. While the latter observation rules out any bias from the H_2 generated at the Au counter electrode in the solution, the former one is an important discovery, since it means that Au actually catalyses to some little but non-negligible extent the heterogeneous hydrolysis of BH_4^- in ocp conditions (at least for the electrolyte composition investigated here), in contradiction to the common belief [40, 46]. Such little H_2 generation still proceeds at an equivalent rate as at ocp until the BOR initiates, at *ca.* 0.3 V *vs.* RHE (Figure 3-A). Indeed, the m/z = 2 signal (obtained by subtracting the absolute m/z = 2 signal at ocp to that measured at the end of each potential hold) showed on Figure 3 is always near zero before any consequent BOR current is monitored.

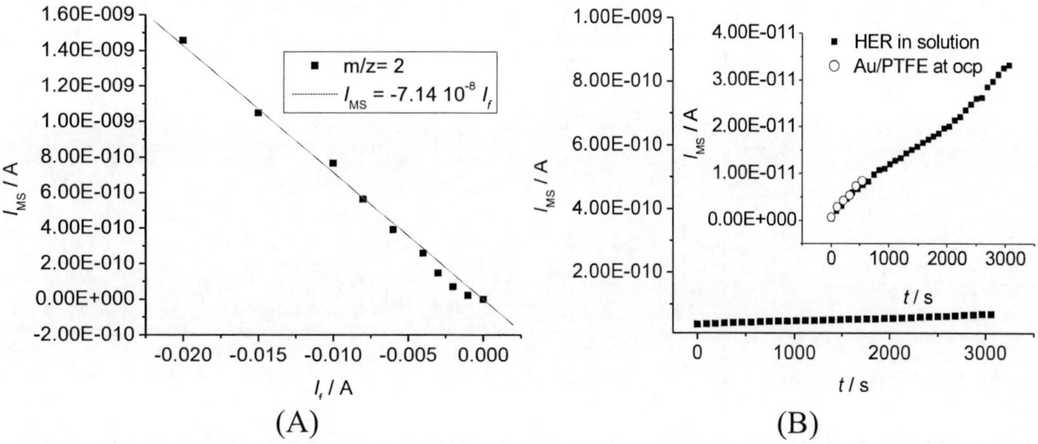

Figure 2. (A) Calibration plot of the *in situ* Mass Spectrometry ionic current (m/z = 2) versus the faradaic current (I_f) during HER experiments on bulk gold in 1 M NaOH + 0.015 M $NaBH_4$ (T = 25°C). The calibration constant of the MS apparatus is in average (based on three different trials) K^* = 1.2 10^{-7} for m/z = 2. (B) Plots relative to HER experiments conducted on an auxiliary Au electrode in the solution at currents in the range -0.5 10^{-4} to -20 10^{-3} A in 0.1 M NaOH (T = 25°C), the Au/PTFE electrode being at ocp (■) and with the Au/PTFE electrode at ocp without HER in solution in 1 M NaOH + 0.015 M $NaBH_4$ (O).

Even more striking are the OLEMS results relative to the steady-state BOR on the Au/PTFE electrode. Figure 3-A makes clear that consequent H_2 generation occurs on gold during BOR, as soon as the Au electrode is polarized positive to its ocp. H_2 generation increases after the BOR onset (*ca.* 0.3 V *vs.* RHE in 0.1 M $NaBH_4$) and reaches a maximum at *ca.* 0.5 V *vs.* RHE. Although of lesser magnitude, H_2 generation is still observed at potentials as high as 0.6 and 0.7 V *vs.* RHE. At the maximum, the MS signal for m/z = 2 amounts *ca.* 7.4 10^{-10} A, which corresponds to an "equivalent HER faradaic current" of *ca.* 10 mA (determined from the calibration plots in Figure 2-A), equal to the real BOR faradaic current at this potential.

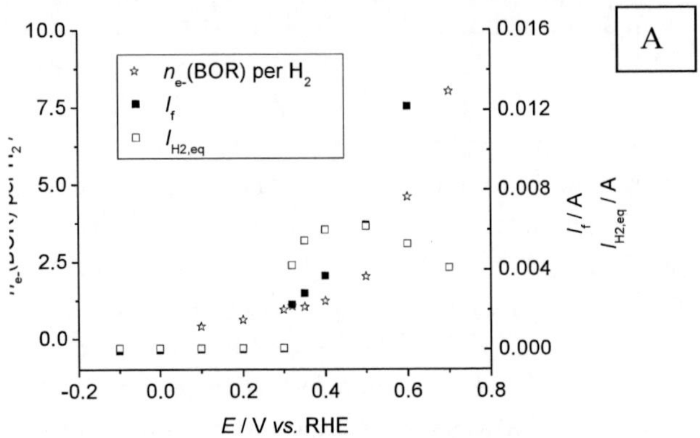

(A)

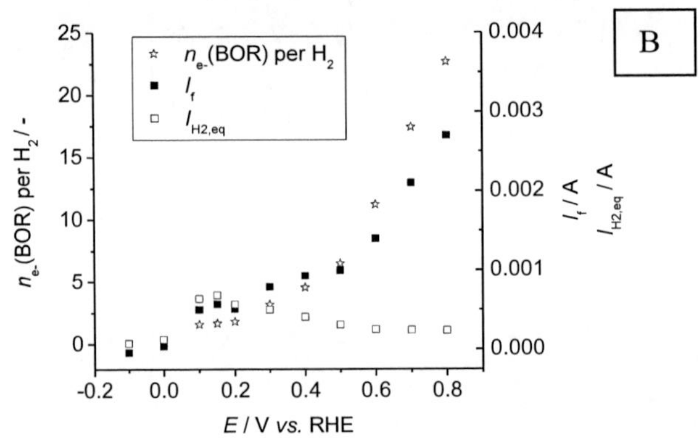

(B)

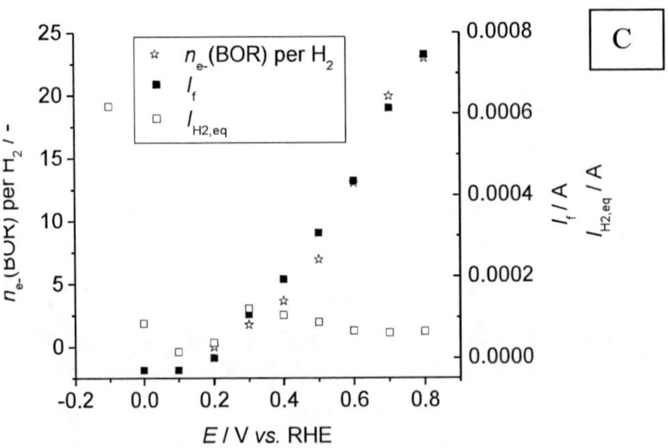

(C)

Figure 3. Number of electrons per H_2 molecule generated by the BOR (✭), faradaic BOR current density (■ I_f) and corresponding equivalent HER current derived from *in situ* mass spectrometry (□ $I_{H2\ eq}$) during steady-state (10 min. per data point) BOR on bulk gold in (A) 1 M NaOH + 0.1 M $NaBH_4$, (B) 1 M NaOH + 0.015 M $NaBH_4$ and, (C) 1 M NaOH + 0.003 M $NaBH_4$ (T = 25°C). Between each steady-state measurement, the MS chamber was vacuumed and the solution was purged with Ar, so as to both remove any trace of dissolved H_2 in the solution and expel the H_2 bubbles trapped at the Au/PTFE working electrode surface.

The same qualitative tendency remains at lower $NaBH_4$ concentration (0.015 M, Figure 3-B and 0.003 M, Figure 3-C), but with the following differences: (i) the BOR and MS onsets are both shifted negative by *ca.* 0.1 V and (ii) the amount of H_2 generated is *ca.* 10 and 40 times smaller at 0.015 and 0.003 M $NaBH_4$ respectively (the steady state current being lower, by a factor *ca.* 4 for 0.015 M $NaBH_4$ and 10 for 0.003 M $NaBH_4$). These results clearly put into evidence that the heterogeneous hydrolysis on gold increases with the borohydride concentration, thereby translating a lower faradaic efficiency for the BOR on gold at high $NaBH_4$ concentration. From Equation 7, assuming that Au is totally inactive for the HOR (which is not so evident, as both Krishnan et al. [93] and Finkelstein et al. [71] found very puzzling results seemingly showing that Au presents a weak but non-negligible HOR activity), one can calculate that the BOR involve *ca.* 6.4 and 7.4 electrons per BH_4^- species in the region of the limiting plateau in hydrodynamic CV performed on RDE (see reference [28, 39]), in agreement with some literature values [28, 34, 39, 40, 93].

In summary, these results show for the first time that, regardless of the solution composition, the amount of H_2 generated during BOR on Au is not zero. The H_2 yield is small but non-negligible at ocp, and then increases sharply above the BOR onset. It reaches a maximum in the potential region of the first BOR wave, slowing down (without stopping) at higher potentials. As a result, the faradaic efficiency for BOR on Au increases monotonically with the electrode potential (up to 0.8 V *vs.* RHE), until *ca.* 6.5 to 7.4 electrons are exchanged per BH_4^- species (for 0.1 M $NaBH_4$ and the more diluted solutions, respectively). So, unlike often stated in the literature, gold promotes the heterogeneous hydrolysis of borohydride, even in strong alkaline medium. This goes in line with some recent results, which indirectly pointed towards the non-negligible catalytic activity of gold electrodes with regards to the heterogeneous catalytic decomposition of borohydride compounds like BH_4^- [71, 93] and NH_3BH_3 [105].

5.2 Evidencing the BOR Intermediates on Gold by FTIR Spectroscopy

Up to now, the amount of literature that provides physical evidences about the nature of the BOR intermediates is extremely scarce. Obviously, *in situ* FTIR spectroscopy is a technique of choice to go beyond this lack. The FTIR spectra of Figure 4-A show five major absorption bands appearing sequentially for increasing potentials during the BOR on a gold electrode. The first two bands appear at 1080 and 1145 cm^{-1} at potentials close to -0.16 V *vs.* RHE. The wave number of the band at 1145 cm^{-1} changes with the electrode potential, which enables us to conclude that it is related to an adsorbed species. At 0.36 V *vs.* RHE a third absorption band is observed at 970 cm^{-1}. The last two significant bands appear at 1325 and 1425 cm^{-1} at potentials close to 0.6 V *vs.* RHE. The corresponding linear BOR voltammogram monitored *in situ* in the FTIR cell is shown on Figure 4-B. Table 3 summarizes these results, and gives the correspondence between these FTIR bands and the most relevant molecular binding at the origin of these vibrations.

Table 2. Assignation of the major bands monitored as a function of the electrode potential in the FTIR spectra of Figure 4-A. The potential *E* displayed in the third column marks the appearance of the corresponding bands.

ν / cm^{-1}	Assignment	*E* / V *vs.* RHE	References
1180	BH_3	-0.16	[106-111]
1080	BH_3	-0.16	[106-111]
970	BH_2	0.36	[112, 113]
1325	BO_2	0.6	[114-119]
1425	O-B-O	0.6	[114-119]

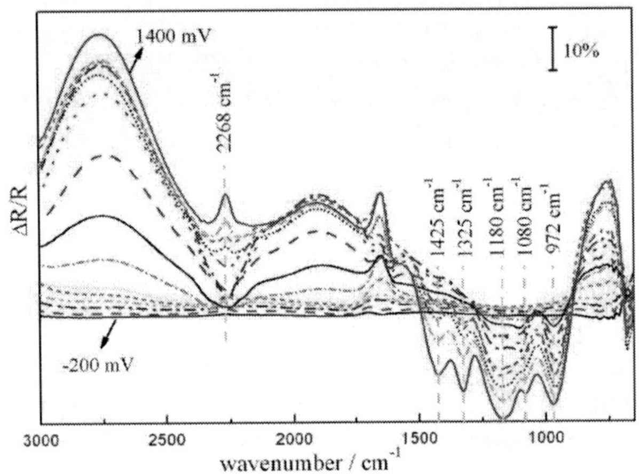

(A)

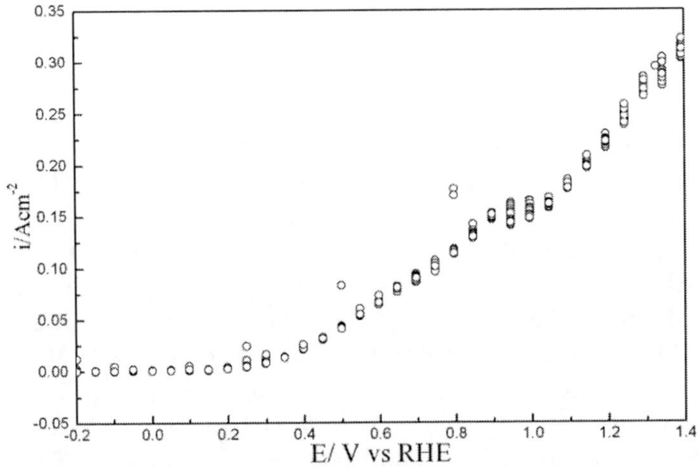

(B)

Figure 4. (A) *in situ* infrared spectra recorded on a polycrystalline gold electrode in 1 M NaOH + 1 M $NaBH_4$ aqueous solutions, at room temperature; R_{Ref} = -0.2 V *vs*. RHE; The FTIR spectra presented here were acquired every 0.1 V from -0.2 to 1.4 V *vs*. RHE. (B) Corresponding polarization curve plotted in the spectro-electrochemical cell in the course of the *in situ* FTIR measurements for BOR on gold from -0.2 to 1.4 V *vs*. RHE.

Table 3. Assignation of the major bands monitored as a function of the electrode potential in the FTIR spectra of Figure 4-A. The potential *E* displayed in the third column marks the appearance of the corresponding bands.

ν / cm^{-1}	Assignment	*E* / V *vs.* RHE	References
1180	BH_3	-0.16	[106-111]
1080	BH_3	-0.16	[106-111]
970	BH_2	0.36	[112, 113]
1325	BO_2	0.6	[114-119]
1425	O-B-O	0.6	[114-119]

In a previous work [94], only the first three bands could be observed, because the infrared measurements were attempted with a CaF_2 window, which induces a cutoff below 1000 cm^{-1}.

The first two bands (at 1080 and between 1145-1190 cm^{-1}) that appear at *ca.* -0.16 V *vs.* RHE during the BOR correspond to symmetric and asymmetric deformation mode in the planar BH_3 molecule (D3h symmetry), respectively. More specifically, the band at 1180 cm^{-1} is reported either to the BH_3 isolated group or included in more complex molecules, whereas that at 1080 cm^{-1} is only reported to the molecules containing BH_3 moieties (e.g., BH_3CO or BH_3NH_3) [106-111]. The appearance of these bands is related to the oxidation or decomposition of BH_4^- into species exhibiting the BH_3 symmetry, like BH_3 or BH_3OH^-. This process starts at very low potential, *ca.* -0.16 V *vs.* RHE, for which no consequent faradaic current is monitored [94]. This consists of an experimental evidence of either the self-poisoning of electrode surface with the BH_3 adsorbed species (they are formed at -0.16 V *vs.* RHE but would not be oxidized on Au at such low potential [94]), or the heterogeneous hydrolysis of BH_4^- on Au at low potential (in agreement with the OLEMS data of section 5.10). The third significant absorption band appears at 970 cm^{-1} at potential close to 0.36 V *vs.* RHE. This absorption is assigned at the wagging mode of B-H bond vibration in BH_2 moieties [112, 113]. So at this potential, the BH_3 (or BH_3OH^-) species formed at -0.16 V *vs.* RHE are oxidized and lose another hydride species, therefore forming a species of symmetry close to that of BH_2. This occurs at a potential close to the BOR onset (see Figure 4-B), showing that the surface of gold electrode is no longer “poisoned” by the adsorbed BH_3 species. Finally, close to 0.6 V *vs.* RHE, the two last absorption bands are observed at 1325 and 1425 cm^{-1}. These bands are assigned to the B-O stretching mode in BO_2 (linear molecule) and to the asymmetric stretching mode in the nearly linear molecule B-O-B [114-119]. Now, these *in situ* FTIR spectra did not put into evidence any absorption bands corresponding to the final product of the BOR (*i.e.* BO_2^- anion) below 0.6 V *vs.* RHE. The B=O absorption bands of isolated BO_2^- in argon matrix should appear at about 1931/2000 cm^{-1} (stretching modes) and 590-600 cm^{-1} (bending modes) for $^{11}B/^{10}B$ respectively [115]. However, the BO_2^- species are prone to polymerize at the electrode surface [94], which might explain the absence of evidence for such bands. Nevertheless, the fact vibrations related to B-O bond have been observed only above 0.6 V *vs.* RHE probably means that the complete oxidation of BH_4^- cannot be considered as quantitative below this potential. Based on the present FTIR results, one can propose the following elementary steps of Equation 8, (C) Equation 9, (E) Equation 10 and (E) Equation 11 for the BOR mechanism:”

$BH_4^- + s + OH^- \rightarrow BH_{3,ads} + H_2O + 2\ e^-$ (E) Equation 8

$BH_{3,ads} + OH^- \rightarrow BH_3OH^-_{ads}$ (C) Equation 9

$BH_3OH^-_{ads} + OH^- \rightarrow BH_2OH_{ads} + H_2O + 2\ e^-$ (E) Equation 10

$BH_2OH_{ads} + 5\ OH^- \rightarrow BO_2^- + 4\ H_2O + 4\ e^-$ (E) Equation 11

The present results agree with the two first steps of the BOR mechanism proposed in reference [94] (see Equation 8 and (C) Equation 9).

The first scenario consists of $BH_{3,ads}$ formation (in a two-electron reaction) at -0.16 V *vs.* RHE, species which presumably poisons the gold electrode at low potential; then, at 0.36 V *vs.* RHE, $BH_3OH^-_{ads}$ is formed in a chemical step, and a this potential this species is oxidized forming BH_2OH ((E) Equation 10). In the second scenario $BH_{3,ads}$ and $BH_3OH^-_{ads}$ are both formed at -0.16 V *vs.* RHE (in electrochemical and chemical steps, respectively) and at 0.36 V *vs.* RHE the $BH_3OH^-_{ads}$ intermediate is electrooxidized in BH_2-containing species in a two electron reaction ((E) Equation 10). One can emphasize that the results from reference [94] did not allow to discriminate the successive steps of the BH_3OH^- oxidation into BO_2^-, because the BH_2 vibration band could not be observed. Both scenarios are consistent with the oxidation of the BH_2OH intermediate into the final product, BO_2^-, above 0.6 V *vs.* RHE. The present FTIR data do not enable to determine the intermediates of this complex four-electron step. In addition, these FTIR data do not rule out the possible heterogeneous catalytic decomposition of BH_4^- on Au below the BOR onset, thereby agreeing with the OLEMS data of section 5.10).

5.3 Towards a Simplified BOR Pathway for Gold Electrodes

Among the few recent studies concerning the BOR, most have been completed on a variety of electrode materials, at various pH and rather high borohydride concentration (usually above 0.01 M, which favors their hydrolysis [21-23, 41]), in a way that it is difficult to unify their conclusions. In consequence, attempting a survey of the BOR in the simplest conditions possible, *i.e.* neither in DBFC nor for electrolyte generating hydrolysis of BH_4^-, is mandatory to enable determining its mechanism. In that frame, the BOR has been studied on gold polycrystalline electrodes, using on purpose a dilute electrolyte (a priori not yielding heterogeneous hydrolysis [21, 39]).

Figure 6 display selected EIS spectra obtained for the BOR on gold during the SPEIS experiments. The EIS spectra chosen are representative of those classically obtained in such experimental conditions. They basically consist of two well separated capacitive loops in the region of the first BOR wave (low potentials), and one ill-defined capacitive loop in the region of the second BOR wave (high potentials). However, the shape of the single capacitive loop measured at high potential suggests it consists of two capacitive loops poorly separated in frequency, as will be discussed below. When the high-frequency and low frequency loops are well separated, the frequencies at their apex (a few 10 Hz versus *ca.* 1 Hz respectively) correspond to classical charge-transfer and mass-transport (diffusion-convection) controlled phenomena respectively, as classically monitored for redox reactions studied on RDE.

Moreover, the frequencies at the apex (*ca.* 0.6 Hz) of the low-frequency loop is not related to the electrode potential, at least in the region of the first BOR wave (for which the two capacitive loops are well-separated, Figure 5), which is expected for mass-transport-limited reactions [120].

As all the EIS spectra from Figure 5 only bare two capacitive loops (at maximum), they can be modeled (Z-fit function in EC-Lab®) with an electrochemical (redox) step E under diffusion-convection control. This naturally consists of an approximation, since a redox E cannot account for the two CV waves experimentally monitored for the BOR on gold [28]. However, and surprisingly, such simple modeling accounts reasonably well for the experimental EIS data, especially for the region of the first BOR wave.

The literature concerning the BOR on gold mainly mentions two reaction pathways. The first one, published by the group of Bard, is based on fast potential scan CV and scanning electrochemical microscopy (SECM) on a gold microdisk: it consists of a reversible ECE (electrochemical – chemical – electrochemical) sequence [46] at low potential (involving 1 electron per E step) followed by an irreversible EC (electrochemical – chemical) sequence involving the remaining 6 electrons to generate the final BOR product, BO_2^-. Mirkin et al. concluded that the first C step is fast [46], which in the quasi-steady state conditions employed here enables simplifying the ECE sequence into a much simpler EE (electrochemical – electrochemical) sequence, overall involving 2 electrons. Nevertheless, even after such simplification, the resulting EECE (electrochemical – electrochemical – chemical – electrochemical) pathway still yields very complex mathematical equations.

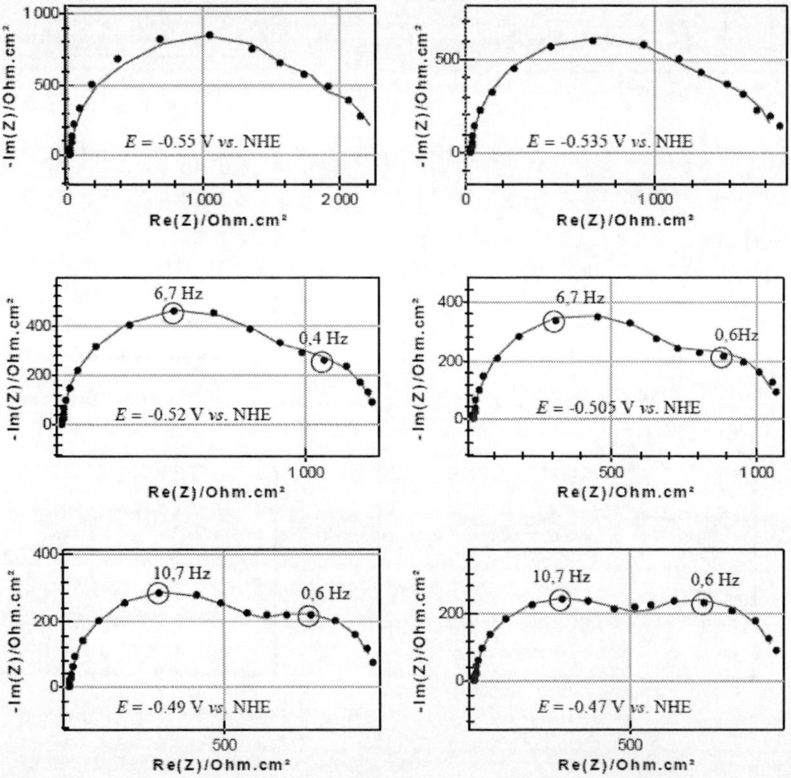

Figure 5. Continued

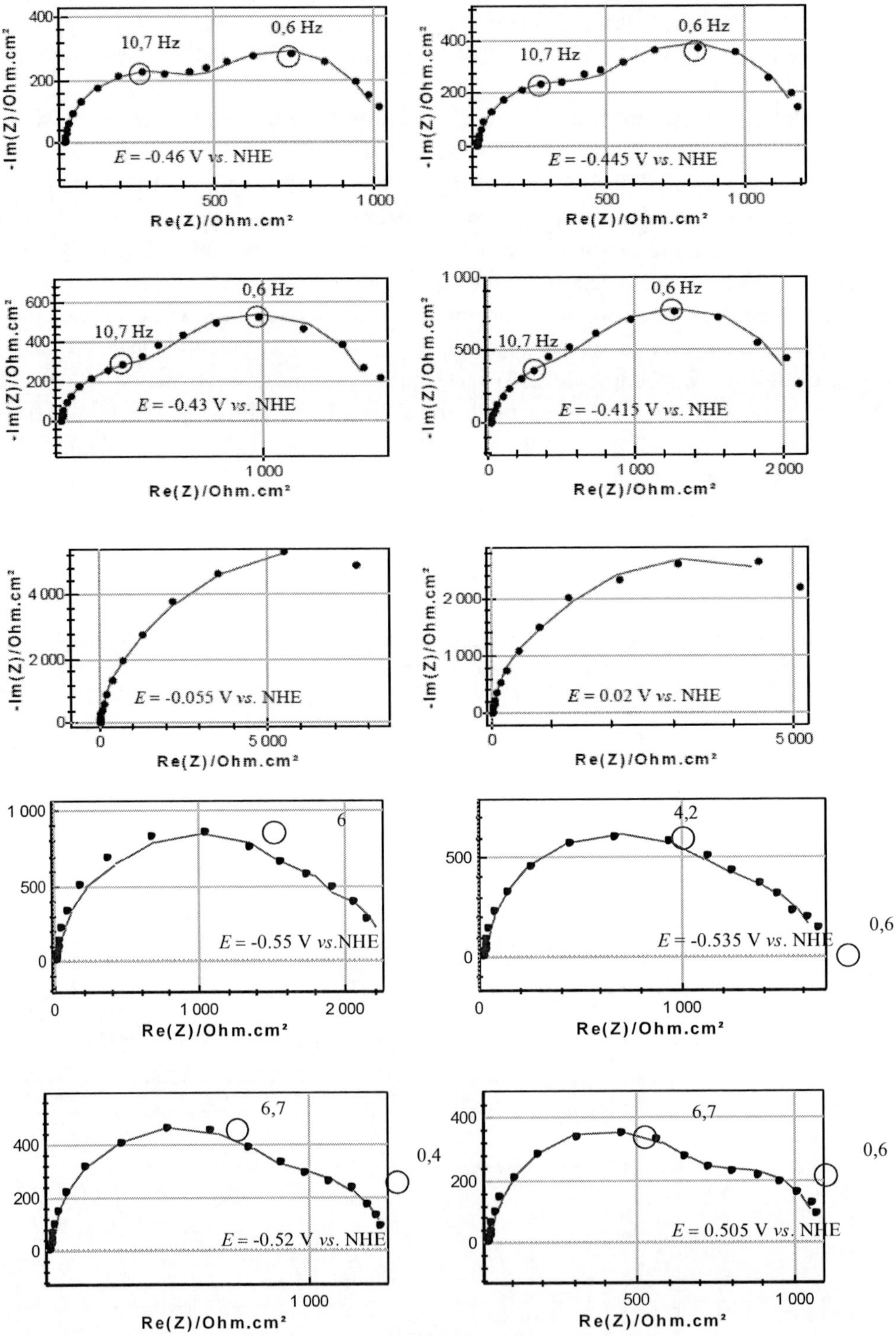

Figure 5. Continued

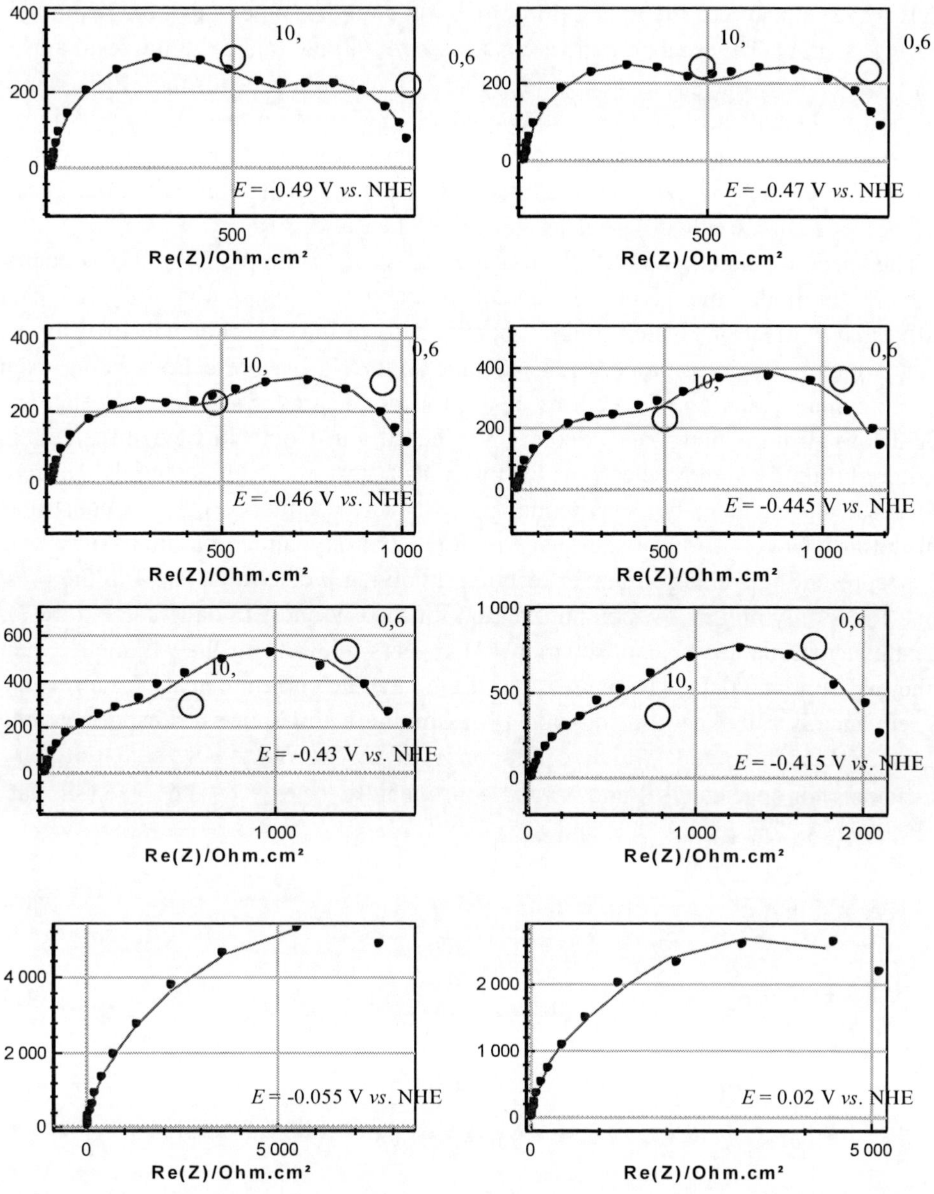

Figure 6 Nyquist representation of selected EIS spectra obtained in SPEIS experiments at representative electrode potentials during BOR measurements on a gold RDE (Ω = 400 rpm) in 0.01 M NaOH + 5 10^{-5} M $NaBH_4$ solution; dots: experimental trace; line: model (based on a redox E, using the Z-fit function in EC-Lab®); only the graphs 1-10, 34 and 39 are displayed among the 50 acquired from $-0.55 < E < 0.1$ V *vs.* NHE. The omitted Nyquist graphs only bare a single non-semi circular capacitive loop, like those for graphs 34 and 39.

The second one, based on slow-scan CV and EIS in hydrodynamic (quasi-steady-state) conditions, consists of an E-step involving 8 electrons (E, at low potentials) in competition with an electrochemical adsorption reaction (EAR, relative to the adsorption of OH^- species, which is the first step of the gold oxide formation at higher potentials), the result being noted: E-EAR [28, 121]. This second pathway is truly over-simple (although the results of Figure 5

Figure 6 show that the E step is nearly sufficient to account for the SPEIS data), because it cannot mathematically describe the two waves observed on the CV. In addition, while the EAR step is necessary to account for gold deactivation monitored at high potentials (see figure 2 above 0.4 V *vs.* NHE in reference [28]), it is not mandatory when one focuses on the region of bare (non oxidized) gold, as considered here.

The shape of the CV and SPEIS traces suggest that a simple EE redox sequence may be sufficient to model the reaction. Lets postulate in first approximation that each E step involves *ca.* 4 electron, which roughly corresponds to the experimental diffusion-convection plateaus monitored in references [28, 48] and Figure 7, *ca.* 8 electrons being exchanged at high potential versus *ca.* 4 electrons at the top of the first wave [4]. It also agrees with the conclusions from Cheng and Scott [49]: based on Koutecki-Levich calculations, they monitored roughly 4 and 8 electrons for the BOR at low and high potentials respectively. The theoretical current density *vs.* potential plot for a EE sequence may be constituted of two well-defined waves in quasi-stationary regime (hydrodynamic conditions) [122, 123], each wave representing a charge-transfer reaction involving n electrons (n = 4 in the present case). If one focus only on the low potential region for the CV and EIS data (see Figure 7), a simple EE pathway (Equation 12 and Equation 13) appears relevant for the parametric identification of the experimental data. The expression of the faradaic current density for a reversible redox EE sequence is calculated in Equation 14, assuming a single mass-transport parameter m for all the species involved, which corresponds to: $D_O = D_R = D_X = D$ (similar diffusion coefficients for species O, R and X) and consequently to: $\delta_O = \delta_R = \delta_X = \delta$ (similar diffusion layer thickness for species O, R and X).

$$\mathrm{R} \leftrightarrow \mathrm{X} + n_1 \mathrm{e}^- \qquad \text{Equation 12}$$

$$\mathrm{X} \leftrightarrow \mathrm{O} + n_2 \mathrm{e}^- \qquad \text{Equation 13}$$

$$i_f = \frac{FmR^* K_{o1}\left((m + K_{o2} + K_{r2})\, n_1 + K_{o2} n_2\right)}{mK_{o2} + (m + K_{r1}) + (m + K_{r2}) + K_{o1}\,(m + K_{o2} + K_{r2})} \qquad \text{Equation 14}$$

K_{r1}, K_{o1}, K_{r1}, K_{o1} and m being given in Equation 15, Equation 16, Equation 17, Equation 18 and Equation 19, respectively:

$$K_{r1} = k_1^\circ \exp\left(-\alpha_{r1}\, n_1 f\left(E - E_1^\circ\right)\right) \qquad \text{Equation 15}$$

$$K_{o1} = k_1^\circ \exp\left(\alpha_{o1}\, n_1 f\left(E - E_1^\circ\right)\right) \qquad \text{Equation 16}$$

4 The OLEMS results of section 5.1 point towards less than 8 electrons exchanged at maximum for BOR on gold in this potential region, but one can neglect in first approximation the heterogeneous hydrolysis of BH_4^- species, because of the low borohydride concentration used here.

$$K_{r2} = k_2^\circ \exp\left(-\alpha_{r2}\, n_2\, f\left(E - E_2^\circ\right)\right)$$ Equation 17

$$K_{o2} = k_2^\circ \exp\left(\alpha_{o2}\, n_2\, f\left(E - E_2^\circ\right)\right)$$ Equation 18

$$m = D/\delta,\ \delta = 1,\ 611\, D^{1/3}\, \nu^{1/6}\, \Omega^{-1/2},\ m = 0,\ 620\, D^{2/3}\, \nu^{-1/6}\, \Omega^{1/2}$$ Equation 19

with: $f = F/(RT)$, $\alpha_{o1} + \alpha_{r1} = 1$, $\alpha_{o2} + \alpha_{r2} = 1$

It is far beyond the scope of the present chapter to go through the complete description of the calculations regarding the EIS modeling. The reader is directed towards a forthcoming paper dedicated to the study of the BOR on gold by EIS for more details.

The FTIR experiments presented in section 0 5.2 brought physical evidence regarding the nature of the BOR intermediates. From these experimental physical findings, or postulated intermediates (from purely electrochemical data [46, 57, 63, 80, 93]), or even compounds isolated from *ab initio* calculations (either regarding the oxidation of borohydride [95] or dimethylamine borane [64, 65], a compound yielding BH_3OH^- in alkaline solution), Equation 12 and Equation 13 could translate into Equation 20 and Equation 21 respectively, each of these steps involving 4 electrons and 4 hydroxide anions [5]:

$BH_4^- + 4\ OH^- \rightarrow BH_2(OH)_2^- + 2\ H_2O + 4\ e^-$ Equation 20

$BH_2(OH)_2^- + 4\ OH^- \rightarrow B(OH)_4^- + 2\ H_2O + 4\ e^-$ Equation 21

These steps are very close to that proposed from the FTIR/electrochemistry coupling (Equation 20 can be considered nearly as the sum of Equation 8, (C) Equation 9 and (E) Equation 10, while Equation 21 and (E) Equation 11 can be can be compared, assuming the species can either be hydrated or not). The parametric identification of the present experimental data using the Mathematica software [124], reveals that the modeled plots are superposed to the experimental traces (Figure 7). The parameter values calculated in the modeling are summarized in Table 4.

It is wise to note that the borohydride diffusivity values (*D*, Table 4) identified from the experimental CV plot assuming 8 electrons are exchanged overall in the reaction ($n = n_1 + n_2 = 8$) are non-negligibly inferior (by a factor *ca.* 2) to those recently measured in reference [48] (*ca.* 3.1 10^{-5} $cm^2\ s^{-1}$ in these conditions). Most of such difference can be attributed to a partial "obstruction" of the gold electrode surface by (i) the presence of hydrogen bubbles *in situ* formed during the BOR (see the OLEMS results of section 5.10) and (ii) boron oxides, which are prone to precipitate upon formation, as detected from *in situ* FTIR coupled to electrochemistry (see [94] and section 5.20). The fact that (iii) a pathway involving overall 8 electrons per BH_4^- species was chosen although *ca.* 7 electrons are actually detected (see section 0) also accounts for this bias.

---

[5] In the present section, the reactions are postulated to occur in liquid phase, although some calculated ([95]G. Rostamikia, M. J. Janik, J. Electrochem. Soc., 156 (2009) B86.[95]) and experimental (see section 5.2 and [94]B. Molina Concha, M. Chatenet, C. Coutanceau, F. Hahn, Electrochem. Commun., 11 (2009) 223. [94]) evidences point towards a reaction in adsorbed phase. This probably shows that the reaction(s) involving such adsorbed intermediates are not rate-determining.

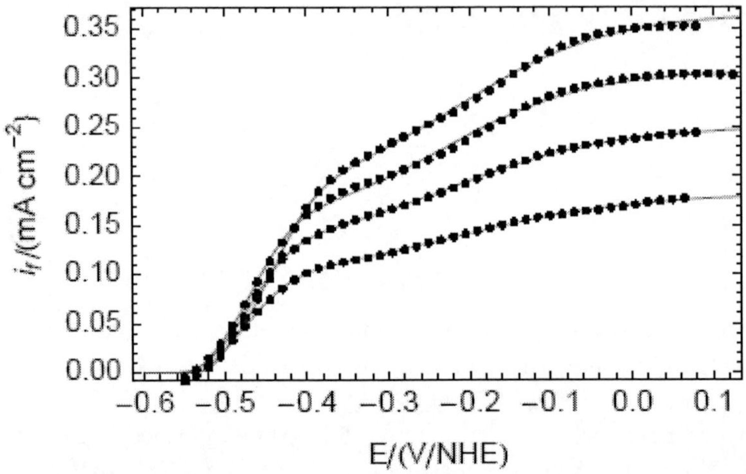

Figure 7. Parametric identification of the SPEIS measurements for the BOR on gold RDE in 0.01 M NaOH + 5 10^{-5} M $NaBH_4$ aqueous solution, dots: experimental data; line: plot resulting from the parametric identification using a reversible redox EE.

Table 4. Parametric identification of the SPEIS measurements for the BOR on gold in 0.01 M NaOH + 5 10^{-5} M $NaBH_4$ aqueous solution (reversible redox EE pathway; $n_1 = n_2 = 4$).

Ω / rpm	$k_1°$ / cm^2 s^{-1}	α_{o1}	$E_1°$ / V *vs.* NHE	$k_2°$ / cm^2 s^{-1}	α_{o2}	$E_1°$ / V *vs.* NHE	m / cm s^{-1}	D / 10^{-5} cm^2 s^{-1}
400	0.0014	0.23	-0.506	0.00040	0.054	-0.515	0.0047	1.24
900	0.0010	0.25	-0.517	0.00035	0.062	-0.520	0.0064	1.08
1600	0.0013	0.24	-0.514	0.00027	0.074	-0.516	0.0080	0.96
2500	0.0016	0.22	-0.599	0.00037	0.070	-0.505	0.0095	0.89

In addition, the standard kinetic constants and symmetry factor extracted from the calculation are abnormally low, which suggest that each of the E step in this redox EE sequence should be divided into additional sub E steps. This enables one to postulate that the BOR on gold can consist of a redox EEEE pathway. In that case, each E step would involve 2 electrons, as would be the case of a sequential electrochemical de-hydridation (electrooxidation) of BH_4^- involving 2 electrons and 2 OH^- anions per each H^- species oxidized. The interested reader is referred to the equations (19) + (20) (respectively electrochemical step + chemical step) and (21) + (22) proposed by Morris et al. for the BOR on Pt [23]. In consequence, the overall pathway could as well correspond to the sequence of Equation 22, Equation 23, Equation 24 and Equation 25. The present experimental findings suggest that the two couples of Equation 22 + Equation 23 and Equation 24 + Equation 25 proceed like the two four electron reaction of Equation 20 and (in other words, the present data neither exhibit a clear separation between Equation 22 and Equation 23 nor between Equation 24 and Equation 25).

$BH_4^- + 2\ OH^- \rightarrow BH_3OH^- + H_2O + 2\ e^-$ Equation 22

$BH_3OH^- + 2\ OH^- \rightarrow BH_2(OH)_2^- + H_2O + 2\ e^-$ Equation 23

$BH_2(OH)_2^- + 2\ OH^- \rightarrow BH(OH)_3^- + H_2O + 2\ e^-$ Equation 24

$BH(OH)_3^- + 2\ OH^- \rightarrow B(OH)_4^- + H_2O + 2\ e$ Equation 25

Some of the reaction intermediates mentioned in Equation 22, Equation 23, Equation 24 and Equation 25 have been detected by FTIR (e.g., BH_3OH^-, BH_2OH, etc.) and are therefore relevant from a physical point of view. In addition, the sequence of Equation 23, Equation 24 and Equation 25 resembles that given by Homma et al. from *ab initio* calculations relative to the oxidation of dimethylamine borane (DMAB), a complex known to spontaneously yield BH_3OH^- (*i.e.* the product of Equation 22) in alkaline solutions [64].

The theoretical equations of the faradaic impedance (Z_f) can be calculated for the redox EE, using the Taylor series of the expression of the faradaic current density [125]. Assuming Butler-Volmer kinetics on a RDE (diffusion-convection) without any coupled chemical reaction and the bulk concentrations for species X and O is zero ($X^* = O^* = 0$) the faradaic impedance actually consists of the sum of four terms [120] Equation 26):

$$Z_f(s) = R_{ct} + Z_O(s) + Z_R(s) + Z_X(s) \qquad \text{Equation 26}$$

with R_{ct} the charge transfer resistance and Z_O, Z_R and Z_X the concentration impedances for the species O, R and X, expressed in the frame of the Nernst approximation (linear concentration profiles between the electrode | electrolyte interface and the outbound of the diffusion layer, assuming a similar diffusion layer thickness δ and diffusion coefficient D for the species O, R and X [6]). For brevity, these expressions will not be detailed in this chapter.

The electrode impedance (Z) can further be obtained from the faradaic impedance, using the double layer capacitance C_{dl} of the gold electrode (Equation 27):

$$Z(s) = \frac{Z_f(s)}{1 + sC_{dl}Z_f(s)} \qquad \text{Equation 27}$$

Using the parameters obtained from the identification of the redox EE mechanism (see Table 4), one can calculate the theoretical Nyquist plots of Figure 9 and further compare them with the experimental graphs of the electrode impedance measured in the same electrolyte solution and at the same electrode potentials.

Comparing

Figure 6 with Figure 9 reveals the near-perfect agreement between the modeled and experimental traces of the electrode impedance (Z). However, while all the Nyquist spectra of

---

6 In the case of the BOR, this hypothesis may not be verified because firstly, one ignores the nature of the reaction intermediates yet and secondly, the diffusion coefficient of most of the possible boron compounds intermediates remain unknown.

Figure 6 (corresponding to the electrode impedance) are composed of two capacitive loops (at maximum), easily fitted with the expression of a redox E (for which only one capacitive loop in faradaic impedance and one wave in stationary hydrodynamic CV will be also monitored), several Nyquist graphs of the faradaic impedance of Figure 9 are composed of two capacitive loops. This latter finding agrees both with the two E steps considered for the modeling and the two waves experimentally monitored in quasi-stationary hydrodynamic CV (see Figure 7).

These results enable to conclude that the EE sequence proposed here is the simplest pathway that perfectly accounts with the experimental electrochemical BOR data. However, it clearly constitutes a simplification of the physical phenomena occurring during the BOR. Firstly, the BOR on gold probably occurs in adsorbed state, as postulated in reference [39], experimentally measured by FTIR spectroscopy [94] (see section 5.20) and lately calculated from first principle theory [95]. Secondly, the two EE steps proposed here (Equation 20 and Equation 21) are far too complex to correspond to elementary steps. They actually should be separated into simpler steps involving mono-electron-transfer and chemical steps (e.g. among those proposed in references [46, 64, 65, 95]). Nevertheless, the fact the experimental behavior of the BOR (in quasi-steady-state conditions in dilute electrolytes) can be modeled using a simple EE pathway probably means that the chemical (C) steps involved in the reaction are fast and thus not rate-limiting, as also supposed by Mirkin et al. [46].

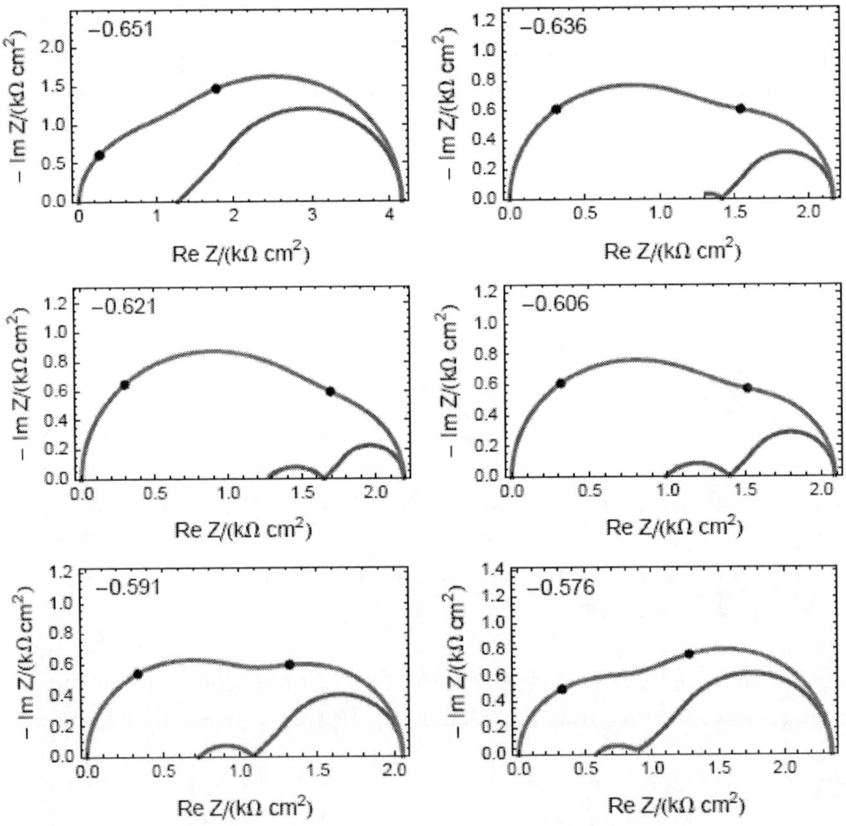

Figure 8. Continued

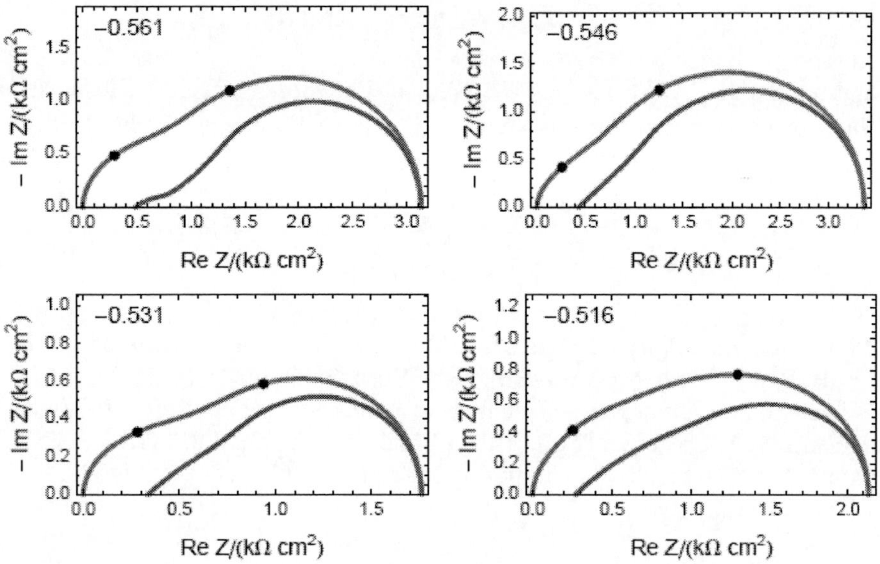

Figure 9. Theoretical faradaic Z_f (—) and electrode Z (—) impedance for a redox EE, as plotted in the Nyquist plane using the parameters determined from the identification of the experimental data. The dots on the electrode impedance (Z) trace represent the "characteristic" frequencies circled in blue on Figure 4.

CONCLUSION

In its first sections, this chapter gives an overview of the potential and limitation of the DBFC. From a selected literature review, it details how (i) the composition of the borohydride fuel, the (ii) membrane and (iii) electrode materials, as well as the cell (iv) geometry and (v) operating conditions drive the DBFC performances. It also emphasizes the marked differences between some basic properties (e.g. faradaic efficiency of the BOR) found in the literature for similar electrode materials. This can be understood in the frame of the real complexity of the borohydride oxidation reaction, but also on ill-adapted operating conditions to fundamentally survey the BOR kinetics and mechanism. This further puts into evidence how mandatory are fundamental studies of the BOR on simple electrode surfaces in simple operating conditions (*i.e.* in conditions opposite to those relevant for power generation in DBFC). This is true from the kinetics and mechanistic point of views and requires both (i) coupled physical techniques to get more insight into the nature of the reaction intermediates and (ii) reaction modeling to propose sufficiently simple reaction pathways that agree with the experimental findings. Such strategy is detailed in the last section on the present chapter, providing insight into the BOR mechanism on gold polycrystalline electrodes. The reaction was particularly studied by on-line electrochemical mass spectrometry (OLEMS) and *in situ* infrared spectroscopy (FTIR), techniques which both enable to detect the nature of some reaction intermediates. From these findings and reaction modeling based on electrochemical impedance spectroscopy and cyclic voltammetry data, a relevant BOR mechanism is finally proposed and discussed.

ACKNOWLEDGMENTS

The authors thank CAPES/COFECUB (Project Ph598/08) and the "Cluster Energie" of Région Rhône-Alpes for funding this study. MC thanks Emmanuel Billy (LEPMI) for the schematic representation of the DBFC.

REFERENCES

[1] G. M. Ehrlich, in *Lithium-ion batteries, Handbook of batteries,* Third ed., D. B. Linden and T. B. Reddy (Ed.), McGraw-Hill, New-York, 2002, pp 35,1 - 35,94.
[2] W. Vielstich, A. Lamm, H. A. Gasteiger, in *Handbook of Fuel Cells,* W. Vielstich, A. Lamm and H. A. Gasteiger (Ed.), Wiley, Chichester, 2003, Vol. 1-4.
[3] E. Guilminot, A. Corcella, F. Charlot, F. Maillard, M. Chatenet, *J. Electrochem. Soc.*, 154 (2007) B96.
[4] M. Chatenet, E. Guilminot, C. Iojoiu, J.-Y. Sanchez, E. Rossinot, F. Maillard, *ECS Trans.,* 11 (2007) 1203.
[5] E. Guilminot, A. Corcella, M. Chatenet, F. Maillard, F. Charlot, G. Berthome, C. Iojoiu, J. Y. Sanchez, E. Rossinot, E. Claude, *J. Electrochem. Soc.*, 154 (2007) B1106.
[6] C. Iojoiu, E. Guilminot, F. Maillard, M. Chatenet, J.-Y. Sanchez, E. Claude, E. Rossinot, *J. Electrochem. Soc.,* 154 (2007) B1115.
[7] J. Xie, D. L. Wood, K. L. More, P. Atanassov, R. L. Borup, *J. Electrochem. Soc.,* 152 (2005) A1011.
[8] J. Xie, D. L. Wood, D. M. Wayne, T. A. Zawodzinski, P. Atanassov, R. L. Borup, *J. Electrochem. Soc.,* 152 (2005) A104.
[9] P. J. Ferreira, G. J. la O', Y. Shao-Horn, D. Morgan, R. Makharia, S. Kocha, H. A. Gasteiger, *J. Electrochem. Soc.,* 152 (2005) A2256.
[10] H. Cheng, K. Scott, K. V. Lovell, J. A. Horsfall, S. C. Waring, *J. Membr. Sci.,* 288 (2007) 168.
[11] H. Cheng, K. Scott, *J. Appl. Electrochem.*, 36 (2006) 1361.
[12] H. Cheng, K. Scott, *J. Power Sources,* 160 (2006) 407.
[13] R. X. Feng, H. Dong, Y. D. Wang, X. P. Ai, Y. L. Cao, H. X. Yang, *Electrochem. Commun.,* 7 (2005) 449.
[14] N. A. Choudhury, R. K. Raman, S. Sampath, A. K. Shukla, *J. Power Sources*, 143 (2005) 1.
[15] R. K. Raman, N. A. Choudhury, A. K. Shukla, *Electrochem. Solid State Lett.*, 7 (2004) A488.
[16] B. H. Liu, Z. P. Li, S. Suda, *J. Power Sources,* 175 (2008) 226.
[17] B. H. Liu, Z. P. Li, *J. Power Sources,* 187 (2009) 291.
[18] R. K. Raman, S. K. Prashant, A. K. Shukla, *J. Power Sources,* 162 (2006) 1073.
[19] F. A. Coowar, G. Vitins, G. O. Mepsted, S. C. Waring, J. A. Horsfall, *J. Power Sources*, 175 (2008) 317.
[20] D. Miliauskas, R. Tarozaite, L. T. Tamasiunaite, *Mater. Sci.-Medzg.*, 14 (2008) 20.
[21] J. A. Gardiner, J. W. Collat, *J. Am. Chem. Soc.*, 87 (1965) 1692.
[22] J. P. Elder, A. H. Hickling, *Trans. Farad. Soc.*, 58 (1962) 1852.
[23] J. H. Morris, H. J. Gysling, D. Reed, *Chem. Rev.*, 85 (1985) 51.
[24] U. B. Demirci, *J. Power Sources, 1*69 (2007) 239.
[25] Z. P. Li, B. H. Liu, K. Arai, S. Suda, *J. Alloys Compd.,* 404-406 (2005) 648.
[26] Z. P. Li, B. H. Liu, J. K. Zhu, S. Suda, *J. Power Sources*, 163 (2006) 555.

[27] B. H. Liu, S. Suda, *J. Power Sources,* 164 (2007) 100.
[28] M. Chatenet, B. Molina-Concha, J.-P. Diard, *Electrochim. Acta,* 54 (2009) 1687.
[29] J. P. I. de Souza, S. L. Queiroz, F. C. Nart, *Quim. Nova,* 23 (2000) 384.
[30] J. P. I. de Souza, S. L. Queiroz, K. Bergamaski, E. R. Gonzalez, F. C. Nart, *J. Phys. Chem. B*, 106 (2002) 9825.
[31] K. Bergamaski, A. L. N. Pinheiro, E. Teixeira-Neto, F. C. Nart, *J. Phys. Chem. B,* 110 (2006) 19271.
[32] T. Iwasita, F. C. Nart, in *Advances in Electrochemical Science and Engineering,* 4th ed., H. Gerischer and C. W. Tobias (Ed.), VCH, Weinheim, 1995.
[33] S. Amendola, P. Onnerud, M. T. Kelly, M. Binder, *Talanta,* 49 (1999) 267.
[34] S. C. Amendola, P. Onnerud, M. T. Kelly, P. J. Petillo, S. L. Sharp-Goldman, M. Binder, *J. Power Sources*, 84 (1999) 130.
[35] S. C. Amendola, S. L. Sharp-Goldman, M. S. Janjua, M. T. Kelly, P. J. Petillo, M. Binder, *J. Power Sources,* 85 (2000) 186.
[36] S. C. Amendola, S. L. Sharp-Goldman, M. S. Janjua, N. C. Spencer, M. T. Kelly, P. J. Petillo, M. Binder, *Int. J. Hydrogen Energy*, 25 (2000) 969.
[37] C. Kim, K.-J. Kim, M. Y. Ha, *J. Power Sources,* 180 (2008) 154.
[38] R. Jamard, A. Latour, J. Salomon, P. Capron, A. Martinent-Beaumont, *J. Power Sources,* 176 (2008) 287.
[39] M. Chatenet, F. Micoud, I. Roche, E. Chainet, *Electrochim. Acta,* 51 (2006) 5459.
[40] E. Gyenge, *Electrochim. Acta,* 49 (2004) 965.
[41] Y. Okinaka, *J. Electrochem. Soc.,* 120 (1973) 739.
[42] S. Suda, in *Aqueous borohydride solutions, Handbook of Fuel Cells Fundamentals, Technology and Applications,* W. Vielstich, H. A. Gasteiger and A. Lamm (Ed.), Wiley, Chichester, 2003, Vol. 3, pp 115-120.
[43] M. M. Kreevoy, R. W. Jacobson, *Ventron Alembic,* 15 (1979) 2.
[44] Z. P. Li, B. H. Liu, K. Arai, K. Asaba, S. Suda, *J. Power Sources,* 126 (2004) 28.
[45] G. Denuault, M. V. Mirkin, A. J. Bard, *J. Electroanal. Chem.,* 308 (1991) 27.
[46] M. V. Mirkin, H. Yang, A. J. Bard, *J. Electrochem. Soc.*, 139 (1992) 2212.
[47] K. Wang, J. Lu, L. Zhuang, *J. Electroanal. Chem.,* 585 (2005) 191.
[48] M. Chatenet, M. B. Molina-Concha, N. El-Kissi, G. Parrour, J.-P. Diard, *Electrochim. Acta,* 54 (2009) 4426.
[49] H. Cheng, K. Scott, *Electrochim. Acta,* 51 (2006) 3429.
[50] G. H. Miley, N. Luo, J. Mather, R. Burton, G. Hawkins, L. Gu, E. Byrd, R. Gimlin, P. J. Shrestha, G. Benavides, J. Laystrom, D. Carroll, *J. Power Sources,* 165 (2007) 509.
[51] B. H. Liu, S. Suda, *J. Alloys Compd.*, 454 (2008) 280.
[52] D. Hua, Y. Hanxi, A. Xinping, C. Chuansin, *Int. J. Hydrogen Energy*, 28 (2003) 1095.
[53] B. H. Liu, Z. P. Li, S. Suda, *J. Alloys Compd.*, 415 (2006) 288.
[54] A. Garron, D. Swierczynski, S. Bennici, A. Auroux, *Int. J. Hydrogen Energy,* 34 (2009) 1185.
[55] B. H. Liu, Z. P. Li, L. L. Chen, *J. Power Sources,* 180 (2008) 530.
[56] V. Simagina, P. Storozhenko, O. Netskina, O. Komova, G. Odegova, T. Samoilenko, A. Gentsler, *Kinetics and Catalysis,* 48 (2007) 168.
[57] L. C. Nagle, J. F. Rohan, *J. Electrochem. Soc.*, 153 (2006) C773.
[58] B. Molina Concha, M. Chatenet, *Electrochim. Acta,* in press 54 (2009) 6119doi:10.1016/j.electacta.2009.05.027.
[59] X. B. Zhang, S. Han, J. M. Yan, H. Shioyama, N. Kuriyama, T. Kobayashi, Q. Xu, *Int. J. Hydrogen Energy,* 34 (2009) 174.
[60] X. B. Zhang, S. Han, J. M. Yan, M. Chandra, H. Shioyarna, K. Yasuda, N. Kuriyama, T. Kobayashi, Q. Xu, *J. Power Sources,* 168 (2007) 167.

[61] A. Sargent, O. A. Sadik, L. J. Matienzo, *J. Electrochem. Soc.,* 148 (2001) C257.
[62] J. I. Martins, M. C. Nunes, *J. Power Sources,* 175 (2008) 244.
[63] L. C. Nagle, J. F. Rohan, *Electrochem. Solid State Lett.*, 8 (2005) C77.
[64] T. Homma, A. Tamaki, H. Nakai, T. Osaka, *J. Electroanal. Chem.*, 559 (2003) 131.
[65] T. Homma, H. Nakai, M. Onishi, T. Osaka, *J. Phys. Chem. B,* 103 (1999) 1774.
[66] O. A. Sadik, H. W. Xu, A. Sargent, *J. Electroanal. Chem.,* 583 (2005) 167.
[67] D. A. Lyttle, E. H. Jensen, W. A. Struck, *Anal. Chem.,* 24 (1952) 1843.
[68] D. M. F. Santos, C. A. C. Sequeira, *J. Electrochem. Soc.*, 156 (2009) F67.
[69] M. V. Mirkin, A. *J. Bard, Anal. Chem.*, 63 (1991) 532.
[70] S. Colominas, J. McLafferty, D. D. Macdonald, *Electrochim. Acta,* 54 (2009) 3575.
[71] D. A. Finkelstein, N. Da Mota, J. L. Cohen, H. D. Abruña, *J. Phys. Chem. C,* accepted 113 (2009) 19700.
[72] M. Chatenet, L. Génies-Bultel, M. Aurousseau, R. Durand, F. Andolfatto, *J. Appl. Electrochem.*, 32 (2002) 1131.
[73] M. Chatenet, M. Aurousseau, R. Durand, F. Andolfatto, J. *Electrochem. Soc.,* 150 (2003) D47.
[74] J. Vondrak, B. Klapste, J. Velicka, M. Sedlarikova, J. Reiter, I. Roche, E. Chainet, J. F. Fauvarque, M. Chatenet, *J. New Mat. Electrochem. Sys.,* 8 (2005) 209.
[75] M. Chatenet, F. Micoud, I. Roche, E. Chainet, *J. Vondrak, Electrochim. Acta*, 51 (2006) 5452.
[76] I. Roche, E. Chainet, M. Chatenet, J. Vondrak, *J. Phys. Chem.* C, 111 (2007) 1434.
[77] M. H. Atwan, C. L. B. Macdonald, D. O. Northwood, E. L. Gyenge, *J. Power Sources,* 158 (2006) 36.
[78] U. B. Demirci, *J. Power Sources*, 172 (2007) 676.
[79] C.-C. Yang, Y. J. Li, S.-J. Chiu, K.-T. Lee, W.-C. Chien, C.-A. Huang, *J. Power Sources,* In Press, Accepted Manuscript.
[80] H. Dong, R. X. Feng, X. P. Ai, Y. L. Cao, H. X. Yang, C. S. Cha, *J. Phys. Chem. B,* 109 (2005) 10896.
[81] E. Gyenge, M. Atwan, D. Northwood, *J. Electrochem. Soc.*, 153 (2006) A150.
[82] B. H. Liu, Z. P. Li, S. Suda, Electrochim. Acta, 49 (2004) 3097.
[83] H. Cheng, K. Scott, J. *Electroanal. Chem.*, 596 (2006) 117.
[84] C. Kim, K.-J. Kim, M. Y. Ha, *J. Power Sources,* 180 (2008) 114.
[85] J. H. Wee, *J. Power Sources,* 161 (2006) 1.
[86] N. Duteanu, G. Vlachogiannopoulos, M. Shivhare, E. Yu, K. Scott, *J. Appl. Electrochem.,* 37 (2007) 1085.
[87] B. H. Liu, Z. P. Li, K. Arai, S. Suda, *Electrochim. Acta,* 50 (2005) 3719.
[88] R. K. Raman, A. K. Shukla, *J. Appl. Electrochem.*, 35 (2005) 1157.
[89] B. Molina Concha, M. *Chatenet, Electrochim. Acta,* in press54 (2009) doi:10.1016/j.electacta.2009.04.074 6130.
[90] A. Schneider, L. Colmenares, Y. E. Seidel, Z. Jusys, B. Wickman, B. Kasemo, R. J. Behm, *Phys. Chem. Chem. Phys.,* 10 (2008) 1931.
[91] F. Helfferich, in *Ion Exchange,* McGraw-Hill, New York, 1962.
[92] M. V. Ivanov, M. V. Tsionskii, *Soviet Electrochemistry,* 25 (1989) 451.
[93] P. Krishnan, T.-H. Yang, S. G. Advani, A. K. Prasad, *J. Power Sources,* 182 (2008) 106.
[94] B. Molina Concha, M. Chatenet, C. Coutanceau, F. Hahn, *Electrochem. Commun.,* 11 (2009) 223.
[95] G. Rostamikia, M. J. Janik, *J. Electrochem. Soc.*, 156 (2009) B86.
[96] R. Tarozaitė, L. Tamašauskaitė Tamašiūnaitė, V. Jasulaitienė, *J. Solid State Electrochem.,* 13 (2009) 721.

[97] M. H. Atwan, D. O. Northwood, E. L. Gyenge, *Int. J. Hydrogen Energy,* 32 (2007) 3116.
[98] E. Sanli, H. Celikkan, B. Z. Uysal, M. L. Aksu, *Int. J. Hydrogen Energy*, 31 (2006) 1920.
[99] E. Sanli, B. Z. Uysal, M. L. Aksu, *Int. J. Hydrogen Energy,* 33 (2008) 2097.
[100] A. Y. Tsivadze, M. R. Tarasevich, V. N. Titova, A. A. Yavich, N. V. Petrova, Dokl. *Phys. Chem.*, 414 (2007) 107.
[101] H. Pan, Y. Chen, C. Wang, X. Wang, C. Chen, Q. Wang, *J. Alloys Compd.,* 293-295 (1999) 680.
[102] J. I. Martins, M. C. Nunes, R. Koch, L. Martins, M. Bazzaoui, *Electrochim. Acta,* 52 (2007) 6443.
[103] B. H. Liu, Z. P. Li, S. Suda, *J. Electrochem. Soc.,* 150 (2003) A398.
[104] E. Gyenge, *Electrochim. Acta,* 49 (2004) 1875.
[105] X.-B. Zhang, S. Han, J.-M. Yan, M. Chandra, H. Shioyama, K. Yasuda, N. Kuriyama, T. Kobayashi, Q. Xu, *J. Power Sources,* 168 (2007) 167.
[106] R. D. Cowan, *J. Chem. Phys.,* 18 (1950) 1101.
[107] J. B. Robb, J. Xiao, J. R. Durig, *J. Mol. Struct.*, 449 (1998) 255.
[108] T. J. Tague, L. Andrews, *J. Am. Chem. Soc.,* 116 (1994) 4970.
[109] T. Shimanouchi, J. Phys. *Chem. Ref. Data,* 6 (1977) 993.
[110] G. W. Bethke, M. K. Wilson, *J. Chem. Phys., 2*6 (1957) 1118.
[111] J. Smith, K. S. Seshadri, D. White, *J. Mol. Spectrosc.*, 45 (1973) 327.
[112] N. Galland, Y. Hannachi, D. V. Lanzisera, L. Andrews, *Chem. Phys.*, 255 (2000) 205.
[113] L. Andrews, X. F. Wang, *J. Am. Chem. Soc.*, 124 (2002) 7280.
[114] L. Andrews, T. R. Burkholder, *J. Phys. Chem.*, 95 (1991) 8554.
[115] T. R. Burkholder, L. Andrews, *J. Chem. Phys.,* 95 (1991) 8697.
[116] L. Andrews, T. R. Burkholder, *J. Chem. Phys*., 97 (1992) 7203.
[117] A. G. Maki, J. B. Burkholder, A. Sinha, C. J. Howard, *J. Mol. Spectrosc.,* 130 (1988) 238.
[118] D. K. Russell, M. Kroll, R. A. Beaudet, *J. Chem. Phys.,* 66 (1977) 1999.
[119] A. Sommer, M. J. Linevsky, D. E. Mann, D. White, J. *Chem. Phys.,* 38 (1963) 87.
[120] J.-P. Diard, C. Montella, B. *Le Gorrec, in Cinétique Electrochimique*, Hermann, Paris, 1996.
[121] M. B. Molina Concha, M. Chatenet, J.-P. D*iard, in Etude du mécanisme d'électrooxydation de borohydrure sur électrode tournante d'or, Proceedings of the XXème forum des impédances électrochimiques*, Paris, 2007, communication matinale #8.
[122] E. Vieil, in *Etude théorique de mécanismes électrochimiques en méthode stationnaire,* Thèse de Doctorat, Université Joseph Fourier, 1979.
[123] A. J. Bard, L. R. Faulkner, in E*lectrochemical Methods*, John Wiley & Sons, Inc., Weinheim, 2001.
[124] S. Wolfram, in T*he Mathematica Book*, 6th ed., Wolfram Media & Cambridge University Press, Champaign-USA, Cambridge-UK, 2007.
[125] G. Parrour, in *Electrooxydation des borohydrures en milieu alcalin sur électrode d'*or, Master thesis, Grenoble-INP, 2008.

In: Boron Hydrides High Potential Hydrogen Storage Material ISBN:978-1-61668-361-0
Editors: Umit B. Demirci and Philippe Miele

Chapter 6

CRYSTAL CHEMISTRY OF LIGHT METAL BOROHYDRIDES

Yaroslav Filinchuk*[1]*, Dmitry Chernyshov and Vladimir Dmitriev
Swiss-Norwegian Beam Lines (SNBL) at the European Synchrotron Radiation Facility (ESRF), BP-220, 38043 Grenoble, France

ABSTRACT

Crystal chemistry of $M(BH_4)_n$, where M is a 2nd-4th period element, is reviewed here. It is shown that except certain cases, the BH_4 group has a nearly ideal tetrahedral geometry. Corrections of the experimentally- determined H-positions, accounting for the displacement of the electron cloud relative to an average nuclear position and for a libration of the BH_4 group, are considered. Recent studies of structural evolution with temperature and pressure are reviewed. Some borohydrides involving less electropositive metals (e.g. Mg and Zn) reveal porous structures and dense interpenetrated frameworks, thus resembling metal-organic frameworks (MOFs). Analysis of phase transitions, and the related changes of the coordination geometries for M atoms and BH_4 groups, suggests that the directional BH_4...M interaction is at the origin of the structural complexity of borohydrides. The ways to influence their stability by chemical modification are discussed.

INTRODUCTION

Borohydrides, also called tetrahydroborates, are largely ionic compounds with a general formula $M(BH_4)_n$, consisting of metal cations M^{n+} and borohydride anions BH_4^-. Due to a high weight percent of hydrogen, they are considered as prospective hydrogen storage materials. Indeed, some borohydrides desorb a large quantity of hydrogen (up to 20.8 wt %), although the decompositon temperatures are usually high. The search for better hydrogen storage materials, with denser structures and lower binding energies, has been hampered by a

[1]Corresponding author (e-mail: Yaroslav.Filinchuk@esrf.fr)

lack of basic knowledge about their structural properties. However, the last few years have seen a remarkably rapid evolution of this field, and a large number of experimental and theoretical studies have recently been carried out, as summarized in several very recent reviews on crystal structures (Filinchuk *et al.*, 2008b), physical properties (Sundqvist and Andersson, 2009) and phase relations (Sundqvist, 2009) of borohydrides at different P-T conditions. Different fundamental aspects of the use of light metal hydrides, including borohydrides, for the storage and production of hydrogen have been comprehensively reviewed by Grochala and Edwards, 2004 and more recently by Orimo *et al.*, 2007 and Züttel *et al.*, 2007. Thermodymanic properties of metal borohydrides with respect to their hydrogen storage applications were briefly reviewed by Soloveichik, 2007. This chapter, which is an update of the recent review by Filinchuk *et al.*, 2008b, has a special focus on crystal structures and crystal chemistry of borohydrides.

Almost all borohydrides are crystalline solids already at room temperature. This gives an advantage of using diffraction methods to study their structure. Diffraction provides an immense ammount of information not only about the structure, but also is more and more frequently used for screening and characterization of new substances, reaction products and intermediates. A dDetailed diffraction study of a promising material at various temperatures and pressures allows to uncover new polymorphs, follow their structural evolution and therefore get a clue to understand (and maybe even alter) the thermodymanic stability. This review will be mostly focused on crystal structures of known metal borohydride phases, also aiming to give a feeling about how much modern diffraction techniques can go beyond a simple structure characterization.

We will review here the experimentally observed structures of light borohydrides $M(BH_4)_n$, where by "light" we assume M to be an element of the 2nd to the early 4th period. Since the most of the theoretically predicted structures remain hypothetical, this chapter is limited to the experimentally observed ones. The known metal borohydride phases and their crystallographic characteristics are summarized in Table 1. The Table, and this short review do not pretend to be exhaustive, so only one (normally the first) work aiming to determine the crystal structuutre using a given diffraction technique is mentioned here.

Table 1. Known phases of light metal borohydrides and their crystallographic characteristics.

Compound	Space group	*a* (Å)	*b* (Å)	*c* (Å)	β (°)	Method[1]	Conditions[2]	Ref.
		7.17858(4)	4.43686(2)	6.80321(4)		XRPD		Soulié *et al.*, 2002
$LiBH_4$	*Pnma*	7.1213(2)	4.4060(1)	6.6744(2)		NPD	3.5K	Hartman *et al.*, 2007
		7.141(5)	4.431(3)	6.748(4)		SCXRD	225K	Filinchuk *et al.*, 2008a
		4.27631(5)		6.94844(8)		XRPD	408K	Soulié *et al.*, 2002
	$P6_3mc$	4.2667(2)		6.9223(8)		NPD	400K	Hartman *et al.*, 2007
		4.3228(10)		7.037(1)		SCXRD	535K	Filinchuk *et al.*, 2008a
	*Ama*2	6.4494(9)	5.307(1)	5.2919(9)		XRPD	2.4 GPa	Filinchuk *et al.*, 2008c
	Fm-3*m*	5.109(2)				XRPD	18.1 GPa	Filinchuk *et al.*, 2008c
$NaBH_4$	*Fm*-3*m*	6.148(1)				NPD		Fischer, Züttel, 2004
		6.1308(1)				SCXRD	200K	Filinchuk, Hagemann, 2008
	$P\text{-}42_1c$	4.332(1)		5.869(1)		NPD	10K	Fischer, Züttel, 2004
	Pnma	7.297(1)	4.1166(5)	5.5692(7)		XRPD	11.2 GPa	Filinchuk *et al.*, 2007
KBH_4	*Fm*-3*m*	6.728(1)				SCXRD		Luck, Schelter, 1999
		6.7306(1)				NPD		Renaudin *et al.*, 2004
NH_4BH_4	*Fm*-3*m*	6.978				XRPD		Karkamkar *et al.*, 2009
	$P4_2/nmc$	4.7004(2)		6.5979(3)		NPD	1.5K	Renaudin *et al.*, 2004
$Be(BH_4)_2$	$I4_1cd$	13.62(1)		9.10(1)		SCXRD		Marynick, Lipscomb, 1972
$Mg(BH_4)_2$	$P6_1$	10.3182(1)		36.9983(5)		XRPD+NPD		Černý *et al.*, 2007
	$P6_122$	10.354(1)		37.055(4)		SCXRD	100K	Filinchuk *et al.*, 2009c2009b
	Fddd	37.072(1)	18.6476(6)	10.9123(3)		XRPD		Her *et al.*, 2007

Table 1. (Continued).

$Ca(BH_4)_2$	*Fddd*	8.791(1)	13.137(1)	7.500(1)		XRPD		Miwa *et al.*, 2006
	*F*2*dd*	8.7759(3)	13.0234(4)	7.4132(2)		XRPD	91K	Filinchuk *et al.*, 2009a
	$P4_2/m$	6.9468(1)		4.3661(1)		XRPD+NPD	480K	Buchter *et al.*, 2008
	P-4 or $P4_2nm$	6.9189(1)		4.3471(1)		XRPD	305K	Filinchuk *et al.*, 2009a
	I-42*d*	5.8446(3)		13.228(1)		XRPD	495K	Filinchuk *et al.*, 2009a
	Pbca	13.0584(8)	8.3881(4)	7.5107(4)		XRPD+NPD	300K	Buchter *et al.*, 2009
$Mn(BH_4)_2$	$P3_112$	10.435(1)		10.835(2)		XRPD		Černý *et al.*, 2009a
$Al(BH_4)_3$	*C*2/*c*	21.917(4)	5.986(1)	21.787(4)	111.90(3)	SCXRD	150K	Aldridge *et al.*, 1997
	$Pna2_1$	18.021(3)	6.138(2)	6.199(1)		SCXRD	195K	Aldridge *et al.*, 1997
$LiK(BH_4)_2$	*Pnma*	7.91337(5)	4.49067(3)	13.8440(1)		XRPD		Nickels *et al.*, 2008
$LiSc(BH_4)_4$	*P*-42*c*	6.07593(6)		12.0338(1)		XRPD		Hagemann *et al.*, 2008
$NaSc(BH_4)_4$	*Cmcm*	8.170(2)	11.875(3)	9.018(2)		XRPD		Černý *et al.*, 2009b2010
$LiZn_2(BH_4)_5$	*Cmca*	8.6244(3)	17.8970(8)	15.4114(8)		XRPD		Ravnsbæk *et al.*, 2009a
$NaZn_2(BH_4)_5$	$P2_1/c$	9.397(2)	16.635(3)	9.136(2)	112.66(2)	XRPD		Ravnsbæk *et al.*, 2009a
$NaZn(BH_4)_3$	$P2_1/c$	8.2714(16)	4.5240(7)	18.757(3)	101.69(1)	XRPD		Ravnsbæk *et al.*, 2009a

[1] SCXRD – single-crystal X-ray or synchrotron diffraction, XRPD – X-ray or synchrotron powder diffraction, NPD – neutron powder diffraction.

[2] Ambient pressure and temperature is assumed if not otherwise specified.

CRYSTAL STRUCTURES

$LiBH_4$

Four $LiBH_4$ phases are known: two at ambient and two at high pressure. The first structural study of the ambient pressure polymorphs was made using synchrotron powder diffraction by Soulié *et al.*, 2002. The low-temperature structure has *Pnma* symmetry. It transforms into a hexagonal wurtzite-like high-temperature phase at ~380K.

Since the originally proposed $P6_3mc$ symmetry of the high-temperature phase has been questioned by theoreticians, it has been revisited by neutron powder diffraction on the triply isotopically substituted $^7Li^{11}BD_4$ (Hartman *et al.*, 2007) and by synchrotron diffraction on single crystals and powder (Filinchuk, Chernyshov, 2007; Filinchuk *et al.*, 2008a). These recent studies confirm the $P6_3mc$ space group symmetry for the high-temperature phase, and reveal large and anisotropic displacements for hydrogen atoms. The observed atomic displacement ellipsoids (Fig. 1b) may indicate, in addition to the thermal vibrations, an orientational disorder (hindered rotations) of the BH_4 group. Large libration amplitudes in the hexagonal phase agree with an apparent shortening of the B-H bonds to ~1.07 Å. An entropy contribution from the disorder is considered as a factor stabilizing the hexagonal structure (Filinchuk, Chernyshov, 2007; Filinchuk *et al.*, 2008a).

Single-crystal investigation of the *Pnma* phase at 225K unambiguously shows that the BH_4 group has a nearly ideal tetrahedral geometry, in contrast to strongly distorted tetrahedra reported in earlier synchrotron diffraction studies (Soulié *et al.*, 2002; Züttel *et al.*, 2003). The observed H-B-H angles, 108.8(9)-109.9(7)°, are very close to the ideal tetrahedral angle of 109.5°, and the B-H bond lengths show a narrow spread from 1.104(11) to 1.131(15) Å. Neutron diffraction at 3.5K also shows that the BH_4 group is very close to the undistorted tetrahedron: B-H distances are 1.208(3)-1.225(6) Å, H-B-H angles – 107.2(3)-111.7(4)°.

The two ambient pressure phases have similar structures (Figs. 1a and 1b), where Li atoms and BH_4 groups are tetrahedrally coordinated. In the orthorhombic phase at 225K, Li-B distances are 2.37-2.57 Å. In the hexagonal phase at 535K, there are three short B-Li contacts of 2.55 Å in the basal plane and one long of 3.00 Å along the *c* axis.

At room temperature and a pressure of 1.2-10 GPa, $LiBH_4$ forms a new phase with pseudo-tetragonal *Ama*2 structure (Filinchuk *et al.*, 2008c). It can be considered as an orthorhombically distorted antistructure of PtS, where Li atoms are tetrahedrally coordinated by BH_4 groups (Li-B distances of 2.35-2.66 Å) and the BH_4 group has a totally new nearly square-planar coordination by four Li atoms (Fig. 1c). Above 10 GPa another $LiBH_4$ phase forms (Filinchuk *et al.*, 2008c). It is isostructural to the cubic $NaBH_4$ (see below); Li atoms and BH_4 groups are octahedrally coordinated (at 18 GPa, the Li-B distances are 2.56 Å). In the *Ama*2 phase, the geometry of the BH_4^- anions was restrained to an ideal tetrahedron with the B-H bond lengths centered at ~1.17 Å, while in the cubic phase at 18.1 GPa the refined B-H distance is 1.08 Å.

The borohydride groups are connected to Li atoms mostly via the tetrahedral edges. However, in the hexagonal phase and in the case of the shortest Li-B contact (2.37 Å) in the *Pnma* phase, the borohydride groups are connected to Li atoms via the tetrahedral faces.

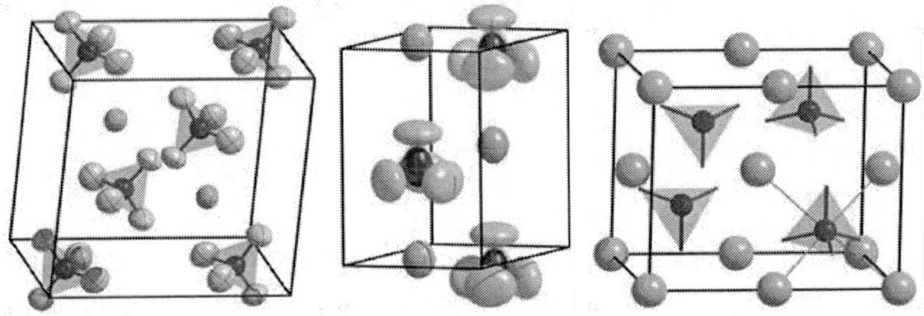

(A) (B) (C)

Figure 1. Crystal structure of LiBH4 phases: Pnma (a), P63mc (b) and Ama2 (c). The cubic Fm-3m structure (not shown here) is identical to the one presented in Fig. 2a. Anisotropic displacement ellipsoids determined from synchrotron diffraction on single crystals (Filinchuk et al., 2008a) are shown for the ambient pressure phases.

$NaBH_4$

Under ambient conditions $NaBH_4$ has a cubic structure, isomorphous to NaCl. From the diffraction data alone, this structure can be desribed in either *Fm*-3*m* or *F*-43*m* space groups. The two models differ only by the absence of the inversion centre in *F*-43*m*. The centrosymmetric model is intrinsically disordered, showing two orientations of the BH_4^- group (Fig. 2a), while the non-centrosymmetric model allows the BH_4^- anion to be fully or partly ordered. A peak of heat capacity at ~190K (Johnston, Hallett, 1953) suggests an order-disorder transition. The estimated entalpy of the transition is consistent with a transformation from the fully orientationally disordered high-temperareture cubic phase to an ordered low-temperature tetragonal phase. Therefore, the cubic phase has to be assumed to have the *Fm*-3*m* space group symmetry (Stockmayer, Stephenson, 1953). Crystal-chemical analysis shows that the shortest, repulsive H…H interactions in the cubic phase favour the disorder: a number of the shortest H…H distances between BH_4^- anions in the disordered *Fm*-3*m* structure is twice smaller than in a hypothetical ordered *F*-43*m* structure (Stockmayer, Stephenson, 1953). Thus, among the two possible models, the disordered *Fm*-3*m* and the ordered *F*-43*m*, only the first agrees both with crystallographic and thermodynamic data, as well as with the crystal-chemical considerations.

The cubic phase was characterized at 293K by neutron diffraction on $NaBD_4$ powder (Fischer, Züttel, 2004) and at 200K by synchrotron diffraction on a $NaBH_4$ single crystal (Filinchuk, Hagemann, 2008). The fully disordered *F*-43*m* model assumed in the first study is equivalent to the conventional *Fm*-3*m* model used in the second one. The ideal tetrahedral geometry of the borohydride group is defined by the site symmetry for the boron atom. The B-D distance at room temperature is ~1.17 Å, and at 200K the apparent B-H bond length is 1.09(2) Å. The latter, being corrected for 0.10 Å increment which takes into account the displacement of the electron cloud relative to an average nuclear position of the H-atom (Filinchuk, Hagemann, 2008), makes 1.19 Å.

On cooling below ~190K or upon compression to ~6 GPa at room temperature a phase with closely related ordered tetragonal structure appears (Kumar, Cornelius, 2005; Sundqvist, Andersson, 2006). Thermal conductivity study allowed to map the P-T boundary between the cubic and tetragonal phases at low temperature (Sundqvist, Andersson, 2006). The tetragonal structure was reported in the P-42_1c space group (Fischer, Züttel, 2004), but as was pointed out (Filinchuk *et al.*, 2007), it deviates insignificantly from the higher symmetry, and thus can also be described in the space group $P4_2/nmc$. Comparing to the cubic phase, the BH_4 groups in the tetragonal phase are ordered in two different orientations (see Figs. 2a and 2b). As a result, a number of the shortest H...H contacts in the tetragonal phase is reduced by one third comparing to the disordered cubic phase, and the network of the repulsive H…H contacts changes from the isotropic three-dimensional (3D) to the two-dimensional (2D) one, oriented in the *ab* plane. Consequently, the *c*/*a* ratio changes discontinuously upon the cubic-to-tetragonal (c-to-t) transition at ~186K from 1 to 0.964 (Filinchuk Babanova *et al.*, 2009b2010); this change is more due to the contraction of the *c*-axis than to the expansion in the basal plane. We also found a small but aubrupt volume drop of 0.40% upon the c-to-t transition at ambient pressure. Both discontinuities confirm that the transition is of the first order.

Neutron powder diffraction at 10K shows that the BD_4 group in the tetragonal phase is nearly ideally tetrahedral and the B-D distance is 1.22 Å (Fischer, Züttel, 2004). Both the cubic and the tetragonal phases are strongly textured when loaded in diamond anvil cells (Filinchuk *et al.*, 2007; Chernyshov *et al.*, 2008).

Above ~9 GPa yet another phase was detected by diffraction and Raman spectroscopy, but both experimental and theoretical efforts originally failed to identify its structure (Kumar, Cornelius, 2005; Araújo *et al.*, 2005). Later its structure has been solved from synchrotron powder diffraction data measured at 11.2 GPa, and it was shown to be of $BaSO_4$ type (Fig. 2c). For the successful solution of the structure, it was essential to model a texture, including one parameter in the global optimization (Filinchuk *et al.*, 2007). The structure was solved with most *a*-axes approximately aligned with the compression direction. Hydrogen atoms were located as a part of a semi-rigid BH_4 group, with B-H distances of ~1.17 Å. A small volume drop of ~1% was found upon the pressure-induced transition from the tetragonal to the orthorhombic phase (Filinchuk *et al.*, 2007).

In all three structures, Na atoms and BH_4 groups are octahedrally coordinated. In the cubic phase at 200K Na-B distances are 3.065 Å, in the tetragonal phase at 180K – 2.976-3.091 Å (Filinchuk Babanova *et al.*, 2009b2010), and in the orthorhombic phase at room temperature and 11.2 GPa – 2.763-2.849 Å.

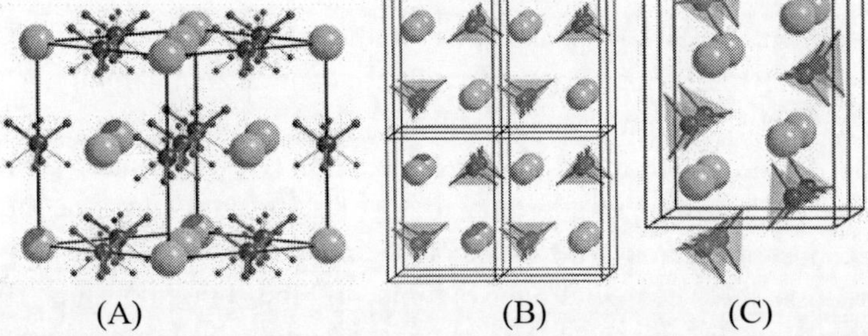

(A) (B) (C)

Figure 2. Crystal structure of NaBH4 phases: Fm-3m (a), P-421c or P42/nmc (b) and Pnma (c). Two orientations of the disordered BH4 group in the cubic phase are shown by bold and thin lines.

KBH_4

This substance shows a similar behaviour to $NaBH_4$. Its cubic phase has been studied at room temperature by X-ray diffraction on single crystals (Luck, Schelter, 1999) and by neutron powder diffraction (Renaudin *et al.*, 2004): in both cases it was described in the space group *Fm-3m*. The cubic phase transforms into a tetragonal one at 65-70K. Its structure is equivalent to the tetragonal $NaBH_4$, but it was refined in the higher symmetry space group, $P4_2/nmc$. Temperature dependence of the unit cell parameters for both phases has been reported by Renaudin *et al.*, 2004. As the cation's size in the cubic MBH_4 increases, as M goes from Na to K and then to Rb and Cs, the unit cell expands and the shortest H...H distances between the neighbouring BH_4 anions increase. In this way, a weaker H...H repulsion in the heavier MBH_4 reduces the temperature of the c-to-t transition. The K-B distance in the cubic KBH_4 is 3.364 Å.

X-ray diffraction on KBH_4 single crystal at 293K reveals an apparent (uncorrected, for an analysis of corrections see below) B-H distance 1.09(1) Å (Luck, Schelter, 1999). The B-D distance in the tetragonal KBD_4, determined by neutron powder diffraction at 1.5K, is 1.205(3) Å, and in the cubic structure at 295K it is 1.196(3) Å (Renaudin *et al.*, 2004). It was shown that the B-D distance in the cubic MBH_4 at 295K increases as M goes from Na to Cs. Besides the ionic size effect, this may be related to the increased anharmonicity in lighter MBH_4 compared to the heavier MBD_4 (Renaudin *et al.*, 2004), which is presumably related to an increased libration amplitudes for the BH_4 group. The latter would lead to an apparent (illusory) shortening of the B-H bonds.

NH_4BH_4

Ammonium borohydride contains ca. 24 wt % of hydrogen, and releases more than 20 wt % hydrogen in three steps at temperatures below 433K (Karkamkar *et al.*, 2009). It has the same cubic structure as Na and K borohydrides at ambient conditions, with a slightly bigger cell parameter than of the potassium analogue (see Table 1). The recently revealed crystal structure has not yet been studied in detail; in particular the hydrogen atom positions have not been experimentally determined.

$Be(BH_4)_2$

The only known $Be(BH_4)_2$ phase was studies by single crystal X-ray diffraction (Marynick, Lipscomb, 1972). Its tetragonal structure contains helical polymeric chains, where the only independent Be cation is coordinated by two bridging borohydride anions (Be-B distances of 2.00 Å) and one terminal borohydride anion (Be-B distance 1.92 Å). It is striking that the therminal BH_4 group is coordinated by the Be atom not via the face but via a tetrahedral edge, just the same way as the bridging BH_4 group. This may be due to the repulsive interaction between hydrogen atoms of the neighbouring BH_4 groups: the corresponding H...H distances (2.24-2.31 Å) are among the shortest known and Be...BH_4 coordination via the tetrahedral face would make these distances even shorter. Be atom has a trigonal-planar environment made by three BH_4 groups, and the bridgind BH_4 group has a linear Be-B-Be geometry. Thus, the low coordination number for the Be atom leads to a reduced dimensionality (1D) of the polymeric structure, where coordination potential of the BH_4 ligands is not fully realized.

The BH_4 groups show slightly deformed tetrahedral geometry with H-B-H angles 104-127°. The apparent (uncorrected) B-H distances fall into the 1.08(2)-1.18(3) Å range at 293K.

$Mg(BH_4)_2$

The most stable polymorph, α-$Mg(BH_4)_2$, has been reported in $P6_1$ space group symmetry by two independent groups (Černý *et al.*, 2007; Her *et al.*, 2007). In one case (Černý *et al.*, 2007) its structure was solved and refined using combination of synchrotron and neutron powder diffraction data. The synchrotron data contributed most of the information, but the use of neutron data allowed to determinate the orientation of the rigid BH_4 tetrahedra. In the other case (Her *et al.*, 2007), the structure was solved from synchrotron powder data alone, however, only positions of Mg and B atoms were reliably determined. Geometry optimization of the experimentally determined hexagonal structure, by using the density functional theory (DFT), suggested that it has a pseudo or maybe even true $P6_122$ symmetry (Dai *et al.*, 2008). The structure has been recently revised in the space group $P6_122$, using 100K single-crystal diffraction data (Filinchuk *et al.*, 2009c2009b). The revision supports the suggestion made in the DFT study. Analysis of the published $P6_1$ models shows that the location of the H-atoms from powder data posed the main problem for the identification of the correct symmetry: wrongly determined orientations of some BH_4 groups hampered a successful detection of the true $P6_122$ symmetry by automatic algorithms.

Pure α-phase transforms into the β-phase above 490K, the latter is quenched (metastable) on cooling, showing an anomalous temperature dependence of the cell parameters (Filinchuk *et al.*, 2009c2009b). Sharp diffraction peaks of β- $Mg(BH_4)_2$ can be described by a relatively simple *Immm* structure. However, all peaks were modeled, assuming an antisite disorder, in the 8 times bigger supercell with *Fddd* symmetry (Her *et al.*, 2007).

An intriguing aspect of the $Mg(BH_4)_2$ structures is their complexity (Fig. 3). The hexagonal structure contains three symmetry independent Mg^{2+} cations and six symmetry independent BH_4^- anions. Mg atoms bridged by the BH_4 groups (Mg-Mg 4.6-4.9 Å) form a framework with a novel 3D topology. It contains a variety of 5-membered $(\text{-Mg-BH}_4\text{-})_5$ rings that dominate in number over 6- and 8-membered ones, and in this way the structure resembles an amorphous state. The orthorhombic structure contains two symmetry independent Mg^{2+} cations and five symmetry independent BH_4^- anions. Mg atoms bridged by the BH_4 groups (Mg-Mg 4.6-4.7 Å) also form a 3D framework, however the latter does not contain 5-membered rings, but only 4- and 6-membered ones.

On the local level, both phases have the same principles of structural organization. Each Mg atom is surrounded by four BH_4 tetrahedra in a strongly deformed tetrahedral environment, while each BH_4 is nearly linearly coordinated by two Mg cations via the opposite tetrahedral edges. However, the Mg-H_2BH_2-Mg fragments are not strictly linear – in the α-phase the Mg-B-Mg angles range from 148 to 170°. At ambient conditions, the Mg-B distances in the hexagonal phase vary within the nearly the same interval, 2.39-2.45 (Filinchuk *et al.*, 2009c2009b) as in the orthorhombic phase, 2.34-2.47 Å (Her *et al.*, 2007).

With respect to the metal-hydrogen bond lengths and coordination geometry for the metal atoms, $Mg(BH_4)_2$ takes an intermediate position between $Be(BH_4)_2$ (flat trigonal coordination by three BH_4 anions) and $LiBH_4$ (nearly ideal tetrahedral coordination by four BH_4 anions). Thus, the ionic metal-to-BH_4 size ratio may be one of the reasons of the high structural complexity of $Mg(BH_4)_2$. Strong distortion of inequivalent Mg polyhedra provides a sufficient flexibility to meet all the crystal-chemical (i.e. energy) requirements: minimization of the repulsive Mg...Mg and H...H interactions, meeting coordination preferences for the BH_4 groups etc.

Accurate single-crystal diffraction data revealed that the H-B-H angles in the Mg-H_2B fragments are in average by 2° more open than for the non-coordinated ones. The DFT model shows the same trend. A detectable distortion of the tetrahedral BH_4 groups can be attributed to the strong polarizing effect of the Mg^{2+} cation or to a partially covalent Mg-BH_4 interaction. Unexpectedly, α-$Mg(BH_4)_2$ contains an unoccupied void, accounting for 6.4% of space in the structure (Filinchuk *et al.*, 2009c2009b). It is large enough (37 $Å^3$) to accommodate a small molecule, such as H_2O (see Fig. 3). The high-temperature β-phase is less dense by ~3%, but contains no unoccupied voids.

Unusual crystal chemistry and the high structural complexity of $Mg(BH_4)_2$ is unprecedented for this class of compounds, and all known attempts to predict $Mg(BH_4)_2$ structures theoretically were so far unsuccessful.

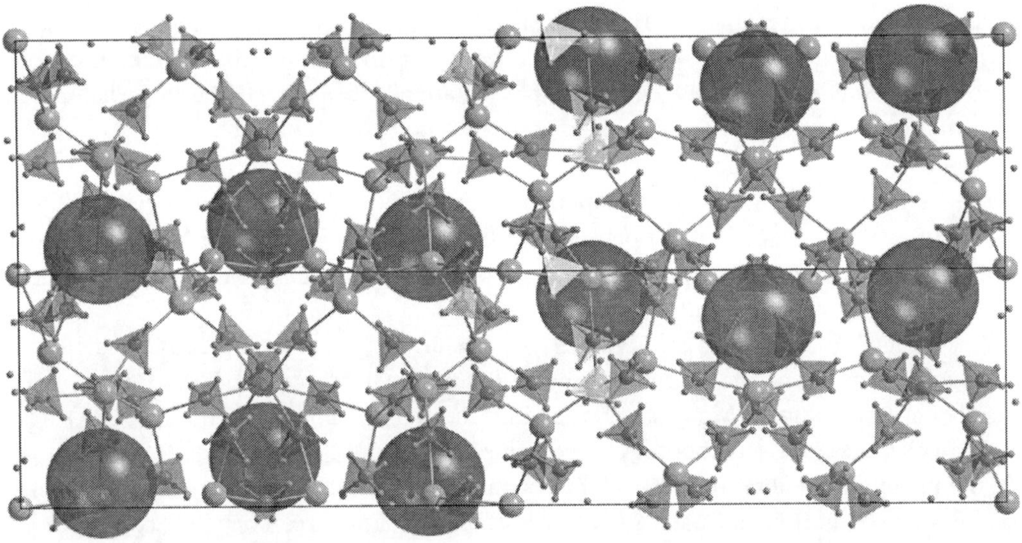

Figure 3. Crystal structure of α-Mg(BH4)2, where the unoccupied voids are shown as large spheres.

$Ca(BH_4)_2$

Riktor *et al.*, 2007 reported on existence of several $Ca(BH_4)_2$ phases at different temperatures. For the α-phase, obtained by desolvation of the tetrahydrofurane complex, Miwa *et al.*, 2006 reported an *Fddd* structure. However, analysis of the high-quality synchrotron powder diffraction data revealed that the actual symmetry of this orthorhombic structure is lower (Filinchuk *et al.*, 2009a). Two noncentrosymmetric subgroups of *Fddd*, namely *F*2*dd* and *Fdd*2, provide a much better fit to the experimental data than the centrosymmetric *Fddd* model, and the *F*2*dd* structure has some advantage over the *Fdd*2 one. The arrangement of the BH_4 groups in the noncentrosymmetric structure forms a simpler ordered pattern than in the *Fddd* model. The structure and symmetry of the orthorhombic phase is maintained from low temperatures up to 495K. However, as the temperature increases, the cell parameters *a* and *c* continuously approach each other, and at ~495K a second order transition into a tetragonal *I*-42*d* α'-phase is taking place (Filinchuk *et al.*, 2009a). The two structures show a group-subgroup relation (index 2). It is important that BH_4 groups are fully ordered in the *F*2*dd* and *I*-42*d* structures (Figs. 4a and 4b), while tetragonal supergroups of *Fddd* impose a disorder of the borohydride group in a hypothetical high-temperature $I4_1/amd$ structure.

Above 450K, the closely related *F*2*dd* and *I*-42*d* phases transform into a completely different β-phase. Determination of its space group symmetry is ambiguous, and its structure was described in two similar models, giving the lowest DFT energies, namely in the space groups $P4_2/m$ (Buchter *et al.*, 2008) and *P*-4 (Filinchuk *et al.*, 2009a), see Fig. 4c. It is likely that the BH_4 group in β-$Ca(BH_4)_2$ is intrinsically disordered, similar to the high-temperature phase of $LiBH_4$ (Filinchuk *et al.*, 2008a), and according to the systematic absences of the diffraction peaks (Filinchuk *et al.*, 2009a) the true symmetry of the disordered structure is $P4_2nm$. The high-temperature phase is by 4-6% (depending on temperature) denser than the *F*2*dd* and *I*-42*d* ones, and it is stable on cooling to room and even to liquid nitrogen temperatures (Filinchuk *et al.*, 2009a). However, Fichtner *et al.* (2008) found that this phase is meta-stable at room temperature, as it slowly transforms back into the orthorhombic phase. Another phase, called γ-$Ca(BH_4)_2$, obtained by wet chemical synthesis, has an orthorhombic structure (Buchter *et al.*, 2009), see Fig. 4d. It is metastable at all temperatures and irreversibly transforms into the β-phase at ~590K. At ambient conditions the γ-phase is slightly (~1% or less) denser than the β-phase.

Crystal structures of all four $Ca(BH_4)_2$ phases (α, α', β and γ) contain calcium cations nearly octahedrally coordinated by six borohydride anions. Ca-B distances are all within the similar range: 2.82-2.97 Å in the *F*2*dd* α-phase at 91K, 2.94-2.98 Å in the *I*-42*d* α'-phase at 495K, 2.92-2.94 Å in the *P*-4 β-phase at 305K and 2.86-2.97 Å in the *Pbca* γ-phase. For the first three structures listed above, the refined geometry of the BH_4^- anions was restrained to an ideal tetrahedral configuration with the B-H bond lengths centered at ~1.17 Å (Filinchuk *et al.*, 2009a), while for the γ-phase the hydrogen atoms were refined independently yielding a distorted tetrahedral geometry with B-H bond length 1.06(3)-1.33(3) Å, at variance with the theoretical calculations suggesting ideal tetrahedral geometry of the BH_4 group (Buchter *et al.*, 2009).

In the *F*2*dd* and *I*-42*d* structures of the α and α'-phases, Ca atoms form a close-packed diamond-like frameworks where the BH_4 groups have a T-shaped coordination. On the contrary, a trigonal-planar coordination of the BH_4 group by Ca atoms is observed in the β and γ-phases.

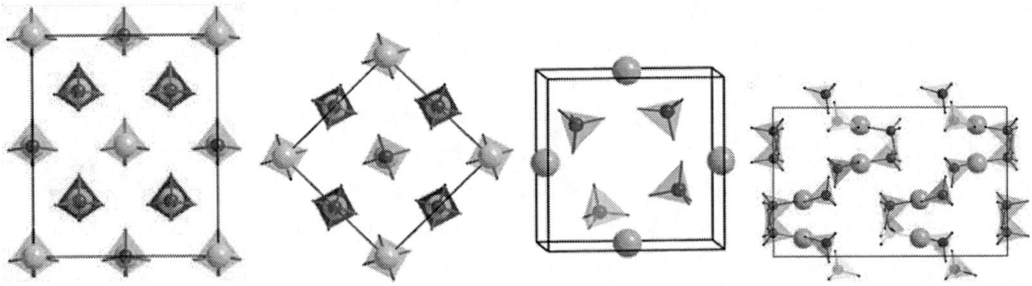

(A) (B) (C) (D)

Figure 4. Crystal structure of Ca(BH4)2 polymorphs: α-phase in F2dd (a), α'-phase in I-42d (after the F2dd → I-42d transition) (b), β-phase in P-4 or P42/m (the true structure is likely disordered with P42nm symmetry) (c) γ-phase in Pbca (d).

$Mn(BH_4)_2$

The first crystal structure of a 3*d*-metal borohydride has been determined only recently for $Mn(BH_4)_2$ by Černý *et al.*, 2009a. The compound forms at ambient conditions in ball-milled mixtures of alkali metal borohydrides and $MnCl_2$, crystallizes with the space group symmetry $P3_112$ and is stable from 90 up to 450K, where the compound melts. Thermal expansion of $Mn(BH_4)_2$ between 90 and 400K is highly anisotropic and very non-uniform.

Similar to magnesium borohydride, both independent metal atoms in $Mn(BH_4)_2$ are surrounded by four BH_4 tetrahedra in deformed tetrahedral coordinations with Mn-B distances ranging 2.39-2.52 Å. Each BH_4 group is approximately linearly coordinated by two Mn atoms (Mn-B-Mn angles vary from 147.4 to 170.0(2)°) via the tetrahedral edges. A single B-H distance of 1.13(2) Å was refined for rigid BH_4 tetrahedra.

The structure of $Mn(BH_4)_2$ shows an interesting similarity to α-$Mg(BH_4)_2$: the two structures are made of similar layers. The layers are stacked along the *c*-axis, and rotated by 120° by the 3_1 axis in $Mn(BH_4)_2$ and by 60° by the 6_1 axis in α-$Mg(BH_4)_2$. Three identical layers are stacked along one unit cell vector *c* in $Mn(BH_4)_2$, six layers are stacked in α-$Mg(BH_4)_2$. In $Mn(BH_4)_2$ the layers are connected directly while in α-$Mg(BH_4)_2$ the layers are connected via a thin intercalated layer which contains one Mg atom per layer per cell. Similar to α-$Mg(BH_4)_2$ (Filinchuk *et al.*, 2009c2009b), the structure of $Mn(BH_4)_2$ is not densely packed and contains isolated voids with the estimated volume of 21 Å^3 each, which occupy in total 6% of the space.

$Al(BH_4)_3$

At ambient conditions, this substance is liquid. However, two crystalline phases are known at low temperatures, with a transition temperature in the 180-195K range. Their structures have been studied by single-crystal X-ray diffraction (Aldridge *et al.*, 1997). Each phase is made up of discrete $Al(BH_4)_3$ units, where Al has a trigonal-planar environment with Al-B distances varying in a very narrow interval, 2.10(2)-2.14(2) Å. All BH_4 groups are coordinated by metal atoms via the tetrahedral edges. Locally, the $Al(BH_4)_3$ structures resemble the one for $Be(BH_4)_2$. However, the shortest H...H distances between neighbouring borohydride anions are longer (an average is over 2.5 Å) than those in $Be(BH_4)_2$ (2.24-2.31 Å).

The BH_4 anions appear somewhat distorted, partly due to experimental inaccuracies; however these distortions correlate with the results of the electron diffraction studies in the gas phase and *ab initio* calculations (Aldridge *et al.*, 1997). For example, the B-H distances are expected to be longer for the H-atoms coordinated by Al. Indeed, the B-H distances in the B-H...Al bridges are 1.12(3)-1.14(4) Å versus 0.99(4)-1.01(4) Å for the terminal B-H bonds. Note, that the experimental distances are much shorter than the theoretical ones (respectively, 1.27 and 1.19 Å), since they have to be corrected for the displacement of the electron cloud relative to an average nuclear position of the H-atom and for the libration shortening of the B-H bond (Filinchuk *et al.*, 2008a).

$LiK(BH_4)_2$

Very recently a number of mixed-cations borohydrides have been obtained and structurally characterized by synchrotron powder diffraction. These compounds are fully stoichiometric, as they do not show statistical occupation of one site by two different cations.

In the first bimetallic borohydride, $LiK(BH_4)_2$ (Nickels *et al.*, 2008), Li atom is tetrahedrally surrounded by BH_4 groups at Li-B distances 2.51-2.61 Å. An increase of the coordination number for K atom from six in KBH_4 to seven in $LiK(BH_4)_2$ increases also the K-B distances from 3.36 Å to 3.40-3.48 Å. Two independent BH_4 anions are coordinated via the tetrahedral edges. One of them has a typical octahedral coordination (Li_2K_4), similar to that in the cubic MBH_4 phases, while the other shows a new type of coordination – square-pyramidal (Li_2K_2 in the base + K in the apical position). The distortion of the BH_4 anions observed in $LiK(BH_4)_2$ is rather related to experimental inaccuracies than to an influence of the polarizing cations suggested by Nickels *et al.*, 2008, for the discussion on the BH_4 geometry see below.

$MSc(BH_4)_4$ (M = Li, Na)

$LiSc(BH_4)_4$ (Hagemann *et al.*, 2008) can be described as a complex containing $[Sc(BH_4)_4]^-$ anion (Borisov, Makhaev, 1988), similar to other structurally characterized anion borohydride complexes, such as $(Ph_4P)_2[Mg(BH_4)_4]$ (Makhaev *et al.*, 2004). Vibrational spectroscopy and DFT calculations indicate the presence of discrete $[Sc(BH_4)_4]^-$ ions. Sc atoms have a distorted tetrahedral coordination by BH_4 groups (Sc-B distances of 2.28 Å), the four BH_4 groups are oriented with a tilted plane of three hydrogen atoms directed to the central Sc ion, resulting in a 8 + 4 coordination. The Li ions are found to be disordered along the *z* axis of the tetragonal cell, thus the coordination of the Li atom is not clearly defined. The borohydride group in $LiSc(BH_4)_4$ was modelled as a rigid body with a common refined B-H distance 1.08 Å.

The structure of $NaSc(BH_4)_4$ (Černý *et al.*, 2009b2010) consists of the similar isolated $[Sc(BH_4)_4]^-$ anions located inside slightly deformed trigonal prisms of Na atoms. Sc atom coordinates the BH_4 groups via the faces, yielding a 12-fold coordination by H atoms. Na is surrounded by six BH_4 tetrahedra in quite regular octahedral coordination with a (6+12)-fold coordination by H atoms. The packing of Na^+ cations and $[Sc(BH_4)_4]^-$ anions in $NaSc(BH_4)_4$ is a deformation variant of the hexagonal NiAs structure type.

$MZn_2(BH_4)_5$ (M = Li, Na)

$LiZn_2(BH_4)_5$ represent a novel type of structure (Ravnsbæk *et al.*, 2009a) which has no distinct analogues among other known inorganic compounds. The structure of $NaZn_2(BH_4)_5$ (Ravnsbæk *et al.*, 2009a) was identified and refined as a monoclinically distorted derivative of the Li-containing analogue.

Two independent Zn atoms in $MZn_2(BH_4)_5$, M = Li and Na, have a trigonal nearly planar coordination (coordination number, CN = 3) by three BH_4 groups, similar to the Be atoms in $Be(BH_4)_2$ (Marynick, Lipscomb, 1972). The Li and Na atoms in these two related structures have a saddle-like coordination (CN = 4), which has not been observed earlier for alkaline atoms in borohydrides. All BH_4 groups in $MZn_2(BH_4)_5$ are linearly coordinated by two metal atoms (the angles in the two structures cover the range 164.5-179.6°), in a similar way as in the $Mg(BH_4)_2$ structures. The BH_4 groups are coordinated via the two opposite tetrahedral edges, bridging either two Zn atoms or one Zn and one M atom. H-atoms were refined as a part of a semi-rigid BH_4 units, with B-H distances close to 1.12 Å and H-B-H angles close to 109.5°.

It is remarkable that $MZn_2(BH_4)_5$ consists of two identical doubly-interpenetrated three-dimensional (3D) frameworks (Fig. 5a), which implies that there are no covalent bonds between them. This type of structural topology is common for the coordination polymers involving organic ligands, also known as metal-organic frameworks (MOFs), but it is observed for the first time in metal hydrides. This suggests directionality and some covalent character of the metal-BH_4 interaction. Indeed, the Zn-H bonds in $LiZn_2(BH_4)_5$ are very short; the refined distances are all below 2 Å, reaching 1.65 Å at the lower limit. Zn-B contacts are also very short, 2.11-2.31 Å. The average Zn-B distance of 2.17 Å in $LiZn_2(BH_4)_5$ is longer than the corresponding 1.97 Å average in $Be(BH_4)_2$ for the trigonally- coordinated Be atom, but considerably shorter than in all other metal borohydrides. Considering only the strongly associated Zn atoms and BH_4 units, the $NaZn_2(BH_4)_5$ compounds contain isolated $[Zn_2(BH_4)_5]^-$ anions (Fig. 5b), counter-balanced by cations M^+, similar to the isolated $[Sc(BH_4)_4]^-$ units in $MSc(BH_4)_4$.

It was recently found that the three-dimensional framework in $Mg(BH_4)_2$ contains empty voids, large enough to accommodate small molecules like H_2O (Filinchuk *et al.*, 2009c2009b). $MZn_2(BH_4)_5$ structures, although not porous, reveal another case of strong and directional metal-BH_4 bonding defining the structural architecture.

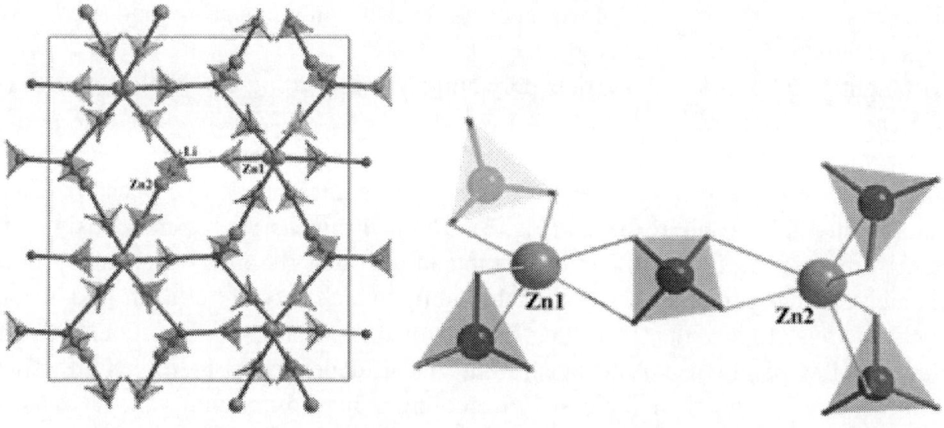

(A) (B)

Figure 5. Crystal structure of LiZn2(BH4)5. The doubly interpenetrated three-dimensional framework is highlighted in blue and red (a). Trigonally coordinated Zn atoms and the BH4 groups are associated into the isolated [Zn2(BH4)5]− anions (b).

$NaZn(BH_4)_3$

This phase shows a new type of structure (Ravnsbæk *et al.*, 2009a), containing a 3D framework made of the metal atoms and the BH_4 groups. The Zn atoms in $NaZn(BH_4)_3$ have a distorted tetrahedral coordination with CN = 4, while the saddle-like coordination of the Na atom by BH_4 groups is similar to the one in $MZn_2(BH_4)_5$. Comparing to $MZn_2(BH_4)_5$, the change of the coordination number from three to four for Zn leads to much longer Zn-B distances, varying from 2.43 to 3.16 Å. The average Zn-B distance of 2.74 Å in $NaZn(BH_4)_3$ is much longer than the 2.42 Å average of the Mg-B distances in the hexagonal $Mg(BH_4)_2$, where the metal atoms also have a tetrahedral coordination by the BH_4 groups (Filinchuk *et al.*, 2009c2009b). Interestingly, while one o the BH_4 groups in $NaZn(BH_4)_3$ exhibits a nearly linear coordination by two metal atoms, the other two show a trigonal-planar coordination.

STRUCTURAL EVOLUTION UNDER NON-AMBIENT CONDITIONS

Diffraction Studies of the Stability Regions and Structure Evolution under External Stimuli

The examples shown above illustrate that powder diffraction provides accurate information about structure of borohydrides. In some of the best examples, its accuracy is comparable to the one of the single crystal diffraction. However, a profound use of this technique is to study structure evolution under variable pressure or temperature. Bulk modulus and coefficients of thermal expansion, together with information on the symmetry and structure, provide the basis for validation of theoretical models. A study of structure evolution with external stimuli, thus, appears to be complementing and verifying the theoretical calculations.

Most light borohydrides show rich polymorphism indicating a complex profile of the potential energy. $LiBH_4$ serves as an illustration of the complex behaviour in a seemingly simple system (Filinchuk *et al.*, 2008a). While the high-temperature hexagonal phase shows a uniform and isotropic temperature expansion, the low-temperature *Pnma* phase reveals non-linear and highly anisotropic response (Fig. 6). The cell dimension *b* continuously contracts on heating from 300K to the transition temperature. Such thermal expansion reflects an anharmonicity of the potential of the crystal binding, which should be taken into account in theoretical models. It is remarkable that the thermal expansion is also anomalous at low temperatures: the parameter *a* deviates from linear dependence below 200K, shows a minimum at ~150K, and then increases on cooling. This observation can be related to a thermodynamic bistability involving the high-pressure *Ama*2 phase. Indeed, according to thermal conductivity measurements at high pressures, the free energy of $LiBH_4$ below 180K and ambient pressure should have two minima (Talyzin *et al.*, 2007). It is remarkable that the thermal expansion in the *a*-direction deviates from the linear behaviour in the same temperature range. A similar bistability and even more pronounced anomalous cell expansion is revealed for the two phases of $Mg(BH_4)_2$ (Filinchuk *et al.*, 2009c2009b), that may be related to the evolution of the free energy profile.

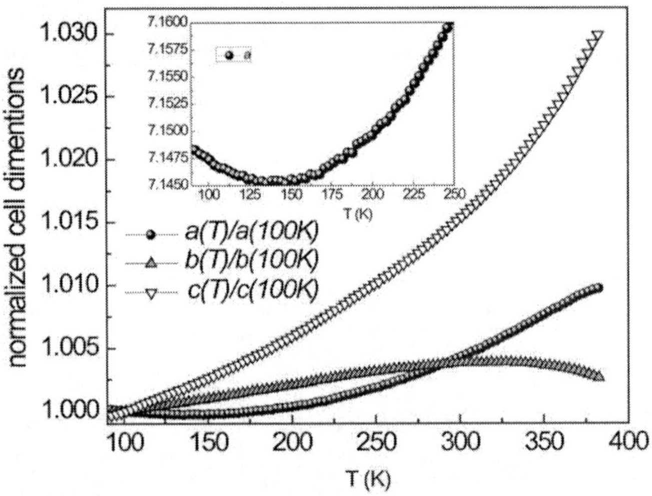

Figure 6. Cell dimensions of the Pnma phase of LiBH4 as a function of temperature (Filinchuk et al., 2008a), scaled to the experimental values at 100K. The low-temperature behaviour of the a parameter is shown in the inset.

The need to consider high-pressure phases while explaining the temperature behaviour of $LiBH_4$ calls for a combined pressure-temperature (P-T) study of the corresponding phase diagram. *In situ* synchrotron diffraction serves as a best probe to map the P-T diagram (Fig. 7), identify the phases and follow their structural evolution (Dmitriev *et al.*, 2008). Such a diagram, with all the supporting information, allows evaluating fundamental thermodynamic and structural properties of $LiBH_4$ and may guide a rational chemical modification of $LiBH_4$. A pressure behaviour of the transition temperature allows to estimate an entropy change on the corresponding transition (Pistorius, 1974), and thus to draw a conclusion about its mechanism. An analysis of symmetry changes and structural deformations, followed by a group theoretical analysis, yields a unified picture of the phase transformations in $LiBH_4$. The results of such study are shortly presented below for $LiBH_4$. A similar investigation of the P-T diagram, followed by a group-theoretical and crystal-chemical analysis has been very recently made for another light hydride, ammonia borane (Filinchuk *et al.*, 2009d2009c).

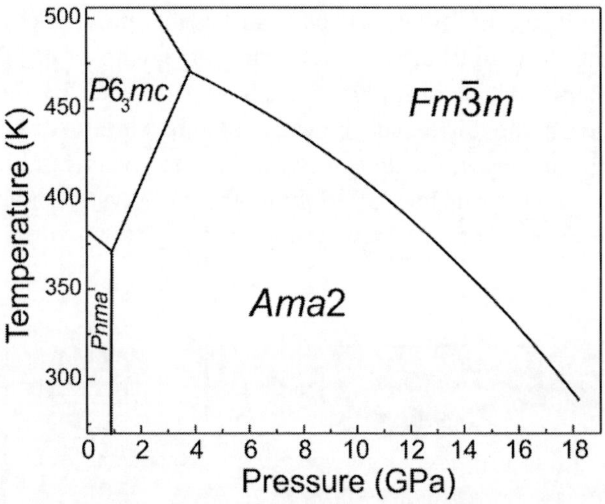

Figure 7. Pressure-temperature phase diagram of LiBH4 from synchrotron diffraction experiments (Dmitriev et al., 2008).

Phenomenological and Crystal-Chemical Analysis of the Mechanisms of the Phase Transitions

An existence of cation-anion layers in all four $LiBH_4$ phases is suggested from the phenomenological analysis of mechanisms of phase transitions (Dmitriev *et al.*, 2008). This conclusion is not trivial from purely geometrical point of view, but it can find a rational explanation considering the experimentally determined structures. A clear evidence for the existence of cation-anion layers is found in the hexagonal phase, where BH_4 tetrahedron has three short Li-B contacts in the *ab* plane and a long one along the *c* axis. In the *Pnma* phase, these layers are corrugated (Fig. 8a) and the structure is less anisotropic. In the high-pressure phases the existence of cation-anion layers is less obvious. However, the layers where Li and BH_4 groups are associated by means of the shorter Li-B contacts can be identified in the (011) plane of the *Ama*2 structure. Due to the high symmetry of the cubic structure different hypothetical layers can be identified there. However, only one type of layers is consistent both with geometrical considerations and with the phenomenological model: these are cation-anion layers situated in the (111) plane, very similar to those found in the hexagonal phase.

Clearly, the formation of layers in the $LiBH_4$ structures is not determined by coordination polyhedra for Li and BH_4 groups, since corresponding coordination numbers and geometries vary with pressure and temperature. However, directional coordination of the BH_4 group by Li atoms indicates geometrical and possibly also electronic preferences of BH_4...M interaction. Interaction of non-spherical BH_4 anions with spherical Li cations results in cation-anions layers, which determine the mechanisms of transitions between polymorphic structures. Directional interaction of the tetrahedral BH_4 with the spherical metal atoms explains the relative complexity of $LiBH_4$ structures and of the P-T phase diagram in comparison with NaCl, where both the cation and the anion are spherical.

It has been shown (Dmitriev *et al.*, 2008) that the order parameter in $LiBH_4$ can be parameterized as a shift of layers formed by Li and BH_4, together with in-layer deformations. The temperature dependence of Li-B distances can be, therefore, taken as a probe of inter- and intralayer deformations. Indeed, it was shown (Hagemann *et al.*, 2009) that the layers defined in the *bc* plane of the orthorhombic phase (Fig. 8a) become more flat as the temperature is increased: the difference between atomic coordinates for Li and B atoms decreases with temperature. Also, the distance between the layers significantly increases with temperature (Fig. 8b). Thus, the thermal evolution of the crystal structure agrees with the phase diagram and description of the order parameters suggested by the Landau theory.

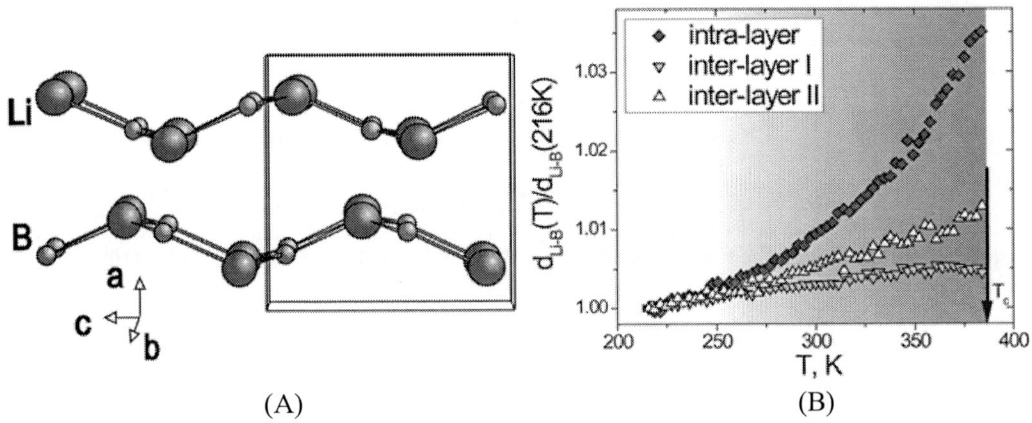

Figure 8. Li-BH4 layers (H atoms omitted) in the orthorhombic phase of LiBH4 (a) and the temperature evolution of Li-B distances within and between the layers (b). The gradient background highlights an increase of lattice anharmonicity, the arrow indicates the temperature of the transition into the hexagonal phase.

Another application of combined crystal-chemical and phenomenological analysis of polymorphic transitions would be to reveal destabilization of borohydrides upon phase transitions, achieved via formation of short H…H distances and deformation of tetrahedral BH_4 anions. A sign of such destabilization was observed in the *Ama*2 phase of $LiBH_4$ (Filinchuk *et al.*, 2008c). A similar destabilization in $LiBH_4$ or similar systems may be achieved during reconstructive phase transitions or melting. Indeed, some hydrogen desorption from $LiBH_4$ occurs upon the transition from *Pnma* to $P6_3mc$ phase and on melting (Mauron *et al.*, 2008). It is suggested (Dmitriev *et al.*, 2008) that other phase boundaries in the P-T phase diagram of pure or chemically modified $LiBH_4$ should be addressed with respect to a possible hydrogen desorption. Chemical modification of borohydrides, for example by exchanging a part of BH_4 anions by halide anions (Mosegaard *et al.*, 2008; Maekawa *et al.*, 2009; Matsuo *et al.*, 2009; Oguchi *et al.*, 2009; Arnbjerg Mosegaard *et al.*, 2009), can be used along with pressure and temperature to investigate their hydrogen desorption properties within the given P-T phase diagram.

CRYSTAL CHEMISTRY

Geometry of the BH_4 Group

Distorted geometry of the BH_4 group in the *Pnma* phase of $LiBH_4$ (Soulié *et al.*, 2002; Züttel *et al.*, 2003), found from the early synchrotron powder diffraction experiments, posed a question whether strongly polarizing Li^+ cation can induce deformations of the covalently bonded and therefore presumably rigid tetrahedral BH_4^- anion.

The recent revision of the $LiBH_4$ structures by single crystal synchrotron diffraction (Filinchuk, Chernyshov, 2007; Filinchuk *et al.*, 2008a) showed that the H-B-H angles are nearly the same as in the ideal tetrahedron (109.5°), while three independent B-H bonds have nearly the same lengths. This showed that the BH_4 group is very close to the undistorted tetrahedron and that the spurious distortions are not related to the nature of the interaction between X-rays and the electron density of light hydrogen atoms. Undistorted geometry of the BH_4 group in $LiBH_4$ was also established by neutron powder diffraction on isotopically-substituted $^7Li^{11}BD_4$ (Hartman *et al.*, 2007).

Synchrotron powder diffraction data obtained using a two-dimensional (2D) image plate detector also define the BH_4 anion as nearly ideally tetrahedral (Filinchuk, Chernyshov, 2007; Filinchuk *et al.*, 2008a). An analysis of the intensity distributions along the diffraction rings suggests that the powder data obtained with a 1D detector, measuring only a small part of Debye rings, suffers from a poor powder average. Intensity integration over the Debye rings recorded with a 2D detector provides much better average over the grains and therefore higher accuracy of the refined structural parameters. It is therefore crucial in the reliable determination of H-atoms on a small-volume powder sample containing relatively large grains. Considerable deviations from the tetrahedral geometry of the BH_4 group, as for instance recently reported by Nickels *et al.*, 2008 from synchrotron powder diffraction data, should be explained by an experimental inaccuracy. We should note, however, that in certain borohydride structures the BH_4 group can indeed be deformed, provided a good structural reason of this unusual behaviour, such as short H...H contacts, exists (Filinchuk *et al.*, 2008c). Despite it is not observed in $LiBH_4$, an interaction of a strongly polarizing cation with BH_4^- can also lead to deformations, $Al(BH_4)_3$ (Aldridge *et al.*, 1997) and α-$Mg(BH_4)_2$ (Filinchuk *et al.*, 2009a) serve as an examples. In any case, such deformation can be reliably detected mainly by diffraction on single crystals or by DFT calculations.

A certain discrepancy exists between the B-H bond lengths determined from X-ray and neutron diffraction experiments. While neutron diffraction at 3.5K gives the refined B-H distances of 1.208(3)-1.225(6) Å (Hartman *et al.*, 2007), the X-ray diffraction at low temperature yields 1.104(11) to 1.131(15) Å (Filinchuk *et al.*, 2008a). A similar difference has been found from accurate diffraction data on other light hydrides. This difference δ_{EL}, illustrated on Fig. 9a, comes from the well-known displacement of the electron cloud (observed by X-ray diffraction) relative to an average nuclear position (seen by diffraction of neutrons). By its nature, this difference is temperature independent.

Furthermore, DFT calculations find the B-H bond lengths still slightly longer than those from neutron diffraction. This second difference δ_{LIB} (Fig. 9b), originates from a geometric effect, caused by a libration of the BH_4 unit, which leads to an underestimation of the experimentally determined distances. An extent of the BH_4 libration is not only temperature-dependent (at higher temperature the motion of the BH_4 groups is more pronounced) but also, via frequencies of the corresponding phonons, is specific to each crystal structure. The extent of this libration shortening can be estimated from atomic displacement parameters determined from an accurate diffraction experiment.

A combined correction of 0.10 Å has been recently suggested from the analysis of the available structural data for $NaBH_4$ and its hydrate at low temperatures (Filinchuk, Hagemann, 2008). It roughly accounts both for the displacement of the electron cloud on H-atoms and for the libration effect observed in ordered structures at 200-300K. Displacement of H-atoms from B along the B-H bonds brings positions of hydrogen atoms determined from X-ray diffraction in agreement with theoretically predicted values.

The two corrections were evaluated separately from synchrotron diffraction data collected on $LiBH_4$ single crystals. It is shown (Filinchuk *et al.*, 2008a) that the displacement of the electron cloud from the nuclear position of H-atoms contributes ~0.08 Å to the apparent shortening of the B-H bonds. The libration correction in the *Pnma* phase at 225K was estimated to be 0.034 Å, while in the high-temperature phase at 535K it is much bigger, ~0.10 Å.

In is interesting to note here that the libration amplitudes for the BH_4 group, along with the related anharmonicities, are larger than for the heavier BD_4. Consequently, the cell parameters for MBH_4 are slightly larger (usually by less than 1% of the cell volume, Renaudin *et al.*, 2004) than those for MBD_4. The different contribution of H and D to the libration correction should also be considered in accurate comparisons of interatomic distances.

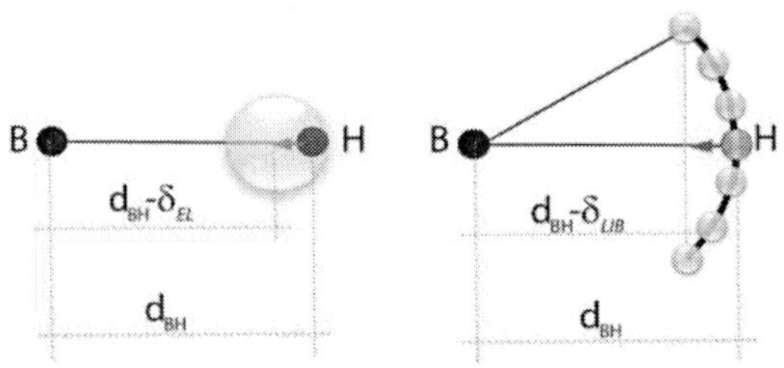

Figure 9. Schematic illustration of systematic differences between B-H distances determined by various methods. A discrepancy δEL between the B-H bond lengths determined from X-ray and neutron diffraction (a) comes from the displacement of the electron cloud (observed by X-ray diffraction) relative to the average nuclear position (seen by diffraction of neutrons). This difference is temperature independent, δEL = 0.08 Å (Filinchuk et al., 2008a). A discrepancy δLIB between DFT calculations and the experimentally determined B-H bond lengths originates also from the geometric effect (Fig. 9b) caused by a libration of the BH4 unit. δLIB is temperature-dependent, specific to each crystal structure, and dependent on the mass of the H-isotope. Note that for the X-ray diffraction data both corrections have to be applied.

Correction of H-positions determined from X-ray diffraction for systematic errors leads to an increased accuracy of this technique applied to low-Z hydride systems. It also allows one to compare directly the results obtained by different diffraction techniques and by theoretical calculations. Making comparisons without applying proper systematic corrections has lead to some odd conclusions (Chater *et al.*, 2006; Siegel *et al.*, 2007; Yang *et al.*, 2007;

Wu *et al.*, 2008), when neutron powder diffraction giving irregular BH_4 groups was considered more accurate than singe crystal X-ray diffraction yielding ideally tetrahedral BH_4 groups. When the corrections are applied, we arrive to the opposite conclusion. Indeed, nonoverlapped three-dimensional information uncovered by the scattering of X-rays on single crystals gives an advantage over the neutron powder diffraction, where all scattering is projected on one dimension.

BH_4...M and BH_4...BH_4 Contacts

Coordination number (CN) for metal atoms increases with their ionic radii. For Be and Al atoms CN = 3 (trigonal-planar coordination by three BH_4 group), for Mg atoms CN = 4 (deformed tetrahedral coordination), for Li atoms at ambient and moderate pressures CN = 4 (tetrahedral coordination), but at higher pressure CN increases to 6 (octahedral coordination). For the largest Na, K and Ca atoms CN = 6 and the coordination is ideal or deformed octahedral. Potassium atom in $LiK(BH_4)_2$ is seven-coordinated, its coordination environment is a capped trigonal prism.

Coordination number for the borohydride anion changes along with $M(BH_4)_n$ shoichiometry. At n = 1, the typical coordination geometries are tetrahedral (ambient pressure phases of $LiBH_4$, $LiK(BH_4)_2$) and octahedral (cubic $LiBH_4$, $NaBH_4$, KBH_4 and $LiK(BH_4)_2$). As n increases, low-connected T-shaped ($Ca(BH_4)_2$) and linear ($Be(BH_4)_2$ and $Mg(BH_4)_2$) geometries occur, and in the structures with low dimensionalities ($Be(BH_4)_2$ and $Al(BH_4)_3$) the BH_4 group can be coordinated even by only one metal atom. Unusual coordination geometries for the BH_4 group are square-planar, found in the *Ama*2 phase of $LiBH_4$, and square-pyramidal in $LiK(BH_4)_2$.

The diversity of coordination geometries for the BH_4^- anion indicates that it behaves as a flexible ligand, adjusting to the requirements (CN) of the metal atom. However, the BH_4...M interaction shows a very directional behaviour. As a rule, the borohydride groups are connected to the metal atoms via the tetrahedral edges. The exceptions are the $MSc(BH_4)_4$ phases and the hexagonal phase of $LiBH_4$, where the BH_4 group is coordinated via the tetrahedral faces, and the shortest Li-B contact (2.37 Å) in the *Pnma* phase of $LiBH_4$, where the borohydride group is coordinated by Li atom also via the face. Coordination via vertices was observed only in some $Ca(BH_4)_2$ structures (Filinchuk *et al.*, 2009a) and in $KZn(BH_4)Cl_2$ (Ravnsbæk *et al.*, 2009b2010). Directional coordination of the BH_4 group by metal atoms clearly indicates non-spherical nature of the BH_4 anion, which has geometrical and possibly also electronic preferences in BH_4...M interaction.

The H...H distances between neighbouring BH_4 groups exceed 2.2 Å, and in most cases they are even longer. The only exception is the *Ama*2 phase of $LiBH_4$, where at rather low pressure of 2.4 GPa a strikingly short H...H contact between adjacent BH_4 anions has been found (1.92 Å in the experimental structure and 1.87 Å in the theoretically optimized model). It was shown that the short H...H interaction distorts the BH_4 anion, and this is likely to decrease the activation energy for hydrogen desorption. It was suggested that the internal pressure in the $LiBH_4$ structure may be tuned by a partial substitution of lithium by larger cations, or substitution of some BH_4 groups by bigger anions (Filinchuk *et al.*, 2008c). The resulting $LiBH_4$-based substance with *Ama*2 structure may show more favourable hydrogen storage properties than pure $LiBH_4$ and may turn out to be useful for hydrogen storage applications.

BH_4...M interaction is also responsible for the formation of doubly-interpenetrated 3D frameworks in $MZn_2(BH_4)_5$ and porous 3D framework in α-$Mg(BH_4)_2$. These structural architectures are common for the coordination polymers involving organic ligands, such as MOFs, but they are observed for the first time in metal hydrides. These observations suggest directionality and some covalent character of the BH_4...M interaction in the borohydrides involving less electropositive metals (e.g. Mg and Zn). This may also explain very complex structures and polymorphism of $Mg(BH_4)_2$.

Chemical Destabilization: Mixed-Cation and Mixed-Anion Borohydrides

Substitution of metal atoms or BH_4 groups by other cations and anions can be considered as a way to alter stability of borohydrides. A few mixed-cations borohydrides have been found recently: $LiK(BH_4)_2$ (Nickels *et al.*, 2008) and $MSc(BH_4)_4$ (Hagemann *et al.*, 2008; Černý *et al.*, 2009b2010). However, hydrogen desorption temperatures for such compounds appear to be high. For $LiK(BH_4)_2$ it is not lower than for the single cation constituents, but merely an intermediate between the two (Nickels *et al.*, 2008).

An apparent linear correlation between Pauling electronegativites and decomposition temperatures for borohydrides has been found (Nakamori *et al.*, 2006). It was pointed out (Ravnsbæk *et al.*, 2009a) that the significant structural diversity and low decomposition temperatures for the novel series of alkali-zinc borohydrides, such as $MZn_2(BH_4)_5$ and $NaZn(BH_4)_3$, may be assigned to the ability to form more covalent bonds between Zn and BH_4 units, as compared to the alkali metal borohydrides. Indeed, the Pauling electronegativity of zinc is higher than those for the alkaline metals and Sc, $Zn(BH_4)_2$ has not yet been isolated in its free form, and its instability may contribute to the lower stability of Zn-based borohydrides. Variation in the ratio between the alkali and transition metal, as well as a use of different metals, thus enables tuning of the properties for hydrogen storage in alkali metal-transition metal-BH_4 systems. Therefore, a variety of novel mixed-cation transition-metal based borohydrides may be discovered in the near future.

The mixed-anion derivatives of metal borohydrides show even more favourable hydrogen storage properties than the individual components; this is well illustrated for the $LiBH_4$-$LiNH_2$ system (Meisner *et al.*, 2006). Currently, two mixed borohydride-amide phases have been structurally characterized: $Li_4(BH_4)(NH_2)_3$ (Filinchuk *et al.*, 2006; Chater *et al.*, 2006) and $Li_2(BH_4)(NH_2)$ (Chater *et al.*, 2007). Very recently an existence of mixed borohydrides-alanates has been reported: $Na_4(BH_4)(AlH_4)_3$ crystallizes in a primitive cubic 2×2×2 supercell of $NaBH_4$ (Smith *et al.*, 2007), $Mg(BH_4)(AlH_4)$ was identified with yet unknown crystal structure (Zhao *et al.*, 2007). Hydrogen desorption properties for the latter class of compounds are not yet reported. Currently, only amide containing mixed-anion derivatives showed some decreasing of the hydrogen evolution temperature comparing to the two hydrogen-containing constituents (Meisner *et al.*, 2006). However, we can see that the mixed-anion derivatives show a bigger improvement of hydrogen storage properties than the mixed-cation ones. This may be partly due to the formation of dihydrogen bonds (such as B-H...H-N) between different anions, which are expected to facilitate hydrogen desorption. A similar modification of light metal borohydrides can be achieved by introducing neutral hydrogen-rich molecules, such as ammonia or water. Such examples include $Mg(BH_4)_2{\cdot}2NH_3$ (Soloveichik *et al.*, 2008) and $NaBH_4{\cdot}2H_2O$ (Filinchuk, Hagemann, 2008); the first substance shows very favourable hydrogen storage properties.

Very recently a mixed-anion borohydride has been obtained (Ravnsbæk *et al.*, 2009b2010), containing isolated $[Zn(BH_4)Cl_2]^-$ anions and octa-coordinated K^+ cations, associated with the anions by weak K...Cl and K...BH_4 interactions. The Zn^{2+} ion has a trigonal planar coordination by two Cl^- and one BH_4^- ion. $KZn(BH_4)Cl_2$ starts to decompose at a significantly lower temperature (383K) than KBH_4, hence this study provides new inspiration for design and preparation of novel materials, which may be suitable for hydrogen storage applications.

All the structures of the borohydride derivatives described above are fully ordered and shoichiometric, without statistical occupation of the cation or anion sites. $LiBH_4$ + LiCl is the first system where a gradual replacement of BH_4 has been found (Mosegaard *et al.*, 2008), showing a partial replacement of the borohydride anions by chloride anions at elevated temperatures (>389K). Comparison of the unit cell volumes for different inorganic salts suggests that the size of anions change according to the sequence $I^- > BH_4^- > Br^- > Cl^-$ (Filinchuk, Hagemann, 2008). This variation gives an efficient tool to tune the unit cell volume and an internal lattice pressure in borohydrides. Larger halide anions, such as iodide, stabilize the high-temperature hexagonal phase of $LiBH_4$ at room temperature, which is interesting as a lithium superionic conductor (Maekawa *et al.*, 2009; Matsuo *et al.*, 2009; Oguchi *et al.*, 2009). Even though smaller halide anions, such as chloride, stabilize the hexagonal phase at significantly lower temperatures, the latter depend on the degree of the anion substitution (Arnbjerg Mosegaard *et al.*, 2009). Heating a $LiBH_4$ + LiCl mixture produces highly chloride-substituted hexagonal lithium borohydride, $Li(BH_4)_{1-x}Cl_x$, with $x \sim 0.42$. At higher temperatures the solubility is higher, and the process of the LiCl entrance into $LiBH_4$ is reversible on heating/cooling (Arnbjerg Mosegaard *et al.*, 2009). The orthorhombic low-temperature $LiBH_4$ phase shows a lower solubility of LiCl than the hexagonal one.

Thus, a partial substitution of the BH_4 group by halide anions (Hal) opens the way to modify metal borohydrides and influence their structural stability. It may possibly allow obtaining high-pressure polymorphs of metal borohydrides at ambient conditions (Filinchuk *et al.*, 2008a). *In-situ* powder diffraction study of the reaction mixtures at variable temperature should be more widely used to probe different $M(BH_4)_n$ + M'Hal combinations.

Chemical Destabilization: Substitution in the BH_4 Group

A note on a possible modification of the BH_4 group is warranted. A partial substitution of hydrogen atoms by fluorine atoms has been achieved in alanates, showing that the complex hydride $Na_3AlH_{6-x}F_x$ is less stable than Na_3AlH_6 (Brinks *et al.*, 2008). Due to the high stability of the covalent B-H bond, a similar chemical modification of the BH_4 anion is much more difficult. Some recent reports suggest that BH_3OH^- exists in solutions (Ruman *et al.*, 2007), and even more interestingly, the first alkoxide-substituted borohydride containing BH_3OR^- anion has been isolated in the solid state and structurally characterized (Gainsford *et al.*, 2009).

We note, however, that amidoboranes, a new class of high-capacity hydrogen storage materials of the general formula $M(BH_3NH_2)_n$ (Xiong *et al.*, 2007), can be considered as modified borohydrides, where one hydrogen atom in BH_4^- has been exchanged for NH_2. Nevertheless, amidoboranes are currently obtained not by modifying borohydrides but from ammonia-borane, which contains the ready-made B-N bond.

REFERENCES

Aldridge, S.; Blake, A. J.; Downs, A. J.; Gould, R. O.; Parsons, S.; Pulham, C. R.: *Some tetrahydroborate derivatives of aluminium: crystal structures of dimethylaluminium tetrahydroborate and the α and β phases of aluminium tris(tetrahydroborate) at low temperature*. J. Chem. Soc., Dalton Trans. (1997) 1007-1012.

Araújo, C. M.; Ahuja, R.; Talyzin, A. V.; Sundqvist, B.: *Pressure-induced structural phase transition in NaBH4*. Phys. Rev. B. 72 (2005) 054125.

Babanova, O.; Soloninin, A.; Stepanov, A.; Skripov, A.; Filinchuk, Y.: *Structural and dynamical properties of NaBH4 and KBH4: NMR and synchrotron X-ray diffraction studies*. J. Phys. Chem. C. 114 (2010) 3712-3718.

Borisov, A. P.; Makhaev, V. D.: *Anion borohydride complexes of scandium, titanium and vanadium with organic cations*. Russ. J. Inorg. Chem. 33 (1988) 3030-3035.

Brinks, H. W.; Fossdal, A.; Hauback, B. C.: *Adjustment of the stability of complex hydrides by anion substitution*. J. Phys. Chem. C. 112 (2008) 5658-5661.

Buchter, F.; Łodziana, Z.; Remhof, A.; Friedrichs, O.; Borgschulte, A.; Mauron, Ph.; Züttel, A.; Sheptyakov, D.; Barkhordarian, G.; Bormann, R.; Chłopek, K.; Fichtner, M.; Sørby, M.; Riktor, M.; Hauback, B.; Orimo, S.: *Structure of Ca(BD4)2 β-Phase from combined neutron and synchrotron X-ray powder diffraction data and density functional calculations*. J. Phys. Chem. B. 112 (2008) 8042-8048.

Buchter, F.; Łodziana, Z.; Remhof, A.; Friedrichs, O.; Borgschulte, A.; Mauron, Ph.; Züttel, A.; Sheptyakov, D.; Palatinus, L.; Chłopek, K.; Fichtner, M.; Barkhordarian, G.; Bormann, R.; Hauback, B. C.: *Structure of the orthorhombic γ-phase and phase transitions of Ca(BD4)2*. J. Phys. Chem. C. (2009) 113 (2009) 17223-17230DOI: 10.1021/jp810521p.

Černý, R.; Filinchuk, Y.; Hagemann, H.; Yvon K.: *Magnesium borohydride: synthesis and crystal structure*. Angew. Chem. Int. Ed. 46 (2007) 5765-5767.

Černý, R.; Penin, N.; Hagemann, H.; Filinchuk, Y.: *The first crystallographic and spectroscopic characterization of a 3d-metal borohydride: Mn(BH4)2*. J. Phys. Chem. C. 113 (2009a) 9003-9007.

Černý, R.; Severa, G.; Ravnsbæk, D.; Filinchuk, Y.; d'Anna, V.; Hagemann, H.; Haase, D.; Jensen, C. M.; Jensen, T. R.: *NaSc(BH4)4: A novel scandium-based borohydride*. J. Phys. Chem. C. (2009b) 114 (2010) 1357-1364submitted.

Chater, P. A.; David, W. I. F.; Johnson, S. R.; Edwards, P. P.; Anderson, P. A.: *Synthesis and crystal structure of Li4BH4(NH2)3*. Chem. Commun. (2006) 2439-2441.

Chater, P. A.; David, W. I. F.; Anderson, P. A.: *Synthesis and structure of the new complex hydride Li2BH4NH2*. Chem. Commun. (2007) 4770-4772.

Chernyshov, D.; Bosak, A.; Dmitriev, V.; Filinchuk, Y.; Hagemann, H.: *Low lying phonons in NaBH4 studied by inelastic scattering of synchrotron radiation.* Phys. Rev. B. 78 (2008) 172104.

Dai, B.; Sholl, D. S.; Johnson, J. K.: *First-principles study of experimental and hypothetical Mg(BH4)2 crystal structures.* J. Phys. Chem. C. 112 (2008) 4391-4395.

Dmitriev, V.; Filinchuk, Y.; Chernyshov, D.; Talyzin, A.; Dzwilewski, A.; Andersson, O.; Sundqvist, B.; Kurnosov, A.: *Pressure-temperature phase diagram of LiBH4: synchrotron x-ray diffraction experiments and theoretical analysis.* Phys. Rev. B. 77 (2008) 174112.

Fichtner, M.; Chłopek, K.; Longhini, M.; Hagemann H.: *Vibrational spectra of Ca(BH4)2.* J. Phys. Chem. C. 112 (2008) 11575-11579.

Filinchuk, Y. E.; Yvon, K.; Meisner, G. P.; Pinkerton, F. E.; Balogh, M. P.: *On the composition and crystal structure of the new quaternary hydride phase Li4BN3H10.* Inorg. Chem. 45 (2006) 1433-1435.

Filinchuk, Y.; Chernyshov, D.: *Looking at hydrogen atoms with X-rays: comprehensive synchrotron diffraction study of LiBH4.* Acta Cryst. A. 63 (2007) s240.

Filinchuk, Y.; Talyzin, A.; Chernyshov, D.; Dmitriev, V.: *High-pressure phase of NaBH4: crystal structure from synchrotron powder diffraction data.* Phys. Rev. B. 76 (2007) 092104.

Filinchuk, Y.; Hagemann, H.: *Structure and properties of NaBH4·2H2O and NaBH4.* Eur. J. Inorg. Chem. (2008) 3127-3133.

Filinchuk, Y.; Chernyshov, D.; Černý, R.: *The lightest borohydride probed by synchrotron diffraction: experiment calls for a new theoretical revision.* J. Phys. Chem. C. 112 (2008a) 10579-10584.

Filinchuk, Y.; Chernyshov, D.; Dmitriev, V.: *Light metal borohydrides: crystal structures and beyond.* Z. Kristallogr. 223 (2008b) 649-659.

Filinchuk, Y.; Chernyshov, D.; Nevidomskyy, A.; Dmitriev, V.: *High-pressure polymorphism as a step towards destabilization of LiBH4.* Angew. Chem. Int. Ed. 47 (2008c) 529-532.

Filinchuk, Y.; Rönnebro, E.; Chandra, D.: *Crystal structures and phase transformations in Ca(BH4)2.* Acta Mater. 57 (2009a) 732-738.

Filinchuk, Y.; Chernyshov, D.; Dmitriev, V.: (2009b) *unpublished results.*

Filinchuk, Y.; Černý, R.; Hagemann, H.: *Insight into Mg(BH4)2 with synchrotron X-ray diffraction: structure revision, crystal chemistry, and anomalous thermal expansion.* Chem. Mater. 21 (2009c2009b) 925-933.

Filinchuk, Y.; Nevidomskyy, A. H.; Chernyshov, D.; Dmitriev, V.: *High-pressure phase and transition phenomena in ammonia borane NH3BH3 from x-ray diffraction, Landau theory, and ab initio calculations.* Phys. Rev. B. 79 (2009d2009c) 214111.

Fischer, P.; Züttel, A.: *Order-disorder phase transition in NaBD4.* Mater. Sci. Forum. 443-444 (2004) 287-290.

Gainsford, J. G.; Kemmitt, T.; Jameson, G. B.; Telfer, S. G. *The first example of a mixed alkoxide hydride of boron: sodium boron isopropoxide trihydride.* Acta Cryst. C. 65 (2009) m180-m181.

Grochala, W.; Edwards, P. P.: *Thermal decomposition of the non-interstitial hydrides for the storage and production of hydrogen.* Chem. Rev. 104 (2004) 1283-1315.

Hagemann, H.; Longhini, M.; Kaminski, J. W.; Wesolowski, T. A.; Černý, R.; Penin, N.; Sorby, M. H.; Hauback, B. C.; Severa, G.; Jensen, *C. M.: LiSc(BH4)4, a new complex salt with discrete Sc(BH4)4– ions*. J. Phys. Chem. A. 112 (2008) 7551-7555.

Hagemann, H.; Filinchuk, Y.; Chernyshov, D.; van Beek, W.: *Lattice anharmonicity and structural evolution of LiBH4: an insight from Raman and X-ray diffraction experiments*. Phase Trans. 82 (2009) 344-355.

Hartman, M. R.; Rush, J. J.; Udovic, T. J.; Bowman, R. C. Jr.; Hwang, S. J.: *Structure and vibrational dynamics of isotopically labeled lithium borohydride using neutron diffraction and spectroscopy*. J. Solid State Chem. 180 (2007) 1298-1305.

Her, J.-H.; Stephens, P. W.; Gao, Y.; Soloveichik, G. L.; Rijssenbeek, J.; Andrus, M.; Zhao, J.-C.: *Structure of unsolvated magnesium borohydride Mg(BH4)2*. Acta Cryst. B. 63 (2007) 561-568.

Johnston, L.; Hallett, N. C.: *Low temperature heat capacities of inorganic solids. XIV. Heat capacity of sodium borohydride from 15-300°K*. J. Am. Chem. Soc. 75 (1953) 1467-1468.

Karkamkar, A.; Kathmann, S. M.; Schenter, G. K.; Heldebrant, D. J.; Hess, N.; Gutowski, M.; Autrey, T.: *Thermodynamic and structural investigations of ammonium borohydride, a solid with a highest content of thermodynamically and kinetically accessible hydrogen*. Chem. Mater. (2009) 75 (1953) 1467-1468DOI 10.1021/cm902385c.

Kumar, R. S.; Cornelius, A. L.: *Structural transitions in NaBH4 under pressure*. Appl. Phys. Lett. 87 (2005) 261916.

Luck, R. L.; Schelter, E. J.: *Potassium borohydride*. Acta Cryst. C. 55 (1999) IUC9900151.

Makhaev, V. D.; Borisov, A. P.; Antsyshkina, A. S.; Sadikov, G. G.: *Synthesis and properties of new magnesium tetrahydroborate complexes: the crystal structure of (Ph4P)2[Mg(BH4)4]*. Russ. J. Inorg. Chem. 49 (2004) 323-330.

Marynick, D. S.; Lipscomb, W. N.: *Crystal structure of beryllium borohydride*. Inorg. Chem. 11 (1972) 820-823.

Maekawa, H.; Matsuo, M.; Takamura, H.; Ando, M.; Noda, Y.; Karahashi, T.; Orimo, S.-i.: *Halide-stabilized LiBH4, a room-temperature lithium fast-ion conductor*. J. Am. Chem. Soc. 131 (2009) 894-895.

Matsuo, M.; Takamura, H.; Maekawa, H.; Li, H.-W.; Orimo, S.-i.: *Stabilization of lithium superionic conduction phase and enhancement of conductivity of LiBH4 by LiCl addition*. Appl Phys. Lett. 94 (2009) 084103.

Mauron, P.; Buchter, F.; Friedrichs, O.; Remhof, A.; Bielmann, M.; Zwicky, C. N.; Zuttel, A.: *Stability and reversibility of LiBH4*. J. Phys. Chem. B. 112 (2008) 906-910.

Meisner, G. P.; Scullin, M. L.; Balogh, M. P.; Pinkerton, F. E.; Meyer, M. S.: *Hydrogen release from mixtures of lithium borohydride and lithium amide: a phase diagram study*. J. Phys. Chem. B. 110 (2006) 4186-4192.

Miwa, K.; Aoki, M.; Noritake, T.; Ohba, N.; Nakamori, Y.; Towata, S.-i.; Züttel, A.; Orimo, S.-i.: *Thermodynamical stability of calcium borohydride Ca(BH4)2*. Phys. Rev. B. 74 (2006) 155122.

Mosegaard, L.; Moeller, B.; Jorgensen, J.-E.; Filinchuk, Y.; Cerenius, Y.; Hanson, J.; Dimasi, E.; Besenbacher, F.; Jensen, T.: *Reactivity of LiBH4: in-situ synchrotron radiation powder X-ray diffraction study*. J. Phys. Chem. C. 112 (2008) 1299-1303.

Arnbjerg, L. M.; Ravnsbæk, D.; Filinchuk, Y.; Vang, R. T.; Cerenius, Y.; Besenbacher, F.; Jørgensen, J.-E.; Jacobsen, H. J.; Mosegaard, L.; Ravnsbæk, D.; Vang, R.; Filinchuk, Y.;

Cerenius, Y.; Hanson, J. C.; Iversen, B. B.; Besenbacher, F.; Jacobsen, H. J.; Jørgensen, J. E.; Jensen, T. R.: *Structure and dynamics for LiBH4 − LiCl solid solutions*. Chem. Mater. 21 (2009) 5772-5782(2009) submitted.

Nakamori, Y.; Miwa, K.; Ninomiay, A.; Li, H.; Ohba, N.; Towata, S.; Züttel, A.; Orimo, S.: *Correlation between thermodynamical stabilities of metal borohydrides and cation electronegativites: first-principles calculations and experiments*. Phys. Rev. B. 74 (2006) 045126.

Nickels, E. A.; Jones, M. O.; David, W. I. F.; Johnson, S. R.; Lowton, R. L.; Sommariva, M.; Edwards, P. P.: *Tuning the decomposition temperature in complex hydrides: synthesis of a mixed alkali metal borohydride*. Angew. Chem. Int. Ed. 47 (2008) 2817-2819.

Talyzin, A. V.; Andersson, O.; Sundqvist, B.; Kurnosov, A.; Dubrovinsky, L.: *High-pressure phase transition in LiBH4*. J. Solid State Chem. 180 (2007) 510-517.

Oguchi, H.; Matsuo, M.; Hummelshøj, J. S.; Vegge, T.; Nørskov, J. K.; Sato, T.; Miura, Y.; Takamura, H.; Maekawa, H.; Orimo, S.: *Experimental and computational studies on structural transitions in the LiBH4-LiI pseudobinary system*. Appl Phys. Lett. 94 (2009) 141912.

Orimo, S.-i.; Nakamori, Y.; Eliseo, J. R.; Züttel, A.; Jensen, C. M.: *Complex hydrides for hydrogen storage*. Chem. Rev. 107 (2007) 4111-4132.

Pistorius, C. W. F. T.: *Melting and polymorphism of LiBH4 to 45 kbar*. Z. Phys. Chem. Neue Folge. 88 (1974) 253-263.

Ravnsbæk, D.; Filinchuk, Y.; Cerenius, Y.; Jakobsen, H. J.; Besenbacher, F.; Skibsted, J.; Jensen T. R.: *A series of mixed-metal borohydrides*. Angew. Chem. Int. Ed. 48 (2009a) 6659-6663.

Ravnsbæk, D. B.; Sørensen, L. H.; Filinchuk, Y.; Reed, D.; Book, D.; Cerenius, Y.; Jakobsen, H. J.; Besenbacher, F.; Skibsted, J.; Jensen, T.R.: *Mixed-anion and mixed-cation borohydride KZn(BH4)Cl2: synthesis, structure and thermal decomposition. Chem. Eur. J. Inorg. Chem.* (2010) 1608-1612(2009b) submitted.

Renaudin, G.; Gomes, S.; Hagemann, H.; Keller, L.; Yvon, K.: *Structural and spectroscopic studies on the alkali borohydrides MBH4 (M = Na, K, Rb, Cs)*. J. Alloys Compd. 375 (2004) 98-106.

Riktor, M. D.; Sørby, M. H.; Chłopek, K.; Fichtner, M.; Buchter, F.; Züttel, A.; Hauback, B. C.: *In situ synchrotron diffraction studies of phase transitions and thermal decomposition of Mg(BH4)2 and Ca(BH4)2*. J. Mater. Chem. 17 (2007) 4939-4942.

Ruman, T.; Kusnierz, A.; Jurkiewicz, A.; Les, A.; Rode, W.: *The synthesis, reactivity and 1H NMR investigation of the hydroxyborohydride anion*. Inorg. Chem. Commun. 10 (2007) 1074-1078.

Siegel, D. J.; Wolverton, C.; Ozoliņš, V.: *Reaction energetics and crystal structure of Li4BN3H10 from first principles*. Phys. Rev. B. 75 (2007) 014101.

Soloveichik, G.: *Metal borohydrides as hydrogen storage materials*. Material Matters (Aldrich). 2. No. 2 (2007) 11-14.

Soloveichik, G.; Her, J.-H.; Stephens, P. W.; Gao, Y.; Rijssenbeek, J.; Andrus, M.; Zhao, J.-C.: *Ammine magnesium borohydride complex as a new material for hydrogen storage: structure and properties of Mg(BH4)2·2NH3*. Inorg. Chem. 47 (2008) 4290-4298.

Soulié, J.-P.; Renaudin, G.; Černý, R.; Yvon, K.: *Lithium boro-hydride LiBH4. I. Crystal structure*. J. Alloys Compd. 346 (2002) 200-205.

Smith, Ch. I.; Jones, M. O.; Johnson, S. R.; David, W. I. F.; Edwards, P. P.: *Mixed borohydride-alanates as potential hydrogen storage materials.* Abstract in ISHE, International symposium on materials issues in a hydrogen economy. Richmond, Virginia. (2007) W32.

Stockmayer, W. H.; Stephenson, C. C.: *The nature of the gradual transition in sodium borohydride.* J. Chem. Phys. 21 (1953) 1311-1312.

Sundqvist, B.: *Pressure-temperature phase relations in complex hydrides.* Solid State Phenomena. 150 (2009) 175-195.

Sundqvist, B.; Andersson, O.: *Low-temperature phase transformation in NaBH4 under pressure.* Phys. Rev. B. 73 (2006) 092102.

Sundqvist, B.; Andersson, O.: *Thermal conductivity and phase diagrams of some potential hydrogen storage materials under pressure.* Int. J. Thermophys. 30 (2009) 1118-1129 (2009) DOI 10.1007/s10765-009-0559-5.

Wu, H.; Zhou, W.; Udovic, T. J.; Rush, J. J.; Yildrim, T.: *Structures and crystal chemistry of Li2BNH6 and Li4BN3H10.* Chem. Mater. 20 (2008) 1245-1247.

Xiong, Z.; Yong, C. K.; Wu, G.; Chen, P.; Shaw, W.; Karkamkar, A.; Autrey, T.; Jones, M. O.; Johnson, S. R.; Edwards, P. P.; David, W. I. F.: *High-capacity hydrogen storage in lithium and sodium amidoboranes.* Nature Mat. 7 (2007) 138-141.

Yang, J. B.; Wang, X. J.; Cai, Q.; Yelon, W. B.; James, W. J.: *Crystal and electronic structures of the complex hydride Li4BN3H10.* J. Appl. Phys. 102 (2007) 033507.

Zhao, J.-C.; Cui, J.; Gao, Y.; Kniajansky, S.; Lemmon, J.; Raber, T.; Rijssenbeek, J.; Rubinsztajn, M.; Soloveichik, G.: GE *Progress within the DOE Metal Hydride Center of Excellence for Hydrogen Storage.* DOE Annual Progress Report. (2007) 370-374.

Züttel, A.; Rentsch, S.; Fischer, P.; Wenger, P.; Sudan, P.; Mauron, Ph.; Emmenegger, Ch.: *Hydrogen storage properties of LiBH4.* J. Alloys Compd. 356-357 (2003) 515-520.

Züttel, A.; Borgschulte, A.; Orimo, S.-i.: *Tetrahydroborates as new hydrogen storage materials.* Scr. Mater. 56 (2007) 823-828.

In: Boron Hydrides High Potential Hydrogen Storage Material ISBN:978-1-61668-361-0
Editors: Umit B. Demirci and Philippe Miele

Chapter 7

AMMONIA BORANE: THERMOLYSIS

P. Gislon and P.P. Prosini
ENEA CR Casaccia, Via Anguillarese 301, 00123 S. Maria di Galeria (Rome) Italy
E-mail :paola.gislon@enea.it

ABSTRACT

Ammonia borane Borazane (NH_3BH_3) also called borazane ammonia borane has been considered, since its synthesis in 1955, a promising compound for hydrogen storage due to its remarkably high hydrogen content (19.6 wt%) and volumetric density (146 g(H_2) l^{-1}); moreover it is non-toxic and stable at ambient temperature. A large amount of scientific literature has, therefore, been produced on ammonia borane dehydrogenation processes. Hydrogen can be produced by hydrolysis or thermolysis of ammonia borane. In the present chapter we focus our attention on thermal decomposition of ammonia borane. Some hints are also given about ammonia borane synthesis and regeneration. Upon Heating heatingammonia borane, a structural modification occurs in ammonia borane at about 70 °C before hydrogen evolution. About 2 moles of hydrogen per mole of ammonia borane (15 wt%) are thermally released in the 110-150 °C temperature range through a two-steps exothermic reaction. However, the decomposition rate is extremely low (several hours). Complete decomposition of ammonia borane takes place at temperatures above 500°C. The decomposition reaction steps have been extensively studied by complementary experimental techniques: differential scanning calorimetry (DSC), thermogravimetric analyses (TGA) and differential thermo-analysis (DTA) at various temperature scan rates followed by isothermal desorption, thermogravimetric analyses (TGA), volumetric and spectroscopic methods. The study of thermodynamic and of the structural and thermodynamic modifications occurring at each step and the identification of the corresponding reaction products allowed a detailed description of the involved chemical reactions. Doping and or dispersing ammonia borane into nanostructures have showned beneficial effects on hydrogen ammonia borane desorption rate. Ammonia borane metal composites have also been investigated as high capacity hydrogen storage materials. Some hints are also given about ammonia borane synthesis and regeneration.

1. Introduction

Hydrogen storage is a key-challenge in the route toward hydrogen utilization as an energy carrier. This aspect is particularly critical for hydrogen utilization for transportation, one of the most pollutant and hydrocarbon consuming sectors. Traditional hydrogen storage methods such as high pressure tank and liquefaction suffer for security and energetic concerns, so solid state hydrogen storage has been explored as a possible alternative. Reversible metal hydrides show low gravimetric hydrogen content and relatively high operative pressure and temperature operative conditions; large high power is requested to guarantee the necessary hydrogen charging and discharging rate. Several families of irreversible chemical hydrides, which are irreversible and need to be regenerated off-board, have also been explored; among them amine and ammonia borane or borazane (NH_3BH_3) [1] have received great interest due to their large capabilities, low molecular weight and high hydrogen content. Several reviews on ammonia borane structural and thermodynamic properties and on its potentiality as hydrogen storage material have been published [2-7].

After the first reports by Shore and Parry [1] and Lippert and Lipscomb [8] on the synthetic route and molecular structural geometry of ammonia borane, the hydrogen storage properties of this material has have been studied in terms of its hydrogen storage properties by the group of Geanangel [9-11] first, and by the Freiberg group in Germany (Wolf, Baitalow, Baumann and co-workers) afterwards [12-15]. They analyzed the differential scanning calorimetry (DSC) and the thermogravimetric analyses (TGA) curves in detail by emphasizing the effect of the temperature scan rate and the pressure variation on the dehydrogenation process. Moreover, during each step of the dehydrogenation process, they followed its the evolution of each step of the dehydrogenation process by splitting spacing out the heat releasing events with isothermal measurements. An extensive and systematic analysis of ammonia boraneAB thermolysis has recently been carried out by the Chemical Hydrogen Storage Center of Excellence partnership: the group of Tom Autrey at the Pacific Northwest National Laboratory (PNNL), the group of Larry Sneddon at University of Pennsylvania, the Los Alamos National Laboratory (LANL) and The Universities of Washington, Alabama and Northern Arizona [16-21]. They have studied ammonia borane dehydrogenation theory and engineering, new ammonia borane compounds and spent products regeneration. They also collaborate with researcher groups in New Zealand, Singapore and United Kingdom.

The reactions bringing to ammonia borane dehydrogenation can be split into the following categories:

- Hydrolysis [22-24];
- Solid state ammonia borane thermal decomposition [12,25,26];
- Ammonia borane solution thermal decomposition [11,27];
- Catalyzed Nnear room temperature decomposition of catalyzed ammonia borane solution in diglyme [28], methanol [24] or other suitable solvents [29-33].

Transition metals that have been proven to have large catalytic effect are iridium [30,32], rhodium [34,35], palladium [36], platinum [36], ruthenium complexes [31], nickel [28, 37], and titanium [38]; aqueous acid solutions also efficiently promote hydrogen evolution [39]. Ammonia borane hydrolysis is the subject of a separate chapter and will not therefore be treated here.

Ammonia borane thermal decomposition has been experimentally studied by complementary techniques: DSC and differential thermo-analysis (DTA) to measure the temperature onset of the dehydrogenation events together with the associated thermal energy, TGA, mass spectrometryic (MS) and volumetric measurements for evaluating the release of hydrogen and for quantification of other volatile by-products, X-ray diffraction (XRD) and neutron (ND) diffraction (ND) and, Fourier transform infrared spectroscopy (FTIR) for reaction product identification [12,14,15]. Monitoring the evolution of boron-containing non-volatile intermediates and products by ^{11}B nuclear magnetic resonance (NMR) spectroscopy, it is possible to deeply describe the nature and kinetics of the chemical reactions involved in the hydrogen release process [26,40]. A comparison between experimental data with density functional theory and ab-initio electronic structure calculations allowed to refine the microscopic dynamics and thermodynamics of the reactions [41,42].

DIn the past years, dehydrogenation of ammonia borane by thermal decomposition suffered is affected by many drawbacks, thus limiting the possible use of ammonia borane as hydrogen storage medium. In particular the long induction period at temperatures lower than 100 °C [14], the low kinetics [15], the formation of foam during solid state decomposition in solid state, the volatile by-products as borazine [12] and the very large energetic cost for ammonia borane regeneration [24,43]. Chemical additives [40], ionic liquids [27], metal catalysts [28,30,32,33,36,37] and different supporting media [44] have been demonstrated to be effective in lowering the decomposition temperature, increasing the rate of hydrogen release and eliminating most of the above mentioned limitations.

2. Ammonia Borane

Ammonia borane is the simplest molecular boron-nitrogen-hydride compound. At normal temperature and pressure it is a white solid, with a specific weight of about 0.75 g cm^{-3}. The melting point of ammonia borane is about 110 °C. Due to its large hydrogen content (19.6 wt %) and its hydrogen density larger than liquid hydrogen (140 kg m^{-3} against 70 kg m^{-3}), it has attracted attention as a hydrogen source. Since the B-N unit is isoelectronic with the C-C unit, the NH_xBH_x compound (X=1-4) can be viewed as inorganic analogues of hydrocarbons. However, the NH_xBH_x compounds are solid rather than gases at ambient temperature, thus showing a larger volumetric hydrogen density than the corresponding hydrocarbons.

Three general procedures for ammonia borane synthesis are described in literature [45], which are given hereafter:

- Condensation of diborane (B_2H_6) in organic ethers followed by treatment with gaseous NH_3. This procedure provides ammonia borane in yields varying from 45- to 60 %. Shore [46] synthesized approximately equimolar amounts of diammoniate of diborane $H_2B(NH_3)_2^+BH_4^-$ (DADB) and AB passing a stream of diborane, diluted in nitrogen, into liquid ammonia at -78 °C. Meyer [47] discovered that ethers accelerate the ammonia borane rate of formation and yield; moreover, the obtained product is thermically more stable.
- Metathesis of NH_4X and MBH_4 in organic solvents. This procedure, the most widely reported in literature, provides ammonia borane in yields ranging from 30- to 85 %. Ammonia borane can be synthesized at room temperature by

reacting sodium or lithium borohydride with an ammonium salt, as ammonium chloride, sulfate or carbonate in diethyl ether or THF or Dioxane [1,24,45,48]:
$NaBH_4 + NH_4Cl \rightarrow NH_3BH_3 + NaCl + H_2$ (in diethyl ether) (1)
Ramachandran and Gagare [24] obtained very high yield, up to 96%, finding that a low concentration of the reaction medium is crucial for the purity and yield of the product. Heldebrant et al. [45] could also attain 99% yield of 99% pure ammonia borane.

- Dissolving $H_2B(NH_3)_2^+BH_4^-$ in glyme solvent containing a trace of diborane [49].

The ammonia borane solid state structure has been determined using both XRD [1,8,50] and ND [51]. The molecule adopts a structure like ethane with a tetragonal unit cell at room temperature and an orthorhombic crystal structure at temperatures lower than -48 °C [52,53]. Since the B-H bond is more ionic than the N-H one, the H atoms bond to boron are hydridic and those attached to nitrogen are protonic; thus ammonia borane exhibits a highly polar nature, with a dative N-B bond oriented along the c-axes of the tetragonal cell [13]. The stability of ammonia borane in the solid state relies on the strong molecular polarity and on the dihydrogen bonds between the boron hydrides and the amine protons of adjacent ammonia borane molecules. The distance between the hydridic hydrogen atoms on boron and the protic hydrogen atoms on nitrogen of adjacent molecule (2.02 Å) is shorter than the Van der Waals distance interaction (2.4 Å) constituting a dihydrogen bond. The B-N distance is 1.58 Å, the B-H and N-H distances are 1.15 and 0.96 Å, respectively. The nature of the bonds between nitrogen, boron and hydrogen atoms in ammonia borane and in every N-H-B compound obtained during ammonia borane thermal dissociation, together with the thermodynamic properties of such compounds, have been calculated with density functional theory calculation by Miranda and Ceder [41].

3. Solid Neat Ammonia Borane Thermolysis

The solid ammonia borane decomposition can be attained through an exothermic three-step reaction. Among the three steps, only the first two, occurring in the 70-150 °C temperature range, have been extensively studied as the last one takes place at temperatures too high for practical use. A typical DSC curve of solid ammonia borane in the 70-190 °C temperature range shows, after an endothermic event around 100°C ascribed to the melting of ammonia borane, two exothermic events, peaked at 110 °C and 150 °C [14]. The dehydrogenation process is very slow, so the temperature scan rates need to be kept under 1 °C min^{-1} to correctly evaluate the energy release; the involved chemical pathways, kinetics and products were found to be deeply influenced by the reaction environment. In the following the three dehydrogenation steps and the preparatory phases leading to hydrogen release, the so- called induction period, are examined.

3.1 The Induction Period

The isothermal kinetics of the dehydrogenation process at $T < T_{melting}$ can adequately be fitted by a sigmoidal curve, evidencing therefore an induction, nucleation and growth mechanism [14]. The first hydrogen equivalent can be released at temperatures as low as 70°C, but after a long induction period of 24 h. A detailed insight by XRD, solid-state NMR and spectroscopic analyses [26,40] clarified that during this induction period a structural modification occurs, associated to heat flow but without any hydrogen release. The optical micrograph images of an ammonia borane single crystal heated at 90°C, taken at different times of an ammonia borane single crystal heated at 90 °C, (Figure 1 from [40]) show a phase boundary moving along the crystal. This new phase is confirmed by the appearance of new peaks in NMR and Raman spectra.

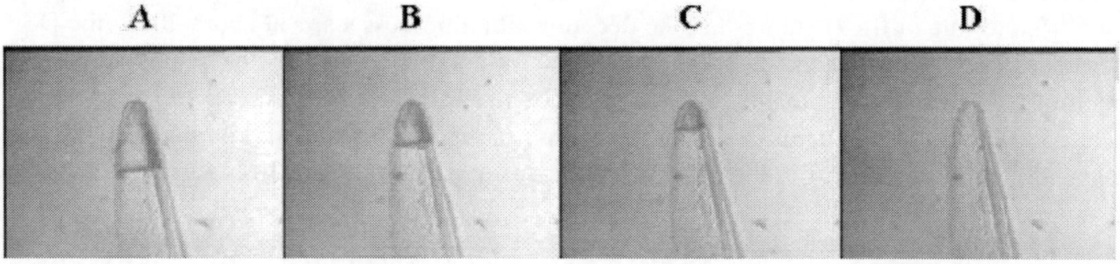

Figure 1. Optical micrograph images of a single crystal of NH3BH3 heated to 90 °C. Reprinted with permission from [40]. Copyright 2009 American Chemical Society

The XRD pattern evidenced the modification by a loss in the crystalline symmetry, associated to the disruption of the dihydrogen bonds. This new phase is amorphous and is more mobile., As seen in the ^{11}B-NMR spectra as heating disrupts the network of hydrogen bonding, permitting greater motion of the molecules, as seen in the ^{11}B-NMR spectra [26,40]. The evolution into a new phase is the reason for the long induction period observed before dehydrogenation starting: if the sample is quickly cooled to room temperature and reheated, the successive induction period is shortened. ^{11}B-NMR experiments carried out at 88 °C [40] revealed, after a 14 min delay from the appearance of the new phase peaks, the appearance of other resonance peaks. These new peaks can be assigned to the -BH_2 cation and BH_4 anion of $H_2B(NH_3)_2^+BH_4^-$; the induction phase can therefore be described by a loss of the alignment along the c-axis of the B-N bond, its breaking and the following isomerization of some ammonia borane molecules in $H_2B(NH_3)_2^+BH_4^-$. A combined pressure composition isotherm (PCI) and ^{11}B-NMR measurement revealed that during the induction period very little hydrogen evolves, even at times at which no ^{11}B-NMR resonance of the starting material is visible any more. During the nucleation phase NMR measurements identified, other than ammonia borane, the presence of $H_2B(NH_3)_2^+BH_4^-$, together with linear and cyclic dimers. The $H_2B(NH_3)_2^+BH_4^-$ is responsible for the reaction nucleation, as evidenced by the H_2 appearance. The release of molecular hydrogen is made easier by the $H_2B(NH_3)_2^+BH_4^-$ formation, as the N-H and/or B-H bonds of $H_2B(NH_3)_2^+BH_4^-$ are more reactive than the corresponding N-H and B-H bonds of ammonia borane. Residual ammonia borane reacts with $H_2B(NH_3)_2^+BH_4^-$ easier than with ammonia borane, as a lower induction period was found for ammonia borane added with $H_2B(NH_3)_2^+BH_4^-$ [26]. It is noticeable that $H_2B(NH_3)_2^+BH_4^-$ formation has been detected also in ammonia borane dehydrogenation when it was in solution both in liquid ammonia and in ionic liquid [27]. The hydrogen release is obtained therefore from a bimolecular reaction between $H_2B(NH_3)_2^+BH_4^-$and ammonia borane.

3.2 The First Reaction Step

The first stage of ammonia borane decomposition starts with a 3-dimensionally growth nucleation [25]. Figure 2 (from ref. [14]) shows two DSC curves obtained at two different temperature scan rates, namely 0.05 ° C min^{-1} and 1 °C min^{-1}. It is well evident how the temperature scan rate influences the result. At temperature rates lower than 1 °C min^{-1} an isothermal decomposition at 90 °C needs 10 h to complete. A DSC measurement at 0.05 °C min^{-1} (solid line in fig.2) shows that the decomposition process starts at about 70°C; the DSC curve reaches a maximum at about 90°C and is completed before the ammonia borane melting point. If a larger temperature scan rate is used, the ammonia borane decomposition, started at 70 °C, is not completed when temperature reaches the melting point at 100 °C, and the endothermic event due to the melting can be observed (dashed line in Figure 2).

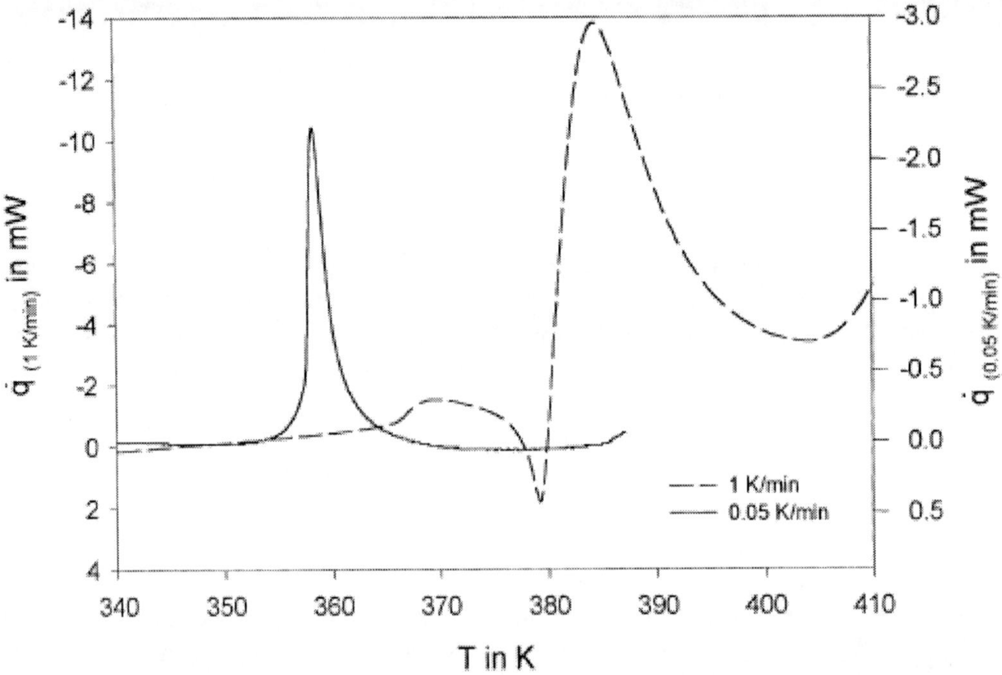

Figure 2. DSC curves of thermal decomposition of BH_3NH_3: (solid line) heating rate 0.05 °C min^{-1}; (dashed line) heating rate 1 °C min^{-1}. Reprinted from [14]. Copyright 2009 with permission from Elsevier

The desorption temperature onset change with temperature scan rate has been explained theoretically by Miranda et al. [41] with the fact that the temperature at which hydrogen is released depends exclusively on the reaction kinetic. XRD allowed to identify the solid product of the first reaction step as polymeric aminoborane, $(BH_2NH_2)_n$. The heat released during this first step has been measured by DSC and resulted to be about -21 kJ mol^{-1} [14]. Helmotz free energy, entropy and enthalpy of reactant and reaction products have been calculated by Miranda and Ceder [41] by first principles calculation within density functional theory in the temperature range up to 320 °C and resulted to be negative in agreement with the exothermic nature of the reaction, as seen by DSC measurements. The activation energy of the first decomposition step has been calculated to be about 117 kJ mol^{-1}. The calculated enthalpy difference was lower than measured, about -5 kJ mol^{-1} [54].

Thermogravimetric analysis associated with DSC revealed an heat release contemporary to the weight loss. In order to identify the gaseous decomposition products, even immediately after their formation, thermo-gravimetric analysis has been associated to FTIR and mass spectrometry [14]. In such a way it was possible to identify also unstable products. Hydrogen is the main evolved gas. Volumetric measurements revealed that 1.1 H_2 mol/mol ammonia borane, corresponding to 7.6 wt%, are released during the decomposition, in good agreement with the formation of $(BH_2NH_2)_n$. The weight loss is slightly larger than what can be expected by the only hydrogen mass; FTIR analysis enabled to recognize traces of monomeric aminoborane and borazine $(BHNH)_3$, not detectable by a volumetric measurement of evolved gas as they condense at room temperature [12]. The borazine formation has been detected after that of monomeric aminoborane [14].

Based on the evolved hydrogen amount and on the presence of $(BH_2NH_2)_n$ as reaction solid product, the reaction was initially described as the following:

$$x\,BH_3NH_3\,(s) \rightarrow (BH_2NH_2)_x\,(s) + x\,H_2\,(g) \tag{2}$$

Recently Heldebrandt et al. [40] analyzing ^{11}B NMR data of ammonia borane heated at 85 °C for 600 min found resonance of boron bonds having reference to $BH(N_2)$, $BH_2(N_2)$, BH_3N and BH_4, indicating that more complex pathways concur to hydrogen release below 100 °C, through intermolecular cross-linking:

$$2\,(NH_2BH_2)_n \rightarrow (NH_2BH\text{-}NHBH_2)_n + n\,H_2\,(g) \tag{3}$$

Stowe et al. [26] deduced from the broadening of the spectral features that going on the reaction, the symmetry around the boron centre decreases. Cross-linking of polyiminoborane for temperature less than 150°C leads to cyclic poly-aminoborane species and ionic polymeric $H_2B(NH_3)_2^+BH_4^-$-like form, as for example $[NH_3BH_2NH_2BH_2NH_2^+][BH_4^-]$, with hydrogen release [26,40]:

$$H_2B(NH_3)_2^+BH_4^- + NH_3BH_3 \rightarrow [NH_3BH_2NH_2BH_2NH_2^+][BH_4^-] + H_2(g) \tag{4}$$

The $H_2B(NH_3)_2^+BH_4^-$-like polymers react again with ammonia borane forming a longer $H_2B(NH_3)_2^+BH_4^-$-like chain. T; the successive steps of the process are sketched in Figure 3.a while in Figure 3.b the starting and final structures are shown.

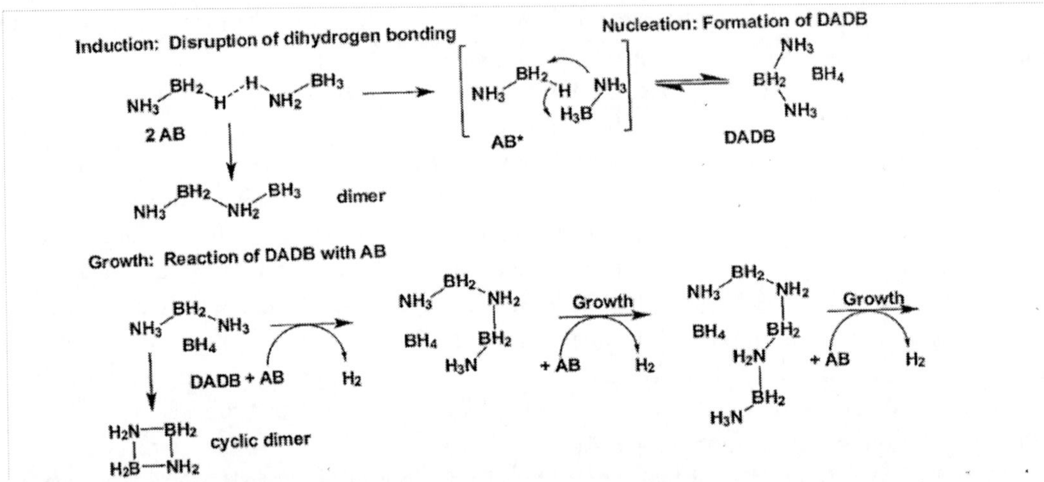

Figure 3.a. The proposed thermal dehydrogenation mechanism of ammonia borane showing discrete induction, nucleation and growth steps leading to hydrogen release. Reproduced from [26] by permission of the PCCP Owner Societies

The DFT (density functional theory)-GIAO (gauge-included atomic orbitals) ^{11}B chemical shift calculations on structures that can result from ammonia borane dehydropolymerization are consistent with the resonance peaks of the BH_2 groups of both cycloborazanes and polyaminoborane, of N-BH-N borons resulting from chain-branched polymeric structures and of polymer-terminating BH_3 groups [11,27]. Both intermolecular and intramolecular reactions are involved during the release of the first equivalent of hydrogen. The first dissociation step is completed at 110 °C.

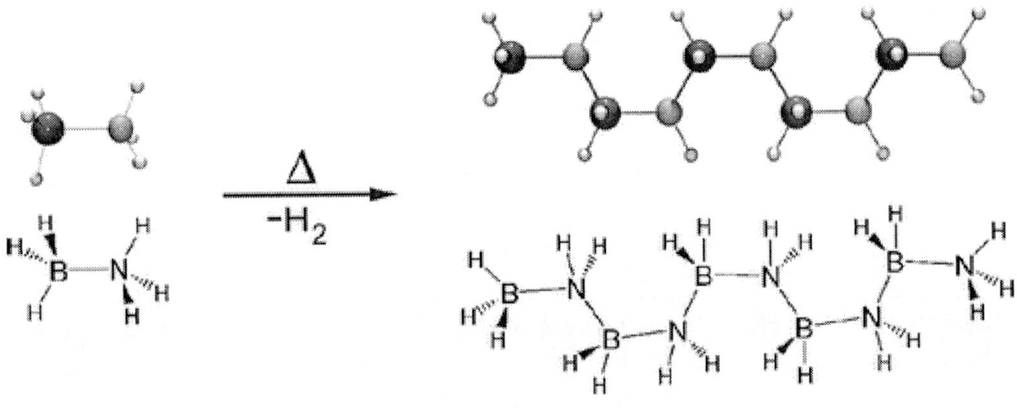

Figure 3.b. A possible structure arising from the dehydropolymerization of ammonia borane. Reprinted with permission from [27]. Copyright 2009 American Chemical Society

3.3 The Second Reaction Step

The second ammonia borane dehydrogenation step, namely the dehydrogenation of $(BH_2NH_2)_n$, formed by ammonia borane decomposition, was studied separately from the first step by Baumann et al. [15]. $(BH_2NH_2)_n$ is a white amorphous solid stable at room temperature. DTA recorded an exothermic event starting at 125 °C, and occurring over a broad temperature range. The reaction enthalpy heavily depends on heating rate, decreasing with increasing heating rate, from -15.4 kJ mol^{-1} at 10 K min^{-1} to -23.9 kJ mol^{-1} at 0.1 K min^{-1}. The released hydrogen is 1.1 mol/mol ammonia borane (7.6 wt%), independent from temperature scan rate. The total weight loss is greater than the hydrogen mass weight, it can reach 20 wt% at 10 K min^{-1} temperature scan rate; this extra-loss increases significantly with increasing temperature scan rate, indicating that different heating rates drive the reaction through different reaction paths, each one characterized by a different reaction heat, in accordance with the found different enthalpy values. The solid residue after the thermal decomposition of $(BH_2NH_2)_n$ is poly-iminoborane such as $(BNH_y)_x$ with $1.2 < y < 2.4$ [15]. It is an amorphous white powder. The composition of the solid product strongly varies in dependence of the decomposition conditions: higher y values have been found at shorter decomposition times. This indicates that the poly-iminoborane structure evolves towards a net-shaped one, with the formation of additional B-N linkages. The net-shaped structure was confirmed by the FTIR spectrum of the solid residue, in which peaks that cannot be attributed only to BH-, NH- and BN- bond were found [15].

Associated TGA, MS and FTIR measurements detected borazine, $(B3HN3H)_3 6$, monomeric aminoborane H_2BNH_2 and small amounts of diborane (B_2H_6) as volatile reaction products. The BH_2NH_2, expected during the first step at about 100°C, should not appear over 130 °C. It is though that it is formed from BHNH evolution through two competitive reactions: trimerization to borazine and hydrogenation to (BH_2NH_2) [12]. The second dehydrogenation step can therefore be described by:

$$(BH_2NH_2)_x\ (s) \rightarrow (BHNH)_x\ (s) + x\ H_2\ (g)(\qquad (5)$$

Calculations from ab initio electronic structure theory foreseen for the two-step reaction pathway $x(BH_3NH_3) \rightarrow (BH_2NH_2)_x \rightarrow (BHNH)_x$ a total released heat of 40 kJ mol^{-1} [42], in agreement with the experimental value measured for the faster decomposition. If a sufficiently low temperature rate is used (0.15 °C min^{-1}), the second DSC peak completely runs out at 160°C.

The effect of hydrogen pressure over the reaction has been checked by Baitalow et al. [13] with a transitiometer. They did not note any influence on reaction kinetics and on product formation up to 600 bar hydrogen pressure.

3.4 The Third Reaction Step

A complete dissociation, with boron nitride BN formation, occurs only above 500 °C, following:

$$(BHNH)_x\ (s) \rightarrow x\ BN\ (s) + x\ H_2\ (g)(\qquad (6)$$

The high stability of boron nitride and the extremely high dissociation temperature inhibit the use of this last step for practical use [9].

4. Thermolysis of Neat Ammonia Borane in Solution

Solutions of ammonia borane have been studied as in principle they could drive ammonia borane thermolysis in a more controllable way. Different volatile organic solvents have been tested. Wang and Geanangel [11] studied the thermolysis of ammonia borane when it was diluted in solvents with boiling point higher than 80 °C: acetonitrile, glyme, diglyme, THF, 2-methylTHF and pyridine were found able to be used for ammonia boraneAB dehydrogenation. The obtained reaction products, both solid and volatile, were identified by ^{11}B-NMR spectroscopy.

In etheral solvents, like glyme, diglyme, tetrahydrofuran, and 2-methyl tetrahydrofuran, ammonia borane decomposition leads to the formation of products suggesting hydrogen-loss. No NH_2BH_2 resonance peak was detected, but its oligomers were observed: cyclotriborazane $(NH_2BH_2)_3$ and/or $(NH_2BH_2)_2$. This can be explained by the fact that NH_2BH_2 molecule is unstable at room temperature and prefers to exist as cyclic oligomers or polymers rather than single molecule [55]. Borazine was also revealed, later than $(NH_2BH_2)_x$. The relative concentrations of ammonia borane, $(NH_2BH_2)_x$ and borazine change with time in such a way that a sequential occurrence of reaction (3) and (5) is likely to happen. Ammonia borane dehydrogenation temperature strongly depends on the kind of ether used. Ammonia borane in other solvent exhibits a really different decomposition pathway: thermolysis of ammonia borane in acetonitrile gives as product ethylamine-borane, indicating that the hydrogen loss of ammonia borane occurs with an hydroboration or hydrogenation of solvent. In any case at the reflux temperature of the used solvent a very low hydrogen evolution was observed.

Bluhm et al. [27] obtained interesting results on hydrogen yield and releasing rate when ammonia borane was diluted (1:1) into the 1-butyl-3-methylmidazolium chloride (bmimCl) ionic liquid. A comparison of ammonia borane dehydrogenation in the solid state versus in bmimCl solvent solution is summarized in Figure 4.

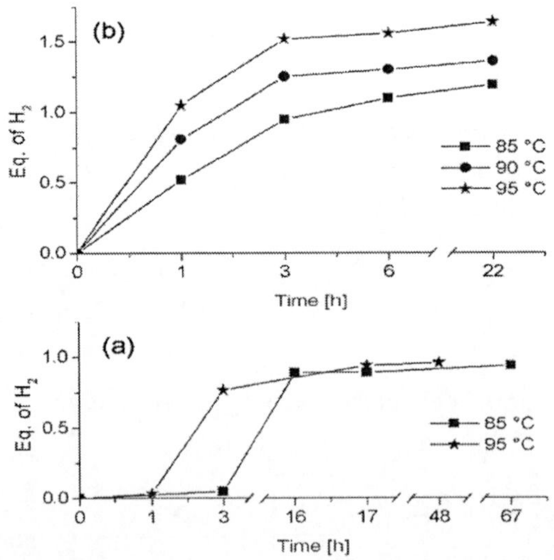

Figure 4. Summary of H_2 release at different temperatures (a) from solid ammonia borane and (b) from equal weight mixtures of bmimCl and ammonia borane. Reprinted with permission from [27]. Copyright 2009 American Chemical Society

Differently from the solid-state dehydrogenation, the dehydrogenation of ammonia borane in bmimCl solution shows no induction period and both the total hydrogen released and its releasing rate are higher. At 95 °C dehydrogenation of solid AB gave 0.9 equiv. of H_2 after 48 h, against 1.6 equiv. of H_2 after only 22 h from the ammonia borane solution. This last H_2 amount corresponds to 5.4 wt% over the weight of AB + bmimCl. ^{11}B NMR spectra of the reaction products showed the presence of $H_2B(NH_3)_2^+BH_4^-$ after the first hour at 85 °C suggesting that the activating effect of the ionic liquid may be related to its ability to induce the $H_2B(NH_3)_2^+BH_4^-$ formation. The larger extent of dehydrogenation has been evidenced by the total disappearing of the polyaminoborane at longer times, resulting in the production of unsaturated residues, containing unsaturated polyborazylene-like structures [18].

Larger H_2 loss rate, namely 2.3 equiv. in 15 min, corresponding to 10.2 mat-wt% from a 80:20 wt% mixture of AB/BmimCl at 120 °C have been obtained at University of Pennsylvania [18]. At the same uUniversity, other ionic liquids have been tested and they demonstrated to have similar improving effects in ammonia borane dehydrogenation. The best performances were obtained with edmin-EtSO_4 and bmim-BF_4: 2.1equiv. (7 wt%) hydrogen released at 85 °C in 22 h [17]. Sneddon reported also that also strong nitrogen bases such as Proton Sponge, react with ammonia borane in ionic liquids producing the H_2NBH_3- anion and favoring the H_2 release [17].

5. Catalyzed Ammonia Borane Thermolysis

5.1 Catalyzed Solid State Ammonia Borane

The observation that $H_2B(NH_3)_2^+BH_4^-$ catalyzes the hydrogen release suggested Heldebrant et al. [40], to dope ammonia borane with $H_2B(NH_3)_2^+BH_4^-$, NH_4Cl and $NaBH_4$. They noted by isothermal gas chromatography that both $H_2B(NH_3)_2^+BH_4^-$ (5 wt%) and NH_4Cl (5 wt%) significantly reduce the induction period (Figure 5) and argued that the $[(NH_3)_2BH_2]^+$ cation, present in both compounds, could promote the dehydrogenation reaction.

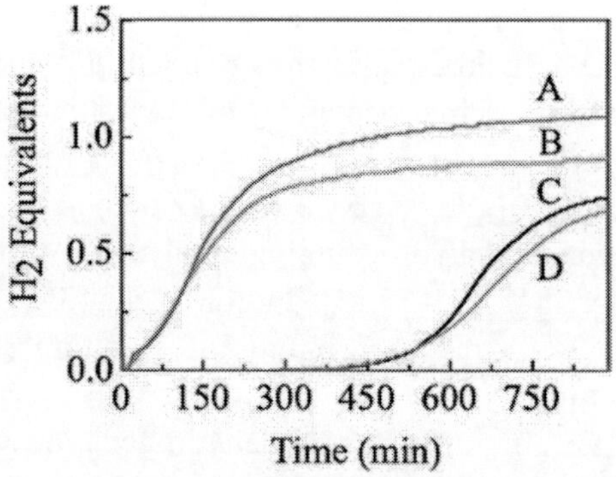

Figure 5. Effect of additives on the induction period for H_2 release from solid ammonia borane at 80 °C. (A) NH_3BH_3:NH_4Cl (5 wt%), (B) NH_3BH_3:DADB (5 wt%), (C) NH_3BH_3, (D) NH_3BH_3:$NaBH_4$ (5 wt%). Reprinted with permission from [40]. Copyright 2009 American Chemical Society

An increment of hydrogen release kinetics with respect to pure ammonia borane has been observed in ammonia borane doped with platinum [56]. A mixture of ammonia borane and hydrogen hexachloroplatinate (IV) hydrate were was milled in a hydrogen atmosphere in 100:1 and 100:2 mol ratio. The doped samples exhibit a shorter induction period with respect to undoped sample and a lower dehydrogenation temperature onset: the 1 mol% Pt doped sample showed some hydrogen evolution at 73 °C after 8 min, whereas the undoped sample started the dehydrogenation process at 82 °C only after 67 min. The 2 mol% Pt doped sample shows even faster dehydration kinetics.

Very recently, Cheng et al. [37] proposed a new kind of catalyst kind, nest-like $Ni_{1-x}Pt_x$ (x=0.03, 0.06, 0.09 and 0.12) submicrometer-sized hollow spheres, which exhibit favorable catalytic activities for both the hydrolysis and the thermolysis of ammonia borane. The NiPt catalyst offers advantages as it acts as well as pure platinum but it is economically convenient; moreover it can be recovered and re-used. TG and DTG curves of $Ni_{0.88}Pt_{0.12}$ catalyzed (1.8 wt% Pt loading) ammonia borane shows a large mass loss of 24.8 wt % in a narrow temperature range centered at 100 °C, and no further mass loss at higher temperatures (up to 300 °C). FTIR spectra of thermally decomposed $Ni_{0.88}Pt_{0.12}$ catalyzed ammonia borane show that the B-H modes and the N-H bending mode typical of unheated ammonia borane, and in lower intensity of thermally decomposed neat ammonia borane, have almost disappeared. That result has been interpreted by Cheng et al. [37] supposing that the catalyzed thermolysis of ammonia borane leads to the formation of $(NBH_x)_n$ at 120 °C, along with the release of H_2 and other volatile boric compounds. This assumption is supported by a inductively coupled plasma emission spectroscopic analysis of the obtained product, indicating an elemental composition of B:N:H=1:1:1.4.

Decreasing of decomposition temperature in the first dehydratation step has also been observed by Helary et al. [57] by adding ammonia borane with ammonium nitrate (NH_4NO_3/ammonia borane 10wt%).

5.2 Catalyzed Ammonia Borane in Solution

The dDehydrogenation rate of ammonia borane in solution is so low that several groups looked for catalysts able to promote the ammonia borane dehydrogenation at lower temperature, speeding up the dehydrocoupling reaction while keeping a large hydrogen yield.

Jaska et al. [31] reported on dehydrocoupling of ammonia borane (and ammonia borane adducts) solution in diglyme and tetraglyme at 45 °C catalyzed by rhodium complexes (0.6mol% of $[Rh(1,5\text{-cod}(\mu\text{-Cl})]_2$, at very low rate and yield; the decomposition gave borazine. Amine-borane adducts dehydrocoupling catalyzed by rhodium and other transition metals has been also explored [31,33-35,38]. Fulton et al. [35] studied the structure of the rhodium acting as catalyst by X-ray absorption fine structure (EXAFS) and concluded it converts into a cluster structure.

An efficient catalysis for ammonia borane dehydrogenation was obtained by Denney et al. at University of Washington [30] by adding 1 mol% of the iridium pincer complex $(POCOP)Ir(H)_2$ into a solution (0.5 M) of ammonia borane in THF. One equivalent of hydrogen was released in 4 min.

Keaton et al. [28] found a 20-fold increase in dehydrogenation rate, with respect to non-catalyzed ammonia borane, for a solution (25 wt%) of ammonia borane in diglyme catalyzed with a N-heterocyclic carbene Ni complex (Ni-NHC) at 60 °C.

Recently Shresta et al. [36] at LANL checked the heterogeneous catalytic activity of supported precious metals to dehydrogenate 2-methoxyethyl ether ammonia borane solution (0.33 M) at 70 °C. They found that platinum has a higher catalytic activity compared to ruthenium and palladium; with 2 mol% platinum/ammonia borane, 1.6 equiv. of H_2 were released at an average rate of 0.04 g s^{-1} kg_{NH3BH3}^{-1} at 70 °C. At 110 °C, the Pt catalyst ($Pt:NH_3BH_3$ 1%) enhances the dehydrogenation rate with respect to pure ammonia borane solution: more than 2.1 mol equiv. of H_2 were released in 120 min.

The product of the catalyzed reaction is thought to be polyborazylene. The Pt activity decreases with cycling, but efficient catalyst regeneration was achieved by heating Pt at 150 °C for 2-4 h.

The ammonia boraneAB dehydrocoupling can also be catalyzed by small amounts of strong Lewis or Brønsted acids. Stephens et al. [39] at LANL and University of Alabama added ammonia borane with different acids at different loading and monitored the released hydrogen amount and releasing rate at 60 °C. They found that cyclic products arise from the reaction, similarly to what happen with ammonia borane in ionic liquid. Table 1 summarizes their results.

Table 1. Reactivity of Ammonia-Borane with acids.

Acid	Loading [mol%]	Conc. of 1 [M]	T [°C]	t [h]	H_2 [equiv]
$B(C_6F_5)_3$	25	0.14[a]	60	24	0.6
$B(C_6F_5)_3$	0.5	2.6[b]	60	20	1.1
$HOSO_2CF_3$	25	0.13[a]	60	18	0.8
$HOSO_2CF_3$	0.5	6.2[c]	60	18	1.3
HCl	0.5	2.9[c]	60	20	1.2

[a] Reaction in glyme. [b] Reaction in tetraglyme. [c] Reaction in diglyme.

6. Ammonia Borane Compounds Thermolysis

6.1 Solid State Ammonia Borane Compounds Thermolysis

Substituting one H atom in the NH_3 group of ammonia borane by one alkali metal atom modifies the ammonia borane molecular electronic energy levels and polarity, the intramolecular distances and the intermolecular binding energies, the dihydrogen bonding; consequently the dehydrogenation enthalpy reaction enthalpy should change. Following this idea, Xiong et al. [58] tested the effect of Li and Na atom substitution. Li-ammonia borane and Na-ammonia borane were prepared by ball milling 1:1 molar ratio of ammonia borane and the corresponding alkali-metal hydrides. The formation of $LiNH_2BH_3$ and $NaNH_2BH_3$ was supposed by hydrogen evolution during ball milling and confirmed by high resolution X-ray powder diffraction data. The B-N binding in alkali amidoborane is stronger than in ammonia borane, and the molecule assumes a $M^+(NH_2BH_3)^-$ ionic character. The hydrogen release occurs in only one step, differently from ammonia borane. It starts at 92 °C for $LiNH_2BH_3$ and slightly earlier for $NaNH_2BH_3$; the obtained hydrogen yield is 10.9 wt% and 7.5 wt%, respectively, corresponding to 2.0 moles H_2. The DSC and MS evolution for hydrogen and borazine measurement results for Li-ammonia borane are shown in Figure 6 together with the pure ammonia borane traces. It is evident that Li-ammonia borane has faster decomposition kinetics in comparison to pure ammonia borane. From DSC measurements the heat released was calculated to be -3 kJ mol^{-1} and -5 kJ mol^{-1} for $LiNH_2BH_3$ and $NaNH_2BH_3$, respectively; both greatly lower than that ofthe -20 kJ mol^{-1} measured for pure ammonia borane dehydrogenation. Differently from ammonia borane dehydrogenation, solid LiNBH and NaNBH are the only reaction products; no other gases except hydrogen evolve during thermolysis. In particular, no borazine has been detected.

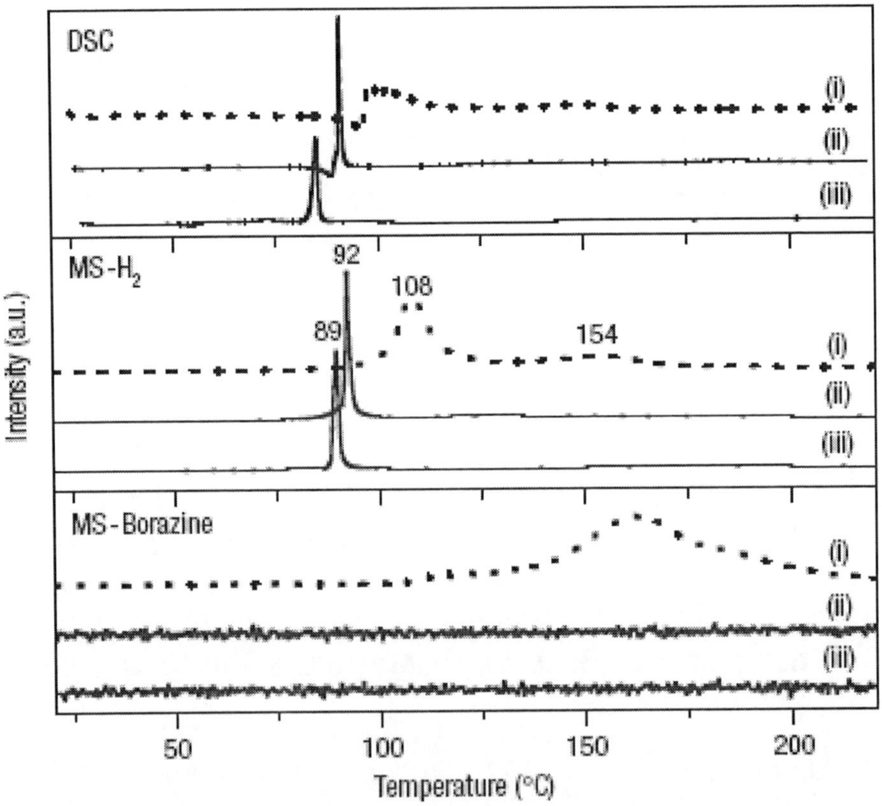

Figure 6. (i) Post-milled ammonia borane; (ii) Li amidoborane sample; (iii) Na amidoborane sample. MS: mass spectrometry. Reprinted by permission from Macmillan Publishers Ltd: Nature Materials [58] copyright (2009)

Diyabalange et al. at Los Alamos National Lab [59] studied calcium amidotrihydroborate $Ca(NH_2BH_3)_2$ in which a new covalent bond between the nitrogen atom of ammonia borane and calcium has been formed. They found that $Ca(NH_2BH_3)_2$ dehydrogenation occurs without any foam production, as it happens in ammonia borane decomposition. It is endothermic rather than exothermic as in ammonia borane, and H_2 is spread over much wider temperature and time ranges. Gas chromatography experiments revealed that 1.1 equiv. of H_2 are released from $Ca(NH_2BH_3)_2$ at 120 °C, 2.4 equiv. at 150 °C and 3.6 equiv. at 170 °C. Four equivalents, corresponding to 7.2 wt%, can be expected from the combination of the four protic ($-NH_2-$) and four of the six hydridic ($-BH_3$) hydrogen atoms.

Differently from ammonia borane dehydrogenation, both the rate and volume of H_2 vary with temperature during $Ca(NH_2BH_3)_2$ dehydrogenation. After the first H_2 equivalent loss the product was identified as $Ca(NBH)_2$ while after the second step at 170 °C $CaN_2B_2H_2C_{0.2}$ was found as residual after the second step.

Bowden et al. [60] studied the thermal decomposition of methylamine borane, $CH_3NH_2BH_3$, an ammonia borane compound whose structure is similar to the ammonia borane one. They observed a large mass loss of at around 100-130 °C, the onset temperature and mass loss being dependent on temperature scan rate. DSC curve shows an exothermic peak, attributed to the hydrogen release event, followed by an endothermic one. The exothermic reaction enthalpy is -25 kJ mol^{-1}, which is similar to the first dehydrogenation step for of ammonia borane. The endothermic peak was attributed to product volatilization, coherent with a mass loss significantly higher than what could be ascribed to hydrogen evolution. A second hydrogen release step is observed at about 190 °C. Although some possibilities have been outlined, no unique and clear evidence of the reaction mechanisms involved in $CH_3BH_2BH_3$ dehydrogenation has been recognized.

Very recently, Graham et al. [61] synthesized a new hybrid, liquid, material by direct reaction of lithium amide with ammonia borane in 1:1 molar ratio. A small amount of H_2 (0.1mol H_2/mol ammonia borane) was recorded detected during the hybrid formation. By heating the compound, two hydrogen evolution steps have been recorded: a first centered at 60 °C, and a second, less intense, spread in the 120-250 °C temperature range. Ammonia was also released in the 60-120 °C temperature range. The total H_2 evolved up to 250 °C was about 3.2 mol H_2 per mol of $LiNH_2+NH_3BH_3$, corresponding to 11.9 wt%. At 85 °C a $NH_3BH_3+LiNH_2$(10 mol%) mixture was able to give 9.28 wt% H_2 in 15 h [16]. By ^{11}B-NMR, isotopic mass spectrometry and gas volume measurement results Graham et al. argued the following dehydrogenation pathway: the reaction should start with the formation of a new B-N bond between ammonia borane and $LiNH_2$ and the release of molecular H_2, thus forming the intermediate $LiNHBH_2NH_3$. Further hydrogen release and B-N bond formation leads to a polymeric $[-LiN-BHNH_3-]_n$ compound; the third equivalent of molecular hydrogen released arises from reaction between the remaining B-H and the $-NH_3$ group [61].

Ammonia triborane, $NH_3B_3H_7$, is also considered a viable complex for chemical hydrogen storage with its potential 17.7 wt% hydrogen [62]. The solid state and ionic liquid thermolysis of a mixture of ammonia borane and $NH_3B_3H_7$ is under study [16,17].

Additives have eliminated the formation of foam during ammonia borane dehydrogenation, keeping the pellets of spent material keepingin their original shape [21].

7. Supported Ammonia Borane Thermolysis

Nanostructured scaffolds have been demonstrated to efficiently act in changing the thermodynamics of different hydrateshydrides, making hydrogen release easier: in addition, facilitating the formation of nanostructured hydrides, the scaffold inhibits sintering and agglomeration of the nanostructured hydride during hydrogen cycling, effects that would compromise the sorption kinetics. Improved reaction rates have been shown for sodium alanate $NaAlH_4$ in carbon aerogels, and $NaAlH_4$ supported on carbon nanofibres [63,64].

Looking for an efficient way to increase dehydrogenation kinetics and contemporaneously to decrease the ammonia borane desorption temperature, Autrey et al. [44] at PNNL found that H_2 evolution increases when ammonia borane is absorbed in the porous channels of mesoporous silica structures, namely SBA-15 or MCM-41. The reaction enthalpy considerably reduces to 1.3 kJ mol^{-1} for SBA-15 supported ammonia borane. Also the activation energy decreases to 50 kJ mol^{-1}. Quasi-elastic neutron scattering measurements on isotopically enriched ^{11}B ammonia borane revealed that energy barriers for NH_3 protons in supported ammonia borane are lower than for bulk ammonia borane [54].

Lighter porous supports other than the heavy silica have also been tested by the same group. Looking for a better thermal exchange between ammonia borane and the supporting media, they tested coherent carbon cryogel (CC) [65]. Ammonia borane (24 wt%) confined into the pores of carbon cryogel nanocomposite, showed a unique dehydrogenation event, starting at 60 °C and peaked at 90 °C, a temperature remarkably lower than the 110 °C of the neat ammonia borane (Figure 7). The DTA peak is broader than for neat ammonia borane one, about 20 °C. The reaction, yielding 9 wt% of hydrogen, is greatly more exothermic (120 kJ mol^{-1} against 21 kJ mol^{-1} for neat ammonia borane).

Moreover, no borazine was found in the reaction products; this last feature is of primary importance if the released hydrogen is intended to be used to power a polymer electrolyte membrane fuel cell, due to its the poisoning effect of borazine on the cell membrane. Possible explanations that have been attempted for the ammonia borane-in-carbon cryogel thermal dehydrogenation behavior are related to the ammonia borane nanosize, supported by the evidenced correlation between the pore size and the onset temperature, and to the catalytic effect of the inner carbon surface dangling bonds. A spectroscopic study by ^{11}B NMR allowed to follow the decomposition pathway; it and to detected some decomposition products during the incorporation of ammonia borane into the carbon cryogel matrix at room temperature, thus explaining the observed detrimental in hydrogen yield, from 13 wt% to 9 wt%. The absence of borazine as reaction product has been ascribed to its polymerization into the cryogel pores, as the polyborazylene presence could suggest. Ammonia borane dehydrogenation in carbon cryogel seems to involve different pathways, as after the thermal reaction the carbon cryogel-ammonia borane NMR spectrum after the thermal reaction is different from the solid-state ammonia borane product spectrum. The enhancement in the dehydrogenation kinetics could be due to different contributing factors: hydroxyl groups and/or defects on the carbon cryogel surface, limitation of ammonia borane diffusion and disruption of the Ammonia ammonia borane dihydrogen bonds by surface interactions.

A very recent paper from Neiner et al. [67] reports on kinetics improvement by ammonia borane ball milling with hexagonal boron nitride (nano-BN). The resulting nanostructured mixture, acting both as support and as catalyst, has the advantage to not introduce elements other than boron and nitrogen, making the recovering procedure easier. Increasing nano-BN content, the first dehydrogenation step onset shifts from 114 °C for neat ammonia borane to 70.6 °C for a nano-BN/ammonia borane ratio equal to 4:1. Also the second step occurs at lower temperature with respect to neat ammonia borane, 120.5 °C vs. 150 °C. The heat released during the reaction is also reduced to 2.2 kJ mol^{-1}. The ammonia production after heating at 150 °C is ten times lower for ammonia borane + nano-BN than for neat ammonia borane. These largely positive effects of the nano-BN over ammonia borane dehydrogenation features are somewhat compensated by a partial decrease of the hydrogen yield, from 2.6 equiv. for neat ammonia borane to 1.9 equiv. for of the mixture, and by a larger borazine production, after heating at 150 °C.

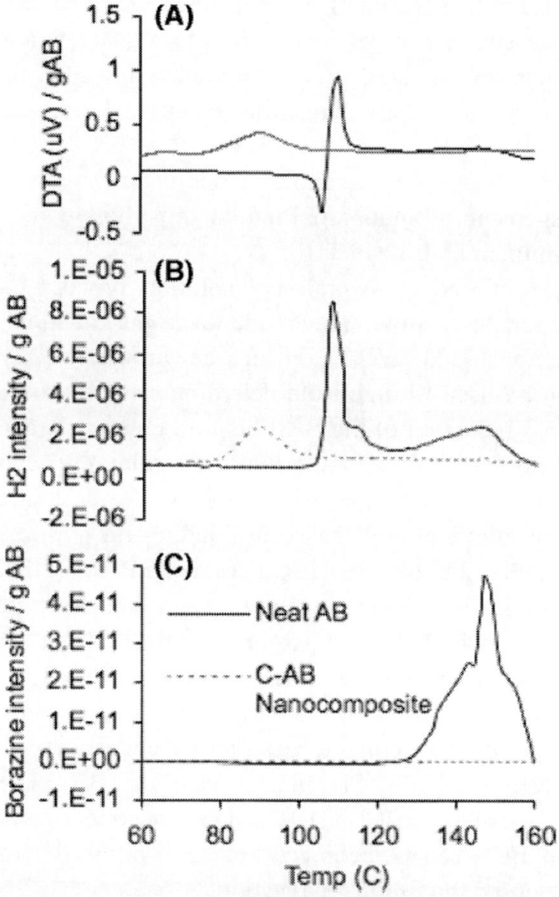

Figure 7. (A) DTA results of the C-NH_3BH_3 nanocomposite and neat ammonia borane with a heating rate of 1 °C min^{-1}; (B) H2 MS readings showing the H_2 release at 90 °C in a C-NH_3BH_3 nanocomposite and at 110 and 150 °C in neat ammonia borane; and (C) the borazine MS readings showing no release of borazine from the C-NH_3BH_3 nanocomposite [66]. Reprinted with permission from [66]. Copyright 2009 American Chemical Society

8. Regeneration of Ammonia Borane Dehydrogenation Products

The regeneration of spent fuel back to ammonia borane is still the main bottleneck in the foreseen use of ammonia borane as a chemical hydrogen storage medium. In the previous paragraphs, it was pointed out that solid state ammonia borane thermolysis leads to the formation of heterogeneous solid BNH-materials, highly polymeric $(BH_2NH_2)_x$, and $(BHNH)_x$.. In addition, smaller quantities of well-defined volatile BNH-compounds were released during ammonia borane pyrolysis, among them monomeric BH_2NH_2 and borazine. Cyclic products were found from ammonia borane thermolysis in glyme and THF solutions.

Polyborazylene-like material cannot be reformed with practical H_2 overpressures. Complex chemical processing pathways are required to regenerate ammonia borane from its dehydrogenation products. NMR, DSC and rat e studies revealed that, whatever the spent fuel, all contain trigonal sp^2 BNH_x structures, thus enabling a common regeneration strategy [18].

Ramachandran and Gagare [24] described a one-pot regeneration of ammonia borane process in which a suspension of ammonium tetramethoxyborate and ammonium chloride is mixed with lithium aluminum hydride in THF.

Tumas et al [68] proposed an overall regeneration process based on five chemical transformations: digestion, activation, reduction, disproportionation and ligand exchange. They individually demonstrated each step in the process. They have showedn that trifluoroacetic acid can be used to digest the dehydrogenated boron-nitrogen product. They and demonstrated the net reduction of the B-Cl bond in chloro catecholborane (ClBCat) with hydrogen mediated by a tin hydride. This tin hydride can be regenerated by decarboxylation of tin formate. The latter can be formed by reaction of formate (itself generable from the hydrogenation of carbon dioxide) and the corresponding tin hydroxide. Diethylaniline was found to drive the redistribution of HBcat (obtained by ClBCat hydrogenation) to (diethylaniline)-BH_3 which reacts quantitatively with ammonia to make ammonia borane.

The procedure reported in Figure 8 has been outlined from Hausdorf et al. [43] for ammonia borane regeneration from different BNH-products. The first regeneration step could be the treatment of BNH-products with a solution of Brønsted acids, e.g. HCl, in organic solvents. This leads to a nearly complete destruction and solubilization of the polymeric material, and to the formation of BCl_3 and, in case of $(BH_2NH_2)_x$ as waste, of a new compound, probably cyclic $(BCl_2NH_2)_3$. The second step, getting B_2H_6 by hydrodechlorination of BCl_3 can be achieved but it's a highly energy consuming reaction. Hausdorf et al. [43] proposed the following reaction:

$$BCl_3.NR_3 + 3\,H_2 + 3\,NR_3 \rightarrow BH_3.NR_3 + 3\,NR_3.HCl \qquad (7)$$

and a recovery of the base NR_3 by thermal decomposition. An appropriate choice of the more favorable base is still to be done [43].

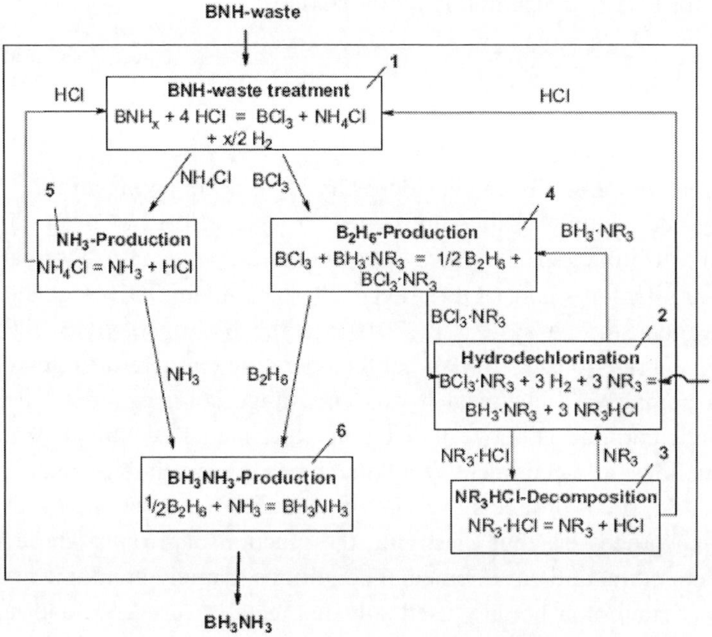

Figure 8. Procedure for the regeneration of ammonia borane from BNH-waste. Reprinted with permission from [43]

The approach of chemical ammonia borane regeneration based on chlorination of decomposition products was studied by Damle et al. [19]. They concluded that the process was excessively energy consuming with respect to U.S. Department of Energy (DOE) targets and, therefore, suspended the related research activities. Also Sneddon et al. [17]] also proposed a general scheme whose key steps are:

- Digestion of the spent fuel with superacidic HX/AlX_3 solutions;
- Coordination of the BX_3 product to a base;
- Reduction of the boron-halide bonds by silanes or stannanes; and
- Exchange of the base by ammonia to regenerate ammonia borane.

They tried HBr/$AlBr_3$ system in CS_2 to digest the BNH-waste, and got good results in the reduction step by using sulfide bases in conjunction with either silane or silane reducing agents.

At PNNL alcohols were found suitable solvents for the digestion step [21]. They also carried on out theoretical studies in order to find the chemical element showing good hydride donor properties and a sufficient acid character for spent fuel digestion, individuating cobalt as a promising candidate [21].

The LANL's approach proposes is the to use of thiacatechol as reagent to digest the spent fuel [69]. For the second regeneration step they found that tin hydride compounds can reduce this thiacatechol digested material to form B-H bonds, and that under appropriate conditions, ammonia can be re-added to this mixture to generate ammonia borane. They are optimizing this procedure for energy consumption minimization.

CONCLUSION

As ammonia borane has been considered by DOE as an applicant for polymer electrolyte membrane fuel cell fuelling in automotive use, a large effort has been done in the last five years aimed to fulfill the gaps that have been evidenced between its thermal dehydrocoupling performances and features and DOE targets [70]. The main DOE targets to be reached by a on-board hydrogen storage system in 2010 are the hydrogen yield (0.06 kg_{H2} kg^{-1}), the hydrogen releasing rate (0.02 g s^{-1} kW^{-1}) at the operating temperatures (lower than 85 °C).

Ammonia borane as a chemical hydrogen storage complex attracted interest since late 70s. A systematic correlation between different experimental technique results and theoretical model computations allowed identifying the mechanisms, intermediate products, residual solid and volatile final products, activation energies and enthalpies associated with its dehydrocoupling processes, thus clarifying the reaction pathways. These were found to be dependent on operating and environmental conditions, namely ammonia borane inthe solid or solution form of ammonia borane, used solvents, catalyst presence and nature, temperature scan rate etc.

The main drawbacks in evidenced by the first experimental results on solid state ammonia borane were the about hydrogen releasing temperature (about 90 °C to get 7.6 wt% hydrogen at 0.0004 g s^{-1}) and even more about the excessively low dehydrogenation rates and the volatile borazine and diborane production.

The hHydrogen yield, acceptable in ammonia borane solid state thermolysis (about 7 wt% at 85 °C), drops to unacceptable low values in case of ammonia borane solutions. Other technological aspects such as ammonia borane foaming during hydrogen releasing were also to be overcome by using calcium complex. The main obstacle towards ammonia borane large scale utilization as hydrogen storage medium was isobviously within the cost and anyway consumed the energeticy cost for the rehydrogenation of the spent fuel to get again ammonia borane.

Ammonia borane has been considered by DOE as an solid H_2 storage method to fuel polymer electrolyte membrane fuel cell for automotive use. A large effort has been done in the last five years aimed to fulfill the gaps between its thermal dehydrocoupling performances and features and DOE targets [70].

Ammonia borane addition with a suitable catalysts, substitution of a hydrogen atom of ammonia borane with a metal atom, dispersion into nanoscale structures or dissolution into solvents allow to reach most of the DOE targets, but not contemporary by on the same material.

Dehydrogenation temperature as low as room temperature and high releasing rate can be attained with catalyzed solutions of ammonia borane; on the other hand, this means to drastically reduce hydrogen gravimetric densities, to introduce expensive and pollutant metals and to make regeneration more difficult. The use of every an ammonia borane solution is not practicable, due to low gravimetric hydrogen content, except for ammonia borane solution in ionic liquids. Ammonia borane diluted in ionic liquid respects DOE 2010 targets both for operative temperature and gravimetric hydrogen density; it exhibits a 8E4g$s^{-1}kg^{-1}$ releasing rate. Good results, even if at 95 °C, were obtained with Li-ammonia borane and Na-ammonia borane compounds. The mixture AB+$LiNH_2$ also seems very promising: it delivered 9.3wt% H_2 in 15h at 85°C. The so far obtained results decreased the marginnarrowed the gap between ammonia borane thermolysis performances and DOE targets. The releasing rates are anyway lower than the 2010 DOE request.

A crucial still open point is the regeneration of the spent fuel. Several proposed methods, each of them composed by several steps, has stillneeds to be technologically validated technologically.

REFERENCES

[1] Shore, S.G.; Parry, R.W.; *J. Am. Chem. Soc.* 1955, *77*, 6084-6085.

[2] Gutowski, M;. Autrey, T . *Chemistry World* 2006, *3*(3).

[3] Karkamar, A.; Aardal, C.; Autrey T. *Mater. Matters* 2007, *2*(2), 6-9.

[4] Matus, M;. Anderson, K.; Camaioni, D.M.; Autrey, T.; Dixon, D.A. *J. Phys. Chem. A* 2007, *111*, 4411-4421.

[5] Stephens, F.H.; Pons, V.; Baker, R.T. *Dalton Trans.* 2007, *25*, 2613-2626.

[6] Peng, B; Chen, J. *Energy Environ. Sci.* 2008, *1*(4), 479.

[7] Demirci, U.B.; Miele P. *Energy Environ. Sci.* 2009, 2, 627-637.

[8] Lippert, L.; Lipscomb, W.N. *J. Am. Chem. Soc.* 1956, *78*, 503-504.

[9] Hu, M.G.; Geanangel, R. A.; Wendlandt, W. W. *Thermoch. Acta* 1978, *23*, 249-255.

[10] Sit, V.; Geanangel, R.A.; Wendlandt W.W *Thermoch. Acta* 1987, *113*, 379-382.

[11] Wang, J. S.; Geanangel, R. A. *Inorg. Chim. Acta* 1988, *148*(2) 185-190.

[12] Baitalow, F.; Baumann, J.; Wolf, G.; Janicke-Rossler, K. J.; Leitner, G. *Thermoch. Acta* 2002, *392*, 159-168.

[13] Baitalow, F.; Wolf, G.; Grolier, J. P. E.; Dan, F.; Randzio, S. L. *Thermoch. Acta* 2006, *445*, 121-125.

[14] Wolf, G.; Baumann, J.; Baitalow, F.; Hoffmann, F.P. *Thermoch. Acta* 2000, *343,* 19-25.

[15] Baumann, J.; Baitalow, F.; Wolf, G. *Thermoch. Acta* 2005, *430*, 9-14.

[16] Sneddon, L.G. *DOE Hydrogen Program FY 2006 Annual Progress Report*, 418-421.

[17] Sneddon, L.G. *DOE Hydrogen Program FY 2007 Annual Progress Report*, 488-491.

[18] Sneddon, L.G. *DOE Hydrogen Program FY 2008 Annual Progress Report*, 583-586.

[19] Damle, A. *DOE Hydrogen Program FY 2007 Annual Progress Report*, 460-463.

[20] Damle, A. *DOE Hydrogen Program FY 2008 Annual Progress Report*, 612-614.

[21] Aardahl, C. *DOE Hydrogen Program FY 2008 Annual Progress Report,* 565-569.

[22] Chandra, M.; Xu, Q. *J. Power Sources* 2006, *156*, 190–194.

[23] Yan, J.M.; Zhang, X.B.; Han, S.; Shioyama, H.; Xu, Q. *Angew. Chem., Int. Ed.* 2008, *47*, 2287-2289.

[24] Ramachandran, P.V.; Gagare, P.D. *Inorg. Chem.* 2007, *46* (19), 7810-7817.

[25] Bowden, M.; Autrey, T.; Brown, I.; Ryan, M. *Curr. Appl. Phys.* 2008, *8,* 498-500.

[26] Stowe, A. C.; Shaw, W. J.; Linehan, J. C.; Schmid, B.; Autrey, T. *Phys. Chem. Chem. Phys.* 2007, *9*, 1831-1836.

[27] Bluhm, M. E.; Bradley, M. G.; Butterick, R.; Kusari, U.; Sneddon, L. G. *J. Am. Chem. Soc.* 2006, *128*(24), 7748-9.

[28] Keaton, R. J.; Blacquiere, J.M.; Baker, R. T. *J. Am. Chem. Soc.* 2007, *129* (7), 1844-1845.

[29] Bowden, M.; Kemmitt, T.; Shaw, W.; Hess, N. Linehan, J.; Gutowski, M.; Schmid, B.; Autrey T. Mechanistic Studies of Hydrogen Release from Solid Amine Borane Materials"in *Hydrogen Storage Materials*, edited by J.C.F. Wang, W. Tumas, A. Rougier, M.J. Heben, E. Akiba (Mater. Res. Soc. Symp. Proc. 927E, Warrendale, PA), 2006.
[30] Denney, M. C.; Pons, V.; Hebden, T. J.; D.; Heinekey, M.; Goldberg, K. I. *J. Am. Chem. Soc.* 2006, *128* (37), 12048-12049.
[31] Jaska, C. A.; Temple, K.; Lough, A. J.; Manners, I. *J. Am. Chem. Soc.* 2003, *125* (31), 9424-9434.
[32] Dietrich, B. L.; Goldberg, K. I.; Heinekey, D. M.; Autrey, T.; Linehan J. C. *Inorg. Chem.* 2008, *47* (19), 8583-8585.
[33] Clark, T. J.; Lee, K.; Manners I. *Chem. Eur.* 2006, *12*(34), 8634-8648.
[34] Chen, Y.; Fulton, J. L.; Linehan, J. C.; Autrey T *J. Am. Chem. Soc.* 2005, *127* (10), 3254–3255.
[35] Fulton, J. L.; Linehan, J. C.; Autrey, T.; Balasubramanian, M.; Chen, Y.; Szymczak, N.K. *J. Am. Chem. Soc.* 2007, *129* (39), 11936-11949.
[36] Shrestha, R.P.; Diyabalanage, H.V.K; Semelsberger, T.A.; Ott, K.C; Burrell, A.K. *Int. J. Hydrogen Energy* 2009, *34*(6), 2616-2621.
[37] Cheng, F.; Ma, H.; Li, Y.; Chen, J. *Inorg. Chem.* 2007, *46*(3), 788-794.
[38] Clark, T. J.; Russell, C. A.; Manners, I. *J. Am. Chem. Soc.* 2006, *128* (30), 9582-9583.
[39] Stephens, F. H.; Baker, R. T.; Matus, M. H.; Grant, D.J.; Dixon, D.A. *Angew. Chem., Int. Ed.* 2007, *46*, 746-749.
[40] Heldebrant, D. J.; Karkamkar, A.; Hess, N. J.; Bowden, M.; Rassat, S.; Zheng, F; Rappe, K.; Autrey T. *Chem. Mater.* 2008, *20* (16), 5332–5336.
[41] Miranda, C.; Ceder, G. *J. Chem. Phys.* 2007 *126* (18), 184703.
[42] Dixon, D.A.; Gutowski M. *J. Phys. Chem. A* 2005 (109), 5129-5135.
[43] Gutowska, A.; Li, L.; Shin, Y.; Wang, C. M.; Li, X.S.; Linehan, J. C.; Smith, R. S.; Kay, B. D.; Schmid, B.; Shaw, W.; Gutowski, M. and Autrey T. *Angew. Chem., Int. Ed.* 2005, *4,* 3578-3582.
[44] Hausdorf, S.; Baitalow, F.; Wolf, G.; Mertens F.O.R.L. *Int. J. Hydrogen Energy* 2008, *33*, 608-614.
[45] Heldebrant, D. J.; Karkamkar, A.; Linehan, J. C.; Autrey, T. *Energy & Environmental Science* 2008, *1*, 479-483.
[46] Shore, S. G.; Boddeker, K. W. *Inorg. Chem.* 1964, *3* (6), 914-915.
[47] Mayer, E. *Inorganic Chemistry* 1972, *11*(4), 866-869.
[48] Hu, M.G.; Van Paaschen, J.M.; Geanangel, R.A. *J. Inorg. Nucl. Chem.* 1977, *39* (12), 2147.
[49] Mayer, E. *Inorg. Chem.* 1973, *12* (8), 1954-1955.
[50] Hoon, C.F.; Reynardt, E.C. *J. Phys.C: Solid State Phys.* 1983, *16*(32), 6129-6136.
[51] Klooster, W. T.; Koetzle, T. F; Siegbahn, E. M.; Richardson, T. B; Crabtree, R. H. *J. Am. Chem. Soc.* 1999, *121*, 6337-6343.
[52] Hoon, C.F; Reinhardt, C. *J. Phys C* 1983, *16*, 6129-6136.
[53] Parvanov, V.M.; Schenter, G. K.; Hess, N. J.; Daemen, L. L.; Hartl, M.; Stowe, A. C.; Camaioni, D. M.; Autrey, T. *Dalton Trans.* 2008, *33*, 4514-4522.
[54] Brown, C.; Jacques, T.; Hess, N.J.; Daemen, L.L.; Mamontov, E.; Linehan, J.C.; Stowe, A.C.; Autrey, T. *Physica B Condensed Matter* 2006, *385,* 266-268.
[55] Sugie, M.; Takeo, H.; Matsumura, C. *Chem. Phys. Lett.* 1979, *64*(3), 573-575.
[56] De Benedetto, S.; Carewska, M.; Cento, C.; Gislon, P.; Pasquali, M.; Scaccia, S.; Prosini, P.P. *Thermoch. Acta* 2006, *441*, 184-190.

[57] Helary, J.; Salandre, N.; Saillard, J.; Poullain, D.; Beaucamp, A.; Autissier, D. *Int. J. Hydrogen Energy* 2009, *34*,169-173.

[58] Xiong, Z.; Keong, C. Y.; Wu, G.; Chen, P.; Shaw, W.; Karkamkar, A.; Autrey, T.; Jones, M. O., Johnson, S. R.;. Edwards, P. P.; David, W. I. F *Nature Materials* 2008, *7*(2), 138-141.

[59] Diyabalange, H. V. K.; Shrestha, R. P.; Semelsberger, T. A.; Scott, B. L.; Bowden, M. E.; Davis, B. L.; Burrell, A. K. *Angew. Chem., Int. Ed.* 2007, *46*, 8995-8997.

[60] Bowden, M. E.; Brown, I.W.M.; Gainsford, G. J.; Wong, H. *Inorg. Chim. Acta* 2008, *361*, 2147-2153.

[61] Graham, K.R.; Kemmit, T.; Bowden, M.E. *Energy & Environ. Sci.* 2009, *2*, 706-710.

[62] Yoon, C.W., Sneddon, L.G. *J. Am. Chem Soc.* 2006, *128*(43), 13992-13993.

[63] Baldé, C. P.; Hereijgers, B. P. C.; Bitter, J. H.; de Jong, K. P. *Angew. Chem.* 2006, *118*, 3581-3583.

[64] Stephens, R. D.; Gross, A. F.; Van Atta, S. L; Vajo, J. J.; Pinkerton, F. E *Nanotechnology* 2009, *20*(20), 204018.

[65] Sepehri, S.; Feaver, A.; Shaw, W. J.; Howard, C. J.; Zhang, Q.; Autrey, T.; and Cao G. *J. Phys. Chem. B* 2007, *111,* 14285-14289.

[66] Feaver, A.; Sepehri, S.; Shamberger, P.; Stowe, A.; Autrey, T.; Cao, G. *J. Phys. Chem. B* 2007, *111* (26), 7469-7472.

[67] Neiner, D.; Karkamkar, A.; Linehan, J.; Arey, C. B.; Autrey, T.; Kauzlarich, S. M. *J. Phys. Chem. C* 2009, *113,* 1098-1103.

[68] Tumas, W.; Baker, R.T.; Burrell, A. Thorn, D. *DOE Hydrogen Program* FY 2006 Annual Progress Report 2006, 386-392.

[69] Ott, K.C. *DOE Hydrogen Program FY 2008 Annual Progress Report,* 577-582.

[70] Marder, T.B.; *Angew. Chem., Int. Ed.* 2007, *46*, 8116-8118.

In: Boron Hydrides High Potential Hydrogen Storage Material ISBN:978-1-61668-361-0
Editors: Umit B. Demirci and Philippe Miele

Chapter 8

AMMONIA BORANE: HYDROLYSIS AND ELECTROOXIDATION

Umit B. Demirci[1] and Philippe Miele[2]

Laboratoire des Multimatériaux et Interfaces, UMR 5615 CNRS Université Lyon 1, Université de Lyon, 43 boulevard du 11 Novembre 1918, F-69622 Villeurbanne, France
Tel.: +33 (0)4 72 44 84 03; +33 (0)4 72 43 12 05, Fax: +33 (0)4 72 44 06 18

ABSTRACT

Ammonia borane NH_3BH_3 combines low molecular weight (30.7 g mol^{-1}) and high gravimetric hydrogen storage capacity (19.6 wt%), what makes it more attractive than both sodium borohydride (37.8 g mol^{-1} and 10.8 wt%) and lithium borohydride (21.8 g mol^{-1} and 18.3 wt%). Like sodium borohydride, ammonia borane is a hydrogen source but ammonia borane is maybe the most promising boron hydride for hydrogen storage applications. Stored hydrogen can be released through either thermolysis or hydrolysis. The present chapter deals especially with hydrolysis of ammonia borane. Its basics, state-of-the-art, and potential are largely discussed on the basis of theoretical data and data available in the open literature. To summarize, it is showned that ammonia borane hydrolysis faces the same issues than those encountered by sodium borohydride hydrolysis and that the latter is superior to the former.

Still like sodium borohydride, ammonia borane is an electron source. Both have attracted attention for their own potential to be electrooxidized because they are able to release up to eight and six electrons, respectively, for electricity generation. Electrooxidation of ammonia borane, i.e. direct ammonia borane fuel cell, is the second purpose of this chapter. Its state-of-the-art, advantages and issues are tackled. Finally direct ammonia borane fuel cell is compared to direct borohydride fuel cell. It is stressed that today research devoted to direct borohydride fuel cell develops even if issues hinder its development but research devoted to direct ammonia borane is at its beginning. However it is showed that the former fuel cell is superior to the latter one. And a question is asked: what can be the advantage of developing direct ammonia borane fuel cell when

1E-mails: Umit.Demirci@univ-lyon1.fr;
2Philippe.Miele@univ-lyon1.fr,

it is less effective, less practical (because of inevitable formation of ammonium ion), and more complex than direct borohydride fuel cell?

The present chapter discussesd thus ammonia borane hydrolysis and ammonia borane electrooxidation and can be seen as a continuation of chapter 7 dealing with ammonia borane thermolysis.

1. Introduction

"Ammonia borane [(NH_3BH_3)]: the hydrogen source *par excellence*?" is the question asked by Stephens et al. [1] while reviewing research devoted to this "intriguing molecule for chemical hydrogen storage applications". This question is very interesting because it reflects the current trend that sees ammonia borane as being the solution for the hydrogen storage issue, especially in automotive applications. This trend is amazingly like the recent craze for sodium borohydride ($NaBH_4$).

Sodium borohydride[3], though known as a hydrogen material since the 1940s [2], has attracted particular attention since the late 1990s [3]. It has been very often showed as being the solution we expect for fuel cell applications. Naturally storing 10.8 wt% of hydrogen, the challenge has been to release efficiently 100 % of stored hydrogen through two different routes, thermolysis or hydrolysis, the former soon showing its inefficiency [4]. Efforts have therefore focused on the hydrolysis process. However too many issues (for example, by-product recycling and cost) have hindered its quick development and the number of challenges increased. The technical and economic difficulties sodium borohydride has encountered made the U.S. Department of Energy recommend a no-go for sodium borohydride for on-board vehicle hydrogen storage [5]. Given the state-of-the-art, such decision was expectable [6]. Today hydrolysis of sodium borohydride is rather intended to portable and niche applications.

While recommending the no-go for sodium borohydride, the U.S. Department of Energy insistently remarked that the experience in sodium borohydride should be valorized in research devoted to ammonia borane [5]. In parallel, a survey about its safety aspects and properties of ammonia borane was published, from which the following points have especially stood out [7]:

- High purity ammonia borane can be safely stored in a closed vessel at ambient temperature;
- Ammonia borane is water-soluble;
- Ammonia borane release hydrogen by hydrolysis only in the presence of catalyst.

In other words, the U.S. Department of Energy has made its position towards ammonia borane clear: it is today the hydride showing the highest potential for automotive applications and much many efforts should focus on this hydride. Moreover ammonia borane combines the following advantages that make it suitable for large-scale applications; it is non toxic, environmentally benign and stable.

---

[3] Chapter 4 specifically deals with hydrolysis of sodium borohydride $NaBH_4$.

Ammonia borane combines low molecular weight (30.7 g mol^{-1}) and high gravimetric hydrogen storage capacity (19.6 wt%), what makes it more attractive than sodium borohydride (37.8 g mol^{-1} and 10.8 wt%) and perhaps even more attractive than lithium borohydride[4] (21.8 g mol^{-1} and 18.3 wt%). Today, the craze is around ammonia borane and, inevitably, it is going to become more pronounced (Figure 1). For example, five review papers devoted to ammonia borane as hydrogen storage material have been published the last two years (cited throughout this chapter). The grand challenge is currently to efficiently dehydrogenate ammonia borane. It is expected effective gravimetric hydrogen storage capacities well above 10 wt% in operating conditions meeting the technical targets established by the U.S. Department of Energy for automotive applications[5]. Chemical dehydrogenation of ammonia borane is envisaged through two distinct routes, which are the thermolysis[6] and the hydrolysis. Hydrolysis of ammonia borane is one of the purposes of the chapter.

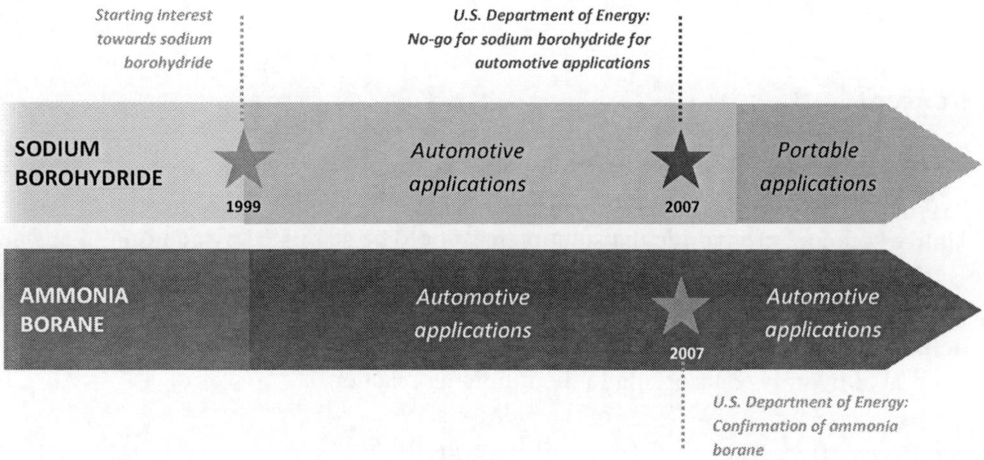

Figure 1. Recent hydrogen-related history of sodium borohydride and ammonia borane.

It has been introduced above that like sodium borohydride ammonia borane is a hydrogen source. Besides, still like sodium borohydride, ammonia borane is an electron source [4]. In the late 1990s, sodium borohydride has also attracted attention for its potential to be electrooxidized because it is able to release up to eight electrons for electricity generation. It has therefore been investigated as direct liquid fuel of polymer membrane electrolyte fuel cell (PEMFC) [8], also called direct borohydride fuel cell (DBFC)[7] [9]. Because sodium borohydride and ammonia borane seem to have a same "destiny", the utilization of ammonia borane as direct liquid fuel of polymer membrane electrolyte fuel cell was expectable. And, indeed, in 2007, for the first time, a paper about direct ammonia borane fuel cell was published [10]. Today, research devoted to direct borohydride fuel cell develops even if issues hinder its development but research devoted to direct ammonia borane is at its early

4 Chapter 3 specifically deals with lithium borohydride LiBH4.
5 Some technical targets for on-board hydrogen storage systems are reported in Chapter 1 (Table 2).
6 Chapter 7 specifically deals with thermolysis of ammonia borane NH3BH3.
7 Chapter 5 specifically deals with direct borohydride fuel cell.

beginning. Electrooxidation of ammonia borane, i.e. direct ammonia borane fuel cell, is the second purpose of the chapter.

Ammonia borane has thus a potential as hydrogen carrier as well as liquid fuel of direct liquid-feed fuel cell (DLFC). Dehydrogenation of ammonia borane is much studied through the thermolysis process, as discussed in chapter 7, but dehydrogenation through hydrolysis is much less studied yet. In parallel, electrooxidation of ammonia borane in fuel cell is a research area still superficially developed. Hence this chapter, as a continuation of chapter 7, deals with ammonia borane, especially focusing on its hydrolysis and its electrooxidation. State-of-the-art, advantages and issues are tackled, and finally ammonia borane as hydrogen carrier and direct ammonia borane fuel cell are compared to sodium borohydride and direct borohydride fuel cell, respectively.

2. Ammonia Borane Hydrolysis

2.1 Concept

Ammonia borane is water-soluble like sodium borohydride but unlike sodium borohydride it is relatively stable to hydrolysis when the solution pH is neutral or basic [11-13]. Unlike sodium borohydride the solution has not to be stabilized by addition of any strong base such as sodium hydroxide. Hydrolysis has therefore to be activated by utilization of a catalyst, which can be an acid or a metal.

In the presence of acid catalyst, ammonia borane and related alkyl-substituted compounds are reported to hydrolyze according to the following reaction:

$$NR_3BH_3 + H^+ + 3\ H_2O \rightarrow B(OH)_3 + 3\ H_2 + NR_3H^+ \quad (1)$$

with R = H or alkyl groups, and so:

$$NH_3BH_3 + H^+ + 3\ H_2O \rightarrow B(OH)_3 + 3\ H_2 + NH_4^+ \quad (2)$$

The kinetics of hydrolysis is first order with respect to ammonia borane and proton concentrations [14,15]. Kelly and Marriott [15] have suggested two mechanisms for the acid-catalyzed hydrolysis, namely:

$$NH_3BH_3 + AH \rightarrow [H_3N(BH_3)(HA)]\ \text{complex} \rightarrow NH_4^+ + A^- + BH_3\ (aq) \rightarrow B(OH)_3 + H_2 \quad (3)$$

the first reaction being slow, the complex being a five-coordinate nitrogen specie (Figure 2, specie a), the third one occurring fast in the presence of water; and,

$$NH_3BH_3 + AH \leftrightarrow [(H_3N)(BH_4)]^+\ \text{complex} + A^- \rightarrow H_3NBH_2^+ + H_2 \rightarrow NH_4^+ + B(OH)_3 + H_2 \quad (4)$$

the first reaction being in equilibrium, the complex being a five-coordinate boron specie (Figure 2, specie b), the third one occurring fast in the presence of water. Analysis of isotope effects and reaction rates led to the conclusion that the mechanism (3) was more likely. The studies cited here dates from the 1960s and the 1970s. To the best of our knowledge, no recent (since the late 1990s) study has reported utilization of acids (mineral such as hydrochloric acid HCl or organic such as formic acid HCOOH) as catalyst for generating hydrogen by hydrolysis of ammonia borane, except one work. Indeed, Xu and co-workers [16] found that solid acids such as cation exchange resins and zeolites exhibit high activities towards hydrolysis of ammonia borane to generate up to 3 hydrogen molecules per ammonia borane molecule at room temperature. Given the parallel with the research devoted to $NaBH_4$, aqueous solution of acids as well as aqueous solution of metal salts will be utilized again when studies envisaging solid storage of ammonia borane will be launched.

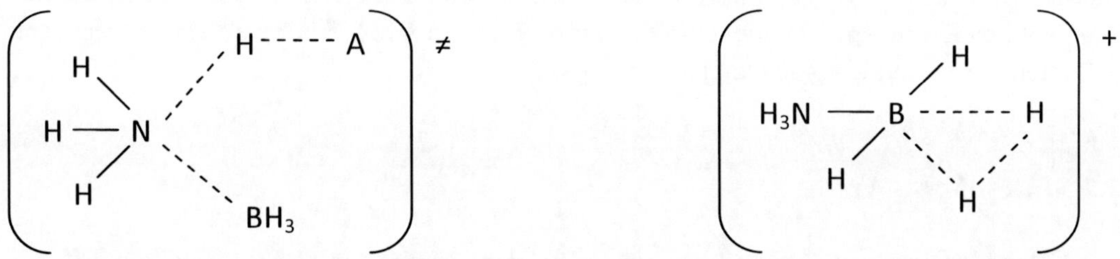

Figure 2. Likely complexes as reaction intermediates in acid-catalyzed ammonia borane hydrolysis [15].

The current research rather focuses on the utilization of heterogeneous transition metal-based catalysts. In the presence of metal catalyst, hydrolysis is reported to ideally achieve according to the following reaction [12,13,17,18]:

$$NH_3BH_3 + 2\ H_2O \rightarrow NH_4^+ + BO_2^- + 3\ H_2 \qquad (5)$$

A recent study has slightly modified this reaction (5) [19]. By analysis of the reaction by-products by ^{11}B nuclear magnetic resonance (NMR), it has been suggested that hydrolysis takes place in excess of water, the borate specie being hydrated[8]:

$$NH_3BH_3 + 4\ H_2O \rightarrow NH_4BO_2.H_2O + 3\ H_2 \qquad (6)$$

Moreover it has been suggested that the dihydrated borate is in equilibrium with different borate species, depending on the concentration of ammonia borane:

$$NH_4B(OH)_4 \leftrightarrow NH_4B_3O_3(OH)_4 \leftrightarrow NH_4B_4O_5(OH)_4 \leftrightarrow [NH_4]B_5O_6(OH)_4(\qquad 7)$$

Whatever the hydrolysis reaction, hydrogen is released with a hydrogen to ammonia borane ratio up to 3. In other words, the formation of the ammonium ion NH_4^+ is a strong

8 The hydrolysis by-product is generally given as $NH_4BO_2.H_2O$ to show that two more water molecules are consumed during the reaction; the by-product is in fact a tetrahydroxyborate compound $NH_4BO_2.H_2O$.

limitation of the effective gravimetric hydrogen storage capacity of the system NH_3BH_3-$2H_2O$ as it decreases as follows:

- Neat NH_3BH_3: capacity of 19.6 wt%;
- System NH_3BH_3-$2H_2O$ without considering the formation of NH_4^+: capacity of 15 wt%;
- System NH_3BH_3-$2H_2O$ while considering the formation of NH_4^+: capacity of 9 wt%.

It is clear that the formation of the ammonium ion is detrimental to the effective storage capacities of ammonia borane. A capacity of 9 wt% is even lower than that of the system $NaBH_4$-$2H_2O$, which stores 10.8 wt%. And, as it will be discussed more in details in section 2.3, the value 9 wt% is decreased more because of the limited solubility of ammonia borane and the formation of hydrated by-products. This asks then a question: is there any relevancy, in an application prospect, to develop the hydrolysis of ammonia borane? The present section will attempt to provide some elements of answer.

2.2 State-of-the-Art

From experience of sodium borohydride hydrolysis, one can easily expect the difficulties hydrolysis of ammonia borane will encounter. The difficulties will then be the ammonia borane challenges. The main ones, difficulties or challenges, are as follows:

- Efficient catalyst, with high hydrogen generation rates, high hydride conversions and high durability for successive utilizations;
- Optimized effective gravimetric (and volumetric) hydrogen storage capacities, meeting the targets set by the U.S. Department of Energy for vehicular applications;
- Recycling of the hydrolysis by-products back into the hydride.

With respect to sodium borohydride, the first point (except the durability issue) has attracted much more attention than the other points. And it appears that, currently, the priority of the studies devoted to ammonia borane hydrolysis is catalytic material, and particularly its efficiency in terms of hydrogen generation rates and hydride conversions.

2.2.1 Transition Metal Catalyst

Xu's group is much involved in studying the catalyzed hydrolysis of ammonia borane as that is characterized by several papers published since 2006 [13]. They first have demonstrated the high catalytic activity of platinum, rhodium and palladium towards hydrolysis of ammonia borane [12]. Simagina et al. [20] have showed that rhodium supported over titanium oxide is more reactive than platinum supported over titanium oxide. In a further study, Xu and co-workers [21] investigated the reactivity of non-noble metals such as iron, cobalt and nickel, supported over various supports such as e.g. alumina and carbon. Cobalt and nickel were found to be equally reactive while iron was inactive. However, further improvements in the synthesis of iron-based catalysts (nanoparticles) led to finding a highly active material that completed the hydrolysis of ammonia borane within only 8 min (against 160 minutes for not optimized iron nanoparticles) [22]. Much more recently, Xu and co-workers [23,24] have developed new catalytic materials. The catalytic abilities of nickel stabilized by poly(N-vinyl-2-pyrrolidone) [23] and hollow nickel-silica nanosphere [24] have been demonstrated. It has been showed that metallic nickel is much more reactive than oxidized nickel for the hydrolysis of ammonia borane, the latter being reduced into the former in the presence of sodium borohydride. Indeed, in both works ammonia borane and sodium borohydride was mixed together with ammonia borane, both catalytically hydrolyzing and generating hydrogen. According to the authors, sodium borohydride is necessary for activating the metal catalyst. Other metals like e.g. ruthenium and gold have also been reported [25,26]. Ramachandran and Gagare [19] and Kalidindi et al. [18] investigated the catalytic abilities of metal chlorides. It has been showed that e.g. ruthenium chloride, cobalt chloride and palladium chloride are potential catalytic accelerators [19]. Cheng et al. [17] have successfully developed nickel-platinum hollow spheres as catalysts of the ammonia borane hydrolysis at ambient temperature. The most catalytically active combination was found to be $Ni_{0.88}$-$Pt_{0.12}$. Platinum- or nickel-based alloys (20 wt% supported over carbon) were also studied by Yao et al. [27]. It has been reported that PtRu > Pt_3Au > PtIr > PtCo > Pt_3Ni > PtCu > PtSn ≈ Pt, in terms of hydrogen generation rates and yields, and that nickel-based alloys are less efficient than the best Pt-based alloys. Kalidindi et al. [18,28] much studied copper, either as nanostructured Cu and Cu@Cu_2O core shell catalysts or as copper chloride. Figure 3 summarizes the metals already investigated in ammonia borane hydrolysis.

	III B	IV B	V B	VI B	VII B	XIII	XIII	XIII	I B	II B	III A	IV A
2											**B [a]**	C
3											Al	Si
4	Sc	**Ti [b]**	V	Cr	Mn	**Fe**	**Co**	**Ni**	**Cu**	Zn	Ga	Ge
5	Y	Zr	Nb	**Mo**	Tc	**Ru**	**Rh**	**Pd**	**Ag**	Cd	In	**Sn [c]**
6	La	Hf	Ta	**W**	Re	Os	**Ir**	**Pt**	**Au**	Hg	Tl	Pb

[a] Metal boride (e.g. nickel boride).
[b] Titanium complex.
[c] PtSn alloy.

Figure 3. Elements investigated in catalyzed hydrolysis of ammonia borane (colored cell).

Table 1 reports some of the highest hydrogen generation rates reported so far. Table 2 shows some of the apparent activation energies reported so far. In our opinion, there is still room to further improve the hydrogen generation rates since rates higher than 100 l min^{-1} $g^{-1}_{catalyst}$ had soon been reported for the catalyzed hydrolysis of sodium borohydride.

Table 1. Catalyzed hydrolysis of ammonia borane: hydrogen generation rate (HGR) for some metal-based catalysts.

Catalyst	Temperature (°C)	HGR (ml min^{-1})	HGR (l min^{-1} $g^{-1}_{catalyst}$)	References
2 wt% Ru-γAl_2O_3	25	110	2.4	[25]
3 wt% Ru-C	25	13	0.8	[26]
2 wt% Rh-γAl_2O_3	25	200	4.4	[25]
1 wt% Rh-TiO_2	40	15	2.5	[20]
2 wt% Pt-γAl_2O_3	25	230	5.0	[25]
1 wt% Th-TiO_2	40	10	0.8	[20]
20 wt% Pt-C	20	1	0.1	[27]
20 wt% PtRu-C	20	9	0.4	[27]
Ni with PVP [a]	25	n.i. [b]	1.1	[23]
Ni-SiO_2 hollow sphere	25	7	1.4	[24]
$Ni_{0.88}$-$Pt_{0.12}$ hollow sphere	20	28	2.3	[17]

[a] PVP as poly(N-vinyl-2-pyrrolidone)

[b] n.i. as not informed.

Table 2. Catalyzed hydrolysis of ammonia borane: apparent activation energies for some metal-based catalysts.

Catalyst	Apparent activation energy (kJ mol^{-1})	References
10 wt% Co-γAl_2O_3	62	[21]
2 wt% Ru-γAl_2O_3	23	[25]
3 wt% Ru-C	76	[26]
2 wt% Rh-γAl_2O_3	21	[25]
2 wt% Pt-γAl_2O_3	21	[25]
$Ni_{0.88}$-$Pt_{0.12}$ hollow sphere	30	[17]
20 wt% NiAg-C	52	[27]

Catalyst durability is crucial for materials intended to multi-use applications. Preliminary studies about some of the catalysts discussed in the previous paragraph have showed promising durabilityies. Cheng et al. [17] have reported that the NiPt hollow sphere catalysts, treated by filtrating and washing, can be repeatedly utilized with no significant deactivation in the hydrolysis experiment. Xu and co-workers [12,25] have reported that no significant deactivation was observed for used noble metal catalysts recycled by washing with water and drying. Kalidindi et al. [28] have showed unchanged reactivity over 8 cycles for the Cu@Cu_2O catalyst although the particle size increased after the cycles of hydrolysis reaction. However, it is important to note here that the hydrolysis conditions were rather far from being severe and favored putting in evidence the catalyst durability. Harsher experimental conditions are required in order to validate the catalysts durability. By harsher conditions, it is meant real operating conditions of the likely technological applications.

Xu and co-workers [12,25] have tentatively described the role of the transition metal in the catalyzed hydrolysis of ammonia borane. Zero-order kinetics has often been reported to characterize the ammonia borane hydrolysis reaction [10,17,21,25], except one study that has reported an order of 0.45 [26]. Hence, it has been suggested that the ammonia borane molecule interacts with a catalytic site of the metal to form some activated complex species, to which attack by a water molecule readily leads to the dissociation of the B–N bonding and the hydrolysis of the resulting BH_3 intermediate. This mechanism has been supported by Kalidindi et al. [28]. In other words, it is suggested an Eley-Rideal mechanism [29] where only one reactant, that is, ammonia borane, adsorbs over the catalysts while the other, that is, water, react without adsorbing. Note that such a mechanism has also been suggested for the hydrolysis of sodium borohydride [30]. However, in a further study, Xu and co-workers [21] have specified the role of water through a scheme (Figure 4), which is quite similar to that suggested by Kojima et al. [31] for the PtRu-$LiCoO_2$-catalyzed hydrolysis of sodium borohydride, which shows that water molecules are adsorbed over the support surface. Such a reaction mechanism is now more in agreement with a Langmuir-Hinshelwood mechanism [29] where both reactants are considered adsorbing over the catalyst surface before reacting. Such a mechanism has also been reported for the hydrolysis of sodium borohydride [32]. A recent kinetic study about Ru-catalyzed hydrolysis of ammonia borane has also supported the Langmuir-Hinshelwood mechanism [26].

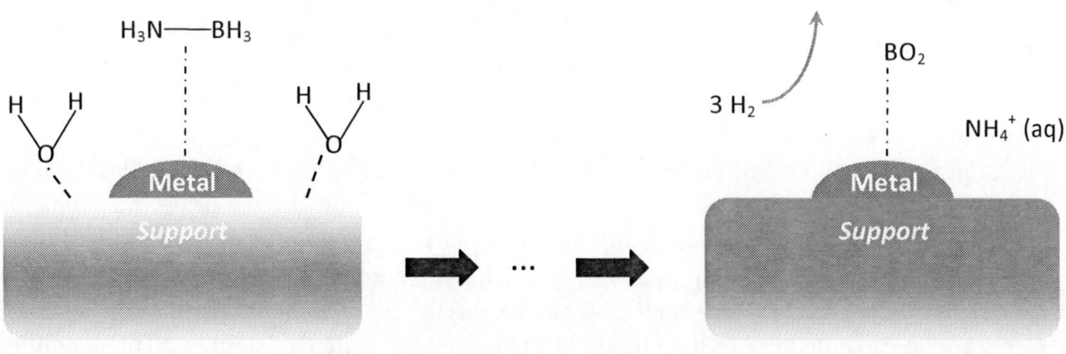

Figure 4. Hydrolysis of ammonia borane over supported transition metal catalysts as suggested by Xu and co-workers [21].

As it has been the case for sodium borohydride, the studies devoted to the catalyzed hydrolysis of ammonia borane mainly focus on the metal-based catalytic material. One may even say that today improving hydrogen generation kinetics is rather favored, and this, at the detriment of improving both effective storage capacities and by-products recycling. These last two aspects are discussed hereafter.

2.2.2 Effective Gravimetric Hydrogen Storage Capacity

As discussed in section 2.1, the occurrence of equation (5) is detrimental to the effective gravimetric hydrogen storage capacity of the system NH_3BH_3-$2H_2O$. No study has shown hydrogen release with a hydrogen to ammonia borane ratio higher than the value 3 (equation (5)).

Xu and co-workers [12,21,25] have reported that upon the completion of hydrolysis the ratio reaches the same value 3, at different ammonia borane concentrations, and such behavior is reported to be independent of the weight of water. Note that this has been confirmed by other studies [10,17,19]. Yao et al. [10] have even reported ratios lower than 3 for Pt-based bimetallic alloys. For example, the 20 wt% PtRu-C catalyst released only 2.8 mol hydrogen per mol ammonia borane. According to Cheng et al. [17], the ratio 3 is a promising result because it corresponds to 8.9 wt% of the reactants NH_3BH_3 + 2 H_2O and the authors have well remarked that solvent water is excluded from the calculation. This is a mistake since the solvent water has also to be on-board stored for any application. Hence, by taking into consideration the excess of water in addition to the reactant water, the gravimetric hydrogen storage capacity is below 1 wt%, which is far from the lowest technical target, that is, of 9 wt%.

As it has been the case for the hydrolysis of sodium borohydride, the optimization of the aqueous solution of ammonia borane is a research priority yet. Screening of catalytic materials conditions the experimental conditions that are in favor of high reactivity. Typically, the effective gravimetric hydrogen storage capacities are below 1 wt%. However, one question arises: is it realistic to think that it would be likely to reach a capacity of 9 wt%? This will be discussed in section 2.3.

2.2.3 Recycling of the Reaction by-Products

Recycling of the reaction by-products is crucial for any chemical as well as, in the present case, ammonia borane because it has the disadvantage of not being a compound storing hydrogen reversibly. Hence, it should be, at least, recyclable, want of being reversible. A target for the regeneration efficiency has been set. It is of 60 % [5].

Concerning sodium borohydride, the recycling of the reaction by-products has been one of the key issues leading to a no-go recommendation for this hydride for vehicular applications [5]. Processes for recycling the (hydrated) metaborates stem from the hydrolysis of sodium borohydride exist but none has been efficient enough in terms of yield (lower than the target 60 %) and cost. Nevertheless, while specifying the reasons for the no-go decision, it was acknowledged that the sodium borohydride recycling challenge is rather difficult and that all of the efforts could be of benefit for the hydrogen storage via ammonia borane. And actually, regeneration of ammonia borane is as crucial as regeneration of sodium borohydride, especially with the prospect of automotive applications [4].

As showed by reactions (1) to (7), ammonia borane hydrolysis generates various borate species and ammonium ion. Recently it has been reported the liberation of ammonia (NH_3) [19], the quantity of ammonia liberated depending on the concentration of ammonia borane and the rate of addition of the catalyst (i.e. an aqueous solution of salt). For hydrogen storage materials intended to generate hydrogen to feed fuel cells, ammonia liberated during hydrolysis can be detrimental to such power technologies [19]. Up to now, several regeneration processes (acidic digestion and reduction, electrochemical, and so on), which depend on the by-products, have been tested. They have been largely reviewed by Stephens et al. [1] and Peng and Chen [33]. None of the suggested regeneration processes has showed efficiency enough. According to Crabtree [34], ammonia borane regeneration is not at all straightforward. An alternative solution has been recently suggested. Instead of recycling the by-products of ammonia borane, it has been proposed that new uses of spent fuel could be a solution but this would defeat the sustainability objectives (depletion of boron deposits) [35].

Actually recycling of ammonia borane by-products have not been deeply investigated yet and there is no doubt that new regeneration routes will be soon found. Today, there are two major criteria: the yield of 60 % and the recycling overall cost. Recyclability of ammonia borane is still one of the main challenges to its practical use [33].

2.3 Ammonia Borane versus Sodium Borohydride

Ammonia borane has often been compared to sodium borohydride. Today the latter is often considered as a reference for the studies about ammonia borane. Recently Demirci and Miele [4] compared both hydrides. On the basis of their respective state-of-the-art, it has been reported the following conclusions:

- Both hydrides suffer from low (not-optimized) hydrogen storage capacities, inefficiency in spent fuel recycling, and catalyst inefficiency in terms of durability;
- The sodium borohydride hydrolysis has a higher potential than the ammonia borane hydrolysis for portable applications (sodium borohydride being commercially available and having a higher effective storage capacity);
- Sodium borohydride is not really competing with ammonia borane; each may be intended to specific applications: *e.g.* sodium borohydride for portable (considering its hydrolysis) and ammonia borane for automotive applications (considering its thermolysis).

To summarize Demirci and Miele's analysis, it appears to be rather irrelevant to develop hydrolysis of ammonia borane in parallel to that of sodium borohydride. A new comparison of the state-of-the-art of both hydrides would not be relevant because of the differences in efforts devoted to each technology. Nevertheless a discussion about the theoretical gravimetric hydrogen storage capacities could be envisaged.

Table 3 shows the theoretical gravimetric hydrogen storage capacities of ammonia borane versus those of sodium borohydride for the neat hydrides and the hydride-water systems. Hydrogen density of neat ammonia borane is indeed higher than that of neat sodium borohydride but when the water weight is integrated to the storage capacity calculations hydrogen density of ammonia borane-water system is lower than that of sodium borohydride-water system. For example, when the reactants amounts are taken in the stoichiometric conditions, the latter system theoretically stores 10.8 wt% versus 9.0 wt% for the former system. And it is the case whatever the water amount taken into consideration. These data support the Demirci and Miele's conclusion given in the previous paragraph [4]. In fact there are two problems with respect to ammonia borane in aqueous solution [1]:

- Formation of ammonia that reduces significantly hydrogen release;
- Volatile ammonia in the hydrogen stream that can reach and poison a fuel cell [1]; and,
- Ammonia borane solubility (23 g / 100 g water at 23 °C [10]) that is lower than that of sodium borohydride (55 g / 100 g water at 23 °C [4]); according to Stephens et al. [1] none hydride (in aqueous solution) can release more than 7 wt% hydrogen.

The consequences of the first problem have been illustrated in Table 3 since, because of the formation of ammonia, ammonia borane only releases half of its hydrogen. In order to circumvent the ammonia problem, it has been suggested the usage of methanol instead of water [28]. Methanolysis of ammonia borane has been reported to follow either reaction (8) [28,36] or reaction (9) [13,19]:

$$NH_3BH_3 + 4\ CH_3OH \rightarrow NH_4B(OCH_3)_4 + 3\ H_2 \quad (8)$$

$$5\ NH_3BH_3 + 22\ CH_3OH \rightarrow [NH_4]B(OCH_3)_4]_5.2CH_3OH + 15\ H_2 \quad (9)$$

Table 3. Theoretical gravimetric hydrogen storage capacities of ammonia borane versus those of sodium borohydride.

	Ammonia borane	Sodium borohydride
Neat hydride	10.8	19.6
Hydride-Water in stoichiometric amounts	10.8	9.0
Hydride-Water if $B(OH)_4^-$ as by-product	7.3	5.8
Hydride-Water at the hydride solubility [a]	6.6	4.3
Hydride-Water at the metaborate solubility [b]	5.0	4.1

[a] Solubility: 55 g / 100 g water for sodium borohydride and 23 g / 100 g water for ammonia borane
[b] Solubility: 33 g / 100 g water

However, it is to specifically note that the theoretical gravimetric hydrogen storage capacity of the system ammonia borane-methanol (according to reaction (8)) is only of 3.8 wt% (4.8 wt% for the system sodium-borohydride-methanol). Hence the system becomes not competitive at all. The second problem tackled above is also a limiting factor against high gravimetric hydrogen storage densities. For example, Cheng et al. [17] have showed that the dissolution of solid ammonia borane at room temperature becomes difficult at concentrations higher than 5 wt%, which happens to be a hurdle in attaining a further higher gravimetric density of hydrogen production.

One has to recognize that the picture drawn here is quite negative and seems to be against ammonia borane hydrolysis. However it is important to note that research on this boron hydride as hydrogen carrier is at its early beginning. These last years, improvements have been showed, especially in terms of hydrogen generation rates by virtue of more reactive and durable catalytic materials. One might expect further improvements. According to Chen et al. [17], there is still much room left to further improve the performance of hydrogen generation. It is true but two question arouse: what can be said about the effective storage capacity? Is there room left to further improve the effective hydrogen density although the theoretical data show values lower than the criteria? Xu and co-workers [37] compared various hydrides and have reported that there is no benefit of using ammonia borane instead of sodium borohydride.

Note that besides the issues discussed above, there are many other issues that can hurdle the hydrides development more or less severely. One can cite, for example, the purity of generated hydrogen, the by-product ($NaBO_2$) recovery, the by-product crystallization, the thermal management, and the refueling of fresh hydride. To this list, another issue can be added. When hydrogen storage capacities, especially the gravimetric one, are calculated, the net storage system as a whole is always neglected. Capacities are assessed on the basis of the hydride and not on the basis of the whole system. This is a mistake since the specifications have been set in relation to the net storage system [4].

3. Ammonia Borane Electrooxidation

3.1 Direct Liquid Fuel Cells

Fuel cells are power technologies of high potential because they offer the possibility to convert chemical energy (found in e.g. hydrogen, methanol, sodium borohydride or ammonia borane) into electrical energy. One way of seeing this technology is that fuel cell stores indirectly electricity. Fuel cells are generally intended to be powered by hydrogen but, especially for the polymer electrolyte membrane fuel cells (PEMFC), it is also possible to use liquid chemicals (precisely aqueous solutions) as fuels instead of hydrogen. This kind of fuel cells is called direct liquid-fed fuel cell (DLFC).

A lotMany of chemicals have been already suggested as potential fuels of direct liquid-fed fuel cells. Most of them are organic but few are inorganic. Those chemicals are ammonia, ammonia borane, dimethoxymethane, dimethyl ether, ethanol, ethylene glycol, formic acid, hydrazine, methanol, 1-methoxy-2-propanol, 1-propanol, 2-propanol, sodium borohydride, tetramethyl orthocarbonate, trimethoxymethane and trioxane [8]. The most common and studied fuel is methanol. Except formic acid, all the fuels display acceptable theoretical energy properties. According to Demirci [8], direct liquid-fed fuel cells fed by dimethyl ether, ethanol, propanol or sodium borohydride are the most promising systems from the energy point of view. However, there are all below polymer electrolyte membrane fuel cell fed with hydrogen, which is able to show energetic capacities more than three times superior to that of direct borohydride fuel cell.

Sodium borohydride and ammonia borane (as aqueous solutions) are thus potential fuels of direct liquid-fed fuel cells. Direct borohydride fuel cell (DBFC) has actually been the object of intense investigations since the early 2000s [9] while direct ammonia borane fuel cell (DABFC) is a more recent technology [4]. The present section discusses basics, state-of-the-art and issues relating to the latter fuel cell.

3.2 Basics

Direct borohydride fuel cell and direct ammonia borane fuel cell consist of a hydride solution fed anode, an oxygen fed cathode and an anion (or cation) exchange membrane. The direct borohydride fuel cell (Figure 5) is based on the borohydride ion electrooxidation and the oxygen electroreduction:

Anode $BH_4^- + 8\ OH^- \rightarrow BO_2^- + 6\ H_2O + 8\ e$ (10)

Cathode $2\ O_2 + 4\ H_2O + 8\ e^- \rightarrow 8\ OH($ 11)

Overall $BH_4^- + 2\ O_2 \rightarrow BO_2^- + 2\ H_2O$ (12)

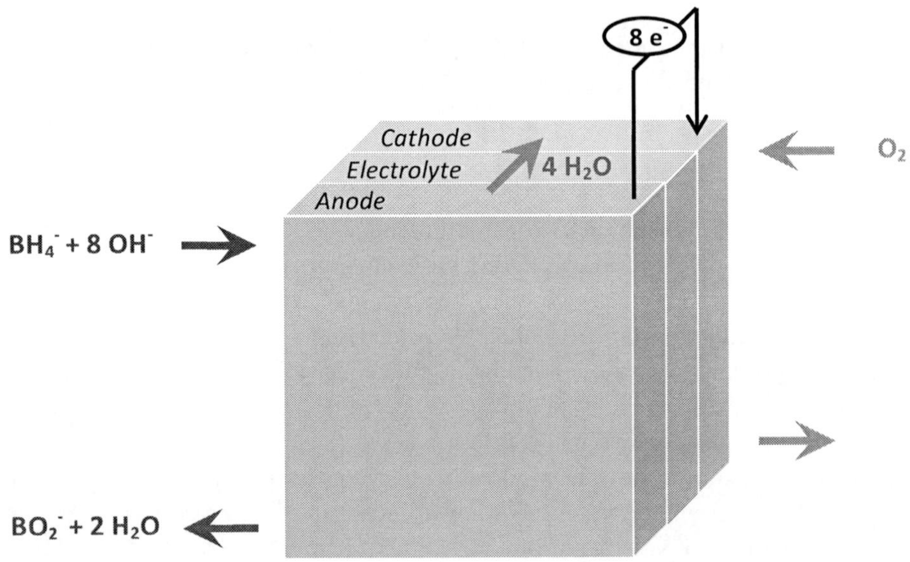

Figure 5. Direct borohydride fuel cell.

The standard potential is 1.64 V versus the standard hydrogen electrode (SHE). The electrooxidation takes place in alkaline medium because sodium borohydride is unstable in acidic medium. The theoretical specific energy of direct borohydride fuel cell (9300 Wh kg^{-1}) is among the highest ones. The pure compound capacity is 5700 Ah kg^{-1}. On the basis of these theoretical data, one may assert that direct borohydride fuel cell is among the best direct liquid-fed fuel cells and that has motivated many research groups for working on its development [8,9].

Direct ammonia borane fuel cell (Figure 6) is based on the ammonia borane electrooxidation and the oxygen electroreduction:

Anode NH3BH3 + 6 OH- → BO2- + NH4+ + 4 H2O + 6 e (13)

Cathode $3/2\ O_2 + 3\ H_2O + 6\ e^- \rightarrow 6\ OH$ (14)

Overall $NH_3BH_3 + 3/2\ O_2 \rightarrow BO_2^- + NH_4^+ + H_2O$ (15)

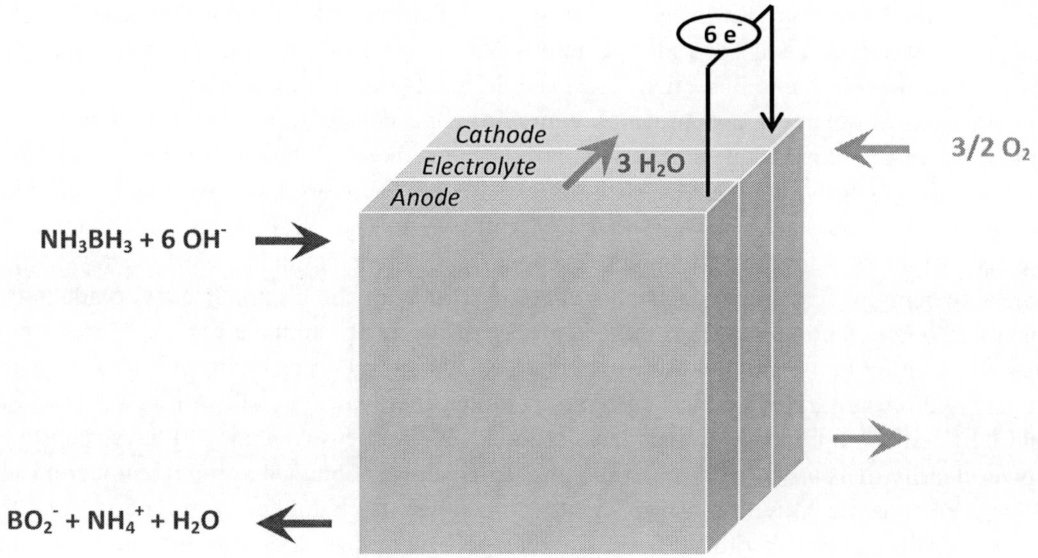

Figure 6. Direct ammonia borane fuel cell.

The standard potential is 1.62 V versus the standard hydrogen electrode (SHE). The electrooxidation takes place in alkaline medium because ammonia borane is unstable in acidic medium. The theoretical specific energy of direct ammonia borane fuel cell (8400 Wh kg^{-1}) is among the highest ones [38]. The pure compound capacity is 5200 Ah kg^{-1} [39]. On the basis of the classification made by Demirci [8], one may assert that direct ammonia borane fuel cell is among the best direct liquid-fed fuel cells but such a comparison is made on the basis of theoretical data. Despite similar potentials (-1.62 versus -1.64 V), a lower number of electrons (6 versus 8) makes direct ammonia borane fuel cell less energetic than direct borohydride fuel cell. It is therefore essential to analyze the real performances of direct ammonia borane fuel cell.

3.3 State-of-the-art and Issues

Direct ammonia borane fuel cell is a very recent technology since, to the best of our knowledge, the first study has been published in 2007 [39]. However, since then, very few studies have been reported throughout the open literature. Actually, there are two groups of researchers involved in this fuel cell, the main one being Xu's group [38-40], that is, that one much involved in the hydrolysis of ammonia borane.

For the first time in 2007, Yao et al. [10] and Xu and co-workers [39] proposed, almost simultaneously, direct ammonia borane fuel cell (or direct borazane fuel cell). They emphasized the six electrons released by ammonia borane electrooxidation and the cell potential comparable to that of direct borohydride fuel cell. Moreover, according to Xu and co-workers [39], ammonia borane for direct ammonia borane fuel cell combines the following advantages: stability, cheapness, availability, solubility in water, environmental safety of the discharge products, and recyclability of the by-products. Nevertheless Demirci and Miele [4] have remarked that this drawn picture entirely benefits the direct ammonia borane fuel cell and that's why they have been more cautious with respect to the cost, the availability, the recyclability (as discussed in section 2.2.3) and the safety of the by-products.

As already reported in a previous paper [4], one can legitimately wonder about the utilization of the expression "state-of-the-art" because direct ammonia borane fuel cell is a new technology for which there is so little investment. There are only five papers reporting performances. These are briefly reported hereafter. Yao et al. [10] screened a number of metals including platinum, gold, nickel, copper and silver as anode electrocatalyst, their objective being to find an anode catalyst not only active for the electrochemical oxidation of ammonia borane but also inert to the chemical hydrolysis of ammonia borane. It was found that silver is the best candidate while platinum shows too much reactivity towards ammonia borane hydrolysis. A fuel cell was constructed and an energy density of 2 Ah g^{-1} at 1 mA cm^{-2} and 0.9 V at room temperature has been reported. Xu and co-workers [39] have reported a power density of 14 mW cm^{-2} (or 93 mW mg^{-1}), which was obtained at room temperature for a fuel cell constructed with a 30 wt% platinum catalyst. Both studies have reported that the open circuit potential is more negative than the hydrogen potential. In a further study, Xu and co-workers [38] obtained a power density of 110 mW cm^{-2} at a current density of 185 mA cm^{-2} and a potential of 0.6 V at 45 °C. Such performances made the authors conclude that the application of direct ammonia borane fuel cell to portable devices may be expected. However it would have been interesting to have an assessment of the energy cost since this fuel cell was constructed with catalyst layers containing 46.6 wt% platinum. Much more recently, Xu and co-workers [40] have suggested the utilization of a gold electrode as anode electrocatalyst. Gold has been proposed as a solution to overcome the unwanted hydrolysis of ammonia borane. The utilization of gold is not a new idea. It has been much largely investigated in the case of direct borohydride fuel cell, for which hydrolysis of borohydride ion is also an unwanted side reaction [9]. By cyclic voltammetry, it was showed that both hydrolysis and electrooxidation of ammonia borane occur over gold electrocatalyst. In a further recent article, Xu and co-workers [41] have reported the synthesis of Fe@Pt core-shell nanoparticles as potential electrocatalyst for ammonia borane electrooxidation, the objective being to decrease the catalyst cost by decreasing the platinum amount while increasing both activity and durability of the catalytic material.

Like direct borohydride fuel cell, direct ammonia borane fuel cell encounters many issues, and unsurprisingly the issues are the same ones [4]. The two main issues that can be discussed now are the occurrence of unwanted hydrolysis and the occurrence of ammonia borane crossover through the fuel cell membrane. Both issues have been recently discussed in a previous paper [4].

Hydrolysis of ammonia borane is expectable with the metal catalysts generally used in fuel cells such as platinum or ruthenium. Yao et al. [10] screened several transition metals because they searched for the metal the most inactive towards this side reaction. It was found that silver is a potential candidate. Xu and co-workers [39] have evidenced the competition between the ammonia borane electrooxidation and the ammonia borane hydrolysis, the latter being put in evidence by cyclic voltammetry. As preliminary information about their work, they [39] have then remarked that the hydrolysis of ammonia borane occurring to some extent in the ammonia borane aqueous solution must be avoided in order to effectively utilize the faradic capacity of ammonia borane. In further studies, Xu and co-workers [38,40] have admitted that there are several problems with direct ammonia borane fuel cell. It has been reported that the hydrogen evolution lowers the cell efficiency and makes the system complicated [38,40]. A gold electrode, known for its low activity towards sodium borohydride hydrolysis, and thiourea, a hydrogen evolution inhibitor, were utilized [40]. The cyclic voltammogram obtained with gold was complex, putting in evidence the competition ammonia borane hydrolysis versus ammonia borane electrooxidation. However, the electrooxidation (or electroreduction) of NH_4^+ was not considered [4] while, for example, a NH_4^+-powered direct liquid-fed fuel cell (*i.e.* based on the NH_4^+ oxidation) has been proposed yet [42]. The utilization of thiourea permitted an effective minimization of the hydrolysis. However Xu and co-workers [40] suggested that the issue of the long-term anode performance, stability and poisoning in the presence of thiourea has to be considered. Note that the anode poisoning in the presence of thiourea had before been discussed by Demirci [43].

Crossover of borohydride ion through the anion exchange membrane makes the development of direct borohydride fuel cell complicated. Crossover significantly reduces the fuel cell performances and makes it less efficient and much expensive [9]. Crossover has also been observed and reported for direct ammonia borane fuel cell. According to Xu and co-workers [40], the crossovering specie is ammonia borane. However, in our opinion, as we have already reported in reference [4], the species crossovering through the cation exchange membrane are certainly the ammonium cation and also, why not, reaction intermediates such as the anion $BH_3(OH)^-$ (since anions may crossover such membrane [9]). The crossover may be reduced or even hindered by increasing the selectivity of the membrane. Nevertheless the utilization of an ion exchange membrane in the direct ammonia borane fuel cell is complicated because of the presence of both cation (ammonium ion) and anions (ammonia borane reaction intermediates) at the anode. Whatever the membrane, there will be crossover.

Now, there are very few papers dealing with direct ammonia borane fuel cell, which is not really enough to discuss relevantly the state-of-the-art of this fuel cell that, or else, appears to be fairly similar to that of direct borohydride fuel cell. They are similar with regard to their theoretical capacities, advantages (*e.g.* aqueous solution as fuel, fuel stability, membrane utilization, and so on) and issues (at least for the ones experimentally put in evidence so far). The side reaction (hydrolysis), the crossover and the cost are common issues.

The only conclusions that can be made are as follows. First, there are real opportunities for performance improvements since research is at its beginning but many more research groups should be involved. The case of direct borohydride fuel cell is a good example. Even after near ten years of research and development efforts it is commonly admitted that direct borohydride fuel cell is still at an initial phase of development. Second, the amount of utilized platinum is high and this is detrimental to the fuel cell cost. Searching for efficient non-noble metals should be a priority as reported by Yao et al. [10] and Xu and co-workers [41]. Third, hydrogen gas evolution and crossover that lower the cell efficiency and make the system complicated are significant issues [39] and the experience in direct borohydride fuel cell has shown that without addressing these problems the fuel cell cannot be technically, commercially viable.

To summarize, it clearly appears that any technological breakthrough in direct borohydride fuel cell should benefit direct ammonia borane fuel cell and accelerate their respective development. The technological backwardness of the latter fuel cell might then be made good by virtue of the know-how and experience developed in the former fuel cell. This is more largely discussed in the next section.

3.4 Direct Ammonia Borane Fuel Cell Vversus Direct Borohydride Ffuel Cell

Experience in direct borohydride fuel cell may benefit direct ammonia borane fuel cell for predicting and above all solving the issues that the latter fuel cell encounters. Direct borohydride fuel cell has encountered the following issues and these ones hinder its development yet [4,9]:

- Cost; for example, Liu et al. [44] have remarked that the electricity cost for the direct borohydride fuel cell is 100 times in comparison with that for the hydrogen-fed fuel cell when only the fuel cost is taken into account;
- The number of electrons utilized per borohydride ion oxidized is less than eight that is the ideal maximum number of electrons;
- Hydrogen gas evolution because of a competition between unwanted sodium borohydride hydrolysis and wanted sodium borohydride electrooxidation; hydrogen gas evolution is particularly detrimental to the fuel cell performance;
- Noble metals with high hydrogen over-potential, such as gold and silver, permit an 8-electron electrooxidation due to lowered hydrolysis but these metals are not effective enough in terms of electrode kinetics and power density [45];
- Borohydride crossover through the electrolyte membrane; this crossover is detrimental to the performance of the cathode electrocatalyst; it is generally suggested that more selective membranes should solve this issue, another solution being to use borohydride-resistant cathode materials such as manganese oxide (MnO_2) [46];

The issues listed here are the main ones. To these, it may be added the following ones: durability of each component of the direct borohydride fuel cell, by-product crystallization at anode after use, stacking loss, direct borohydride fuel cell system design, by-product recycling, and long-term operation [4,47].

The state-of-the-art related to direct ammonia borane fuel cell reveals that the main issues faced by direct borohydride fuel cell are encountered by the former fuel cell. Hydrogen evolution and crossover have been several times reported. Even though cost is not really tackled, it is to recognize that the utilized noble metals and ammonia borane are expensive materials increasing the whole cost of the fuel cell.

Direct ammonia borane fuel cell is a novel technology and in that sense its comparison with direct borohydride fuel cell is made difficult because of the huge difference in research and development efforts. A comparison based on theoretical information is even though possible [4]. In a recent article, it has been reported that:

- Direct borohydride fuel cell is superior to direct ammonia borane fuel cell in terms of theoretical performances (number of electrons involved, standard potential, theoretical specific energy and pure compound capacity) as shown in Table 4;
- With respect to the fuel stability and the by-product features (except ammonium ion), both fuel cells are equivalent;
- The presence of ammonium ion in the anolyte of direct ammonia borane fuel cell has to be clarified in terms of reactivity, recyclability, and crossover.

Table 4. Direct ammonia borane fuel cell versus direct borohydride fuel cell [4].

	DBFC	DABFC
Properties		
Standard electrode potential (V vs. SHE)	1.64	1.62
Number of electrons involved	8	6
Theoretical specific energy (Wh kg^{-1})	9300	8400
Pure compound capacity (Ah kg^{-1})	5700	5200
Others		
Fuel (aqueous solution) stability	Yes $^{\text{NaOH-stabilized}}$	Yes
Cost	High	High
By-products	BO_2^-	BO_2^-, NH_4^+
By-products recyclability	Inefficient	Inefficient
Side reaction	Hydrolysis	Hydrolysis
Crossover	Yes ^{borohydride}	Yes $^{\text{species not clearly specified yet}}$

From these observations, Demirci and Miele [4] have asked a question: what can be the advantage of developing direct ammonia borane fuel cell when it is less effective, less practical (because of ammonium ion), and more complex than direct borohydride fuel cell? It has been answered and concluded that direct ammonia borane fuel cell is an interesting fuel cell but direct borohydride fuel cell appears more attractive and less complicated.

Conclusion

In hydrogen storage, ammonia borane is maybe the most promising boron hydride since it stores near 20 wt% atomic hydrogen. Nevertheless stored hydrogen has to be released and its release is a challenge yet as it is for both sodium borohydride and lithium borohydride. Actually, ammonia borane is quite similar to sodium borohydride. Let us specify that:

- Both can release stored hydrogen either by thermolysis or hydrolysis;
- Both can be considered as liquid fuel of direct liquid-fed fuel cell; and
- Both encounter the same issues, which hinder their development.

Besides its potential as hydrogen storage material, ammonia borane has thus a potential as fuel of fuel cell. Both aspects have been discussed through the present chapter, which has especially focused the hydrolysis of ammonia borane and the direct ammonia borane fuel cell.

Like sodium borohydride hydrolysis, ammonia borane hydrolysis can be accelerated by using a metal-based catalyst. In other words, the hydrogen release can be obtained at controlled rates. However, the amount of hydrogen released is much lower than the theoretical hydrogen density of near 20 wt% because of the formation of ammonia that reduces significantly the hydrogen release and, because of the lower solubility of ammonia borane in relation to that of sodium borohydride. This is maybe the most important issue. In fact it appears that there are many other issues, which are as follows:

- Ammonia borane as well as sodium borohydride suffer from low (not-optimized) hydrogen storage capacities;
- Both suffer from inefficiency in spent fuel recycling; and,
- Both suffer from catalyst inefficiency in terms of durability.

Ammonia borane is often compared to sodium borohydride. Here, it has been showed that the sodium borohydride hydrolysis has a higher potential than the ammonia borane hydrolysis for portable applications. This conclusion is based on the fact that sodium borohydride is commercially available and has a higher effective storage capacity.

Like sodium borohydride, ammonia borane electrooxidizes and this can be valorized in a direct liquid-fed fuel cell. Direct ammonia borane fuel cell has been recently envisaged and few studies have showed that such fuel cell has a certain potential. Nevertheless, like direct borohydride fuel cell, direct ammonia borane fuel cell encounters many issues, and unsurprisingly the issues are the same ones, that is:

- The occurrence of the unwanted hydrolysis; and,
- The occurrence of ammonia borane crossover through the fuel cell membrane.

These issues are the main, most detrimental ones. Direct ammonia borane fuel cell is a novel technology and in that sense its comparison with direct borohydride fuel cell is made difficult because of the huge difference in research and development efforts. Nevertheless from theoretical considerations, it is that direct borohydride fuel cell appears more attractive and less complicated than direct ammonia borane fuel cell.

To summarize both topics developed in the chapter, it has been showned that sodium borohydride is quite superior to ammonia borane in hydrogen storage, hydrogen release through hydrolysis and in fuel cell application. Moreover, sodium borohydride cost is today much lower than that of ammonia borane. Nevertheless, sodium borohydride has no potential is thermolysis while ammonia borane shows high potential for releasing hydrogen through this thermal process[9].

REFERENCES

[1] Stephens, F. H.; Pons, V.; Baker, R. T. *Dalton Trans.* 2007, 2613-2626.
[2] Schlesinger, H. I.; Brown, H. C.; Finholt, A. E.; Gilbreath, J. R.; Hoekstra, H. R., Hyde, E. K. *J. Am. Chem. Soc.* 1953, *75*, 215-219.
[3] Demirci, U. B.; Miele, P. *C. R. Chim.* 2008, doi:10.1016/j.crci.2008.08.002.
[4] Demirci, U. B.; Miele, P. *Energy Environ. Sci.* 2009, doi:10.1039/b900595a.
[5] U.S. Department of Energy Hydrogen Program. (2007). *Independent review. November 2007. Go/no-go recommendation for sodium borohydride for on-board vehicular hydrogen storage.* NREL/MP-150-42220. http://www.hydrogen.energy.gov/.
[6] Demirci, U. B.; Akdim, O.; Miele, P. *Int. J. Hydrogen Energy* 2009, *34*, 2638-2645.
[7] Lane, C. F. *Ammonia-borane and related N-B-H compounds and materials: safety aspects, properties and applications, U.S. Department of Energy Chemical Hydrogen Storage Center.* http://www1.eere.energy.gov/hydrogenandfuelcells/pdfs/nbh_h2_storage_survey.pdf.
[8] Demirci, U. B. *J. Power Sources* 2007, *169*, 239-246.
[9] Demirci, U. B. *J. Power Sources* 2007, *172*, 676-687.
[10] Yao, C.; Yang, H.; Zhuang, L.; Ai, X.; Cao, Y.; Lu, J. *J. Power Sources* 2007, *165*, 125-127.
[11] Storozhenko, P. A.; Svitsyn, R. A.; Ketsko, V. A.; Buryak, A. K.; Ul'yanov, A. V. *Russ. J. Inorg. Chem.* 2005, *50*, 980-985.
[12] Chandra, M.; Xu, Q. *J. Power Sources* 2006, *156*, 190-194.
[13] Umegaki, T.; Yan, J.M.; Zhang, X.B.; Shioyama, H.; Kuriyama, N.; Xu, Q. *Int. J. Hydrogen Energy* 2009, doi:10.1016/j.ijhydene.2009.01.002.
[14] Ryschkewitsch, G. E. *J. Am. Chem. Soc.* 1960, *82*, 3290-3294.
[15] Kelly, H. C.; Marriott, V. B. *Inorg. Chem.* 1979, *18*, 2875-2878.
[16] Chandra, M.; Xu, Q. *J. Power Sources* 2006, *159*, 855-860.
[17] Cheng, F.; Ma, H.; Li, Y.; Chen, J. *Inorg. Chem.* 2007, *46*, 788-794.
[18] Kalindindi, S. B.; Indirani, M.; Jagirdar, B. R. *Inorg. Chem.* 2008, *47*, 7424-7429.
[19] Ramachandran, P. V.; Gagare, P. D. *Inorg. Chem.* 2007, *46*, 7810-7817.
[20] Simagina, V. I.; Storozhenko, P. A.; Netskina, O. V.; Komova, O. V.; Odegova, G. V.; Larichev, Y. V.; Ishchenko, A. V.; Orezova, A. M. *Catal. Today* 2008, doi:10.1016/j.cattod.2008.05.022.
[21] Xu, Q.; Chandra, M. *J. Power Sources* 2006, *163*, 364-370.
[22] Yan, J. M.; Zhang, X. B.; Han, S.; Shioyama, H.; Xu, Q. *Angew. Chem. Int. Ed.* 2008, *47*, 2287-2289.
[23] Umegaki, T.; Yan, J.M.; Zhang, X.B.; Shioyama, H.; Kuriyama, N.; Xu, Q. *Int. J. Hydrogen Energy* 2009, doi:10.1016/j.ijhydene.2009.03.003.

9 For more details, see Chapters 4 and 7.

[24] Umegaki, T.; Yan, J.M.; Zhang, X.B.; Shioyama, H.; Kuriyama, N.; Xu, Q. *J. Power Sources* 2009, doi:10.1016/j.Jpowsour.2009.02.054.
[25] Chandra, M.; Xu, Q. *J. Power Sources* 2007, *168*, 135-142.
[26] Basu, S.; Brockman, A.; Gagare, P.; Zheng, Y.; Ramachandran, P. V.; Delgass, W. N.; Gore, J. P. *J. Power Sources* 2009, *188*, 238-243.
[27] Yao, C. F.; Zhuang, L.; Cao, X. P.; Ai, X. P.; Yang, H. X. *Int. J. Hydrogen Energy* 2008, *33*, 2462-2467.
[28] Kalindindi, S. B.; Sanyal, U.; Jagirdar, B. R. *Phys. Chem. Chem. Phys.* 2008, *10*, 5870-5874.
[29] Somorjai, G. *Introduction to surface chemistry and catalysis*. Wiley-Interscience, John Wiley and Sons: New York, 1994.
[30] Personal communication.
[31] Kojima, Y.; Suzuki, K.I.; Fukumoto, K.; Sasaki, M.; Yamamoto, T.; Kawai, Y.; Hayashi, H. *Int. J. Hydrogen Energy*. 2002, *27*, 1029-1034.
[32] Zhang, J.S.; Delgass, W.N.; Fisher, T.S.; Gore, J.P. *J. Power Sources*. 2007, *164*, 772-781.
[33] Peng, B.; Chen, J. *Energy Environ. Sci.* 2008, *1*, 479-483.
[34] Crabtree, R.H. *Energy Environ. Sci.* 2008, *1*, 134-138.
[35] Marder, T. B. *Angew. Chem. Int. Ed.* 2007, *46*, 8116-8118.
[36] Kalindindi, S. B.; Vernekar, A. A.; Jagirdar, B. R. *Phys. Chem. Chem. Phys.* 2009, *11*, 770-775.
[37] Xu, Q.; Chandra, M. *J. Alloys Compd.* 2007, *447*, 729-732.
[38] Zhang, X. B.; Yan, J. M.; Han, S.; Shioyama, H.; Yasuda, K.; Kuriyama, N.; Xu, Q. *J. Power Sources* 2008, *168*, 167-171.
[39] Zhang, X. B.; Han, S.; Yan, J. M.; Chandra, M.; Shioyama, H.; Yasuda, K.; Kuriyama, N.; Kobayashi, T.; Xu, Q. *J. Power Sources* 2007, *182*, 515-519.
[40] Zhang, X. B.; Han, S.; Yan, J. M.; Shioyama, H.; Kuriyama, N.; Kobayashi, T.; Xu, Q. *Int. J. Hydrogen Energy* 2009, *34*, 174-179.
[41] Zhang, X. B.; Yan, J. M.; Han, S.; Shioyama, H.; Xu, Q. *J. Am. Chem. Soc.* 2009, *131*, 2778-2779.
[42] Maffei, N.; Pelletier, L.; McFarlan, A. *J. Power Sources* 2008, *175*, 221-225.
[43] Demirci, U.B. *Electrochim. Acta* 2007, *52*, 5119-5121.
[44] Li, Z.P.; Liu, B.H.; Arai, K.; Suda, S. *J. Alloys Compd.* 2005, 404-406, 648-652.
[45] Chatenet, M.; Molina-Concha, M.B.; Diard, J.P. *Electrochim. Acta* 2009, *54*, 1687-1693.
[46] Verma, A.; Basu, S. *J. Power Sources* 2005, *145*, 282-285.
[47] Ponce de Leon, C.; Walsh, F.C.; Pletcher, D.; Browning, D.J.; Lakeman, J.B. *J. Power Sources* 2006, *155*, 172-181.

In: Boron Hydrides High Potential Hydrogen Storage Material ISBN:978-1-61668-361-0
Editors: Umit B. Demirci and Philippe Miele

Chapter 9

METAL AMIDOBORANES

Hui Wu,[1,2] Wei Zhou[1,2] and Taner Yildirim[1,3]

[1]NIST Center for Neutron Research, National Institute of Standards and Technology, Gaithersburg, Maryland 20899

[2]Department of Materials Science and Engineering, University of Maryland, College Park, Maryland 20742

[3]Department of Materials Science and Engineering, University of Pennsylvania, Philadelphia, PA 19104, USA

ABSTRACT

Metal amidoboranes, consisting of $[NH_2BH_3]^-$ anion and various metal cations, are a new class of compounds that have recently attracted great attentions due to their intriguing crystal structures and novel chemical and physical properties. Similar to ammonia borane (AB), metal amidoboranes have great potential to meet the stringent gravimetric and volumetric hydrogen storage capacity targets needed for transportation applications. Moreover, compared to AB, metal amidoboranes have been shown to release hydrogen at lower temperatures, and at higher rates and with the release of fewer volatile contaminants. Particularly, several metal amidoboarnes also appear to release hydrogen less exothermically than AB. In this chapter, we review recent research progress on metal amidoborane compounds, including their syntheses, structures, crystal chemistry, and hydrogen storage properties.

INTRODUCTION

Hydrogen is a clean, source-independent energy carrier. One of the most challenging technological barriers to the advancement of hydrogen fuel cell technologies for mobile applications is the storage of hydrogen. Compressed gas and liquid hydrogen storage are not safe for everyday use and unlikely to meet the aggressive US Department of Energy (DOE) targets for on-board H_2 storage systems,[1] while solid state hydrogen storage, where a host material is used as a hydrogen carrier, offers improved volumetric energy densities.

Currently, there are three categories of hydrogen storage materials: high surface area sorbents, metal hydrides, and chemical hydrides. Many recent efforts have been focused on boron-containing materials primarily due to the light constitutional elements and the resulting high hydrogen storage capacities. Several promising hydrogen-rich boron-containing materials have been identified and developed in all three classes of hydrogen storage materials. Some of these B-containing materials have been reviewed in previous chapters, and the most recently developed metal amidoboranes will be discussed in greater details in this chapter.

Metal amidoboranes are a class of metal derivative compounds of the chemical hydride — ammonia borane (AB, H_3NBH_3), with many advantages in improving the hydrogen storage properties of the parent compound. Since the structures and properties of metal amidoboranes are closely related to those of AB, in order to facilitate our discussions, we first briefly summarize the properties, advantages, and disadvantages of ammonia borane as a hydrogen storage material. Ammonia borane has attracted great interest due to its combined low molecular weight (30.7 g mol^{-1}) and high gravimetric hydrogen capacity. [2,3] H_3NBH_3 contains both hydridic B-H and protic N-H bonds and a strong enough B-N bond that hydrogen release from solid AB is favored over dissociation to ammonia and diborane under noncatalytic conditions.[2] Solid AB releases 1 equivalent of H_2 at temperatures up to 110 °C and forms a mixture of products consisting mostly of aminoborane oligomers, $[NH_2BH_2]_n$.[4,5,6,7,8] $[NH_2BH_2]_n$ can further dehydrogenate 1 equivalent of H_2 at 110-200 °C and produce polyiminoboranes, $[NHBH]_n$. [4,5,6,9] While the dehydrogenation reaction in solution was significantly improved by using a variety of catalysts, the hydrogen release from solid AB is still comparably slow and includes a long induction period during which very little hydrogen is released. Also, accompanying the release of hydrogen, volatile toxic species such as borazine form with increasing heating rate[4,5,6,9], thus degrading the purity of hydrogen that can be used for fuel cell systems. In addition, pure AB melts at about 110-114°C, which will cause foaming issues during hydrogen release, leading to additional technical difficulties.[10] Finally, In order to regenerate with reasonable pressure of hydrogen, hydrogen release from a storage material should not be more exothermic than ~4.2 kJ mol^{-1}[1] The necessary cleavage of covalent bond to release hydrogen puts more stringent requirement on reversibility of AB. Pure AB releases hydrogen exothermically at ~25-33.5 kJ mol^{-1} AB, therefore, dehydrogenated AB cannot be directly rehydrogenated under hydrogen pressure. The exothermic dehydrogenation also indicates that hydrogen release of AB is kinetically controlled.[12] As a result, it has been observed that the dehydrogenation temperatures of AB vary largely with heating rate.[5,6,9]

Many investigations on promoting hydrogen release from AB have been done, including solution themolysis,[13] acid-catalyzsed dehydrogenation,[14] anionic dehydropolymerisation,[15] metal-catalyzsed dehydrogenation,[16,17,18] and scaffolded nanostructures.[19,20] Recent studies also show that regenerate H_3NBH_3 is achievable from dehydrogenated BNH_x materials. [11,21,22,23,24] Details about the potential of ammonia borane as a hydrogen storage material are presented in Chapter 7 and 8.

Despite many approaches developed to understanding and tuning the properties of AB to be practically useful in terms of the reduced dehydrogenation temperatures[19] accelerated H_2 release kinetics,[14,15,16,17,18,19,20] and/or minimized borazine release,[15] there is no single approach that can achieve all these improvements simultaneously. Collectively advancing and modifying AB in this regard is metal derivatives of AB—metal amidoboranes, which have been found to exhibit intriguing and advantageous properties over AB as potential hydrogen storage materials.

The recent activities in metal amidoboranes were largely stimulated by the important approach of hydride destabilization in the light-weight complex metal hydrides using destabilization additives to decrease the enthalpy of system hydrogen loss, e.g. $MgH_2/LiBH_4$ mixture,[25,26] or in some cases, to produce a new compound with lower thermal stability, e.g. $Li_4[BH_4][NH_2]_3$ formed[27,28,29,30,31] between $LiNH_2$ and $LiBH_4$. It was hoped that the metal hydride additives could potentially decrease the dehydrogenation enthalpy of AB and accelerate the hydrogen release rate.

Experimentally, Sneddon et al. first noticed that hydrogen release from AB can be promoted using small amounts of LiH or $LiNH_2$,[22,23] and a recent patent also reported the improved properties of AB by addition of LiH or $LiAlH_4$.[32] Such beneficial effect of metal hydride additives were further confirmed by several other groups with variety of additives including LiH, NaH, CaH_2, and $LiNH_2$.[33,34,34,36,37] It is believed that the addition of these metal hydrides deprotonates AB, produces $NH_2BH_3^-$ anions, and form metal amidoborane.

Though the formation of metal amidoboranes has been noticed in solutions decades ago,[38,39] not until recently have these compounds been evidenced in solid state by NMR and X-ray diffraction (XRD) and has their potential as a hydrogen storage material been explored.[34,35] Since metal amidoboranes for hydrogen storage are a relatively new research direction and what have been achieved is still at the very early stage, it is not surprising there is only limited number of these compounds with their crystal structures being determined so far. In addition, there is no direct experimental evidence (e.g. structural information on the intermediates and dehydrogenation products) to elucidate their dehydrogenation mechanism and no effective routes reported thus far to regenerate the dehydrogenated product back to amidoborane, making it difficult to directly compare the reaction enthalpies and barriers between metal amidoboranes and AB through first-principles reaction path calculations and thus promote further studies to achieve the possible reversibility of these materials. Finally, besides expanding the variety of metal amidoboranes, the role of catalysts/additives in controlling dehydrogenation, and possibly rehydrogenation needs to be investigated.

Besides forming metal amidoboranes, amidoborane ligand $NH_2BH_3^-$ and its related ligands $(R)NH\text{-}BH_2(R')^-$ can serve as components in many metalorganics.[40,41,42,43] The structures and properties of these derivatives is relevant to the thriving field of catalytic hydrogen generation from AB[44,45] and the early stage of dehydrogenation mechanism of metal amidoboranes.[40] More recently $NH_2BH_3^-$ was also reported as an important substitute to the ethyl anion $CH_2CH_3^-$ to enhance the agostic interaction between the metal center and the C-H bond and form more stable metal organics complexes.[46] For example, the agostic interaction is supposed to affect the stereoselective olefin polymerization at zirconocene centers, but it is typically too fleeting to characterize in a persistently stable complex. By replacing the $CH_2CH_3^-$ with $NH_2BH_3^-$, several Zr-amidoborane complexes can be formed, which mirror the structural and spectroscopic properties of their isoelectronic β-agostic ethyl analogues and exhibit new stereoisomers configurations that have not been characterized in the chemistry of the less robust β-agostic derivatives. These research fields about amidoborane derivatives have been expanded quickly and bring considerable insight to metal organic chemistry.

Overall, metal amidoboranes have attracted significant attentions worldwide for their intriguing potentials for hydrogen storage, and their many unknown aspects will be hopefully revealed and understood. Below, we will discuss the synthesis, structures, and hydrogen storage properties of this new family of compounds in greater details.

SYNTHESES

Metal amidoboranes can be prepared through both solid state synthesis and/or wet chemical synthesis techniques. In the first reports of formation of metal amidoboranes, $NaNH_2BH_3$ was formed by reacting $(CH_3)_2OBH_3$ and Na in the liquid ammonia,[38] and a reaction between ammonia borane and n-butyllithium in tetrahydrofuran produced $LiNH_2BH_3$.[39] Recently, it was found that amidoboranes containing cations with strong ionicity such as $LiNH_2BH_3$, $NaNH_2BH_3$ and $Ca(NH_2BH_3)_2$ can be formed by ball milling mixtures of metal hydrides and AB powders.[34,35] The H^- anion in these hydrides serves as a strong Lewis base and can easily deprotonate AB, generate H_2 molecule and form metal amidoboranes. The mechanical milling is a facile and direct technique to synthesis metal amidoboranes, however, it is difficult to obtain single-phase amidoboranes through such method because i) homogeneously mixing in the solid state cannot be achieved in the atomic scale, resulting in an incomplete reaction or local compositional heterogeneity, ii) other reaction mechanisms such as head-to-tail dimerization may lead to the formation of adducts that are off the stoichiometry of the initial mixtures.

Xiong et al. reported that ball milling 1:1 ratio LiH/AB mixture at 200 rpm can produce pure $LiNH_2BH_3$,[34] however, a later study using the same synthesis condition found that in the mixture of LiH/AB noticeable amount of secondary phase $LiNH_2BH_3 \cdot NH_3BH_3$ was formed (Figure 1),[35] which was also present in the Xiong et al.'s XRD pattern of $LiNH_2BH_3$ but was not addressed at the time. Studies also found that the XRD pattern on the milled LiH/NH_3BH_3 powders varies with milling time and rate,[32,35] indicating more phase changes with further reactions between the precursors and/or products.

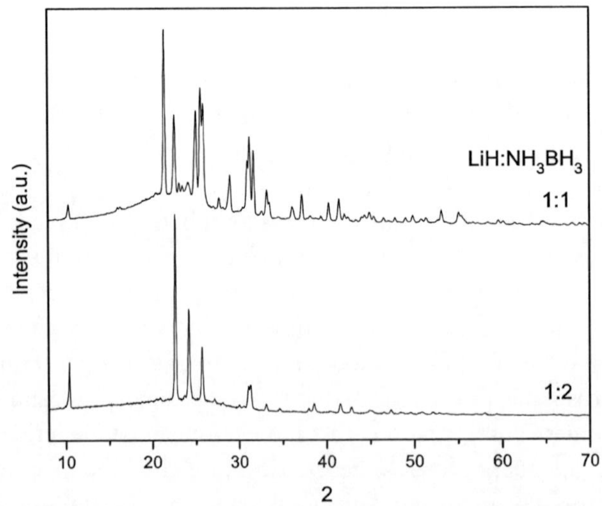

Figure 1. XRD patterns of a) $LiNH_2BH_3$ formed after ball milling 1:1 ratio of LiH/NH_3BH_3 with the presence of a secondary phase $LiNH_2BH_3 \cdot NH_3BH_3$. b) $LiNH_2BH_3 \cdot NH_3BH_3$ formed after ball milling 1:2 ratio of LiH/NH_3BH_3.

$Ca(NH_2BH_3)_2$ can be formed through ball milling CaH_2 and NH_3BH_3 without appearance of any other secondary off-stoichiometric phases, but small amount of un-reacted precursors sustain even with extended milling time (Figure 2), indicating deficient reactions between precursors.

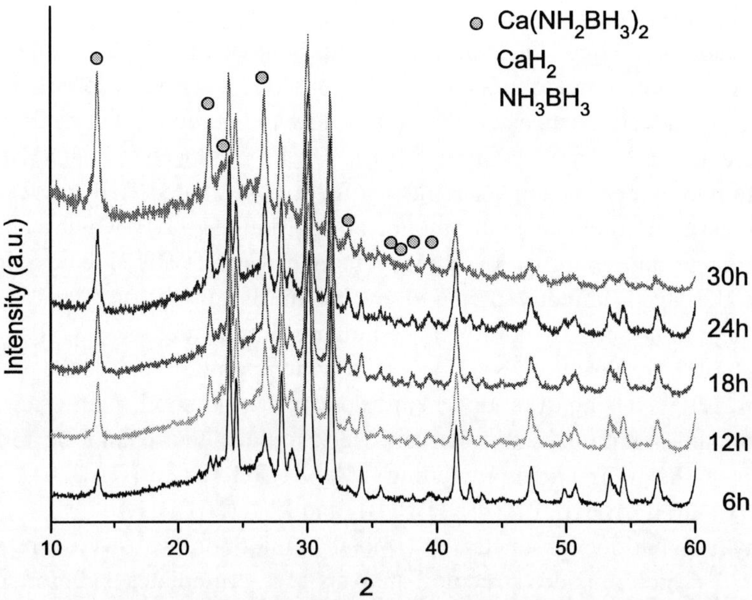

Figure 2. XRD patterns of the formation of $Ca(NH_2BH_3)_2$ with extended ball milling time at 350 rpm ball mill rate. Adapted from [35] with permission. Copyright 2008 American Chemical Society.

$NaNH_2BH_3$ was claimed to form via ball milling NaH/NH_3BH_3 without NaH left,[34] whereas a recent detailed study revealed that the product initially reported by Xiong et al. was actually contaminated with residual NaH.[47] Besides, the later study indicated that significant amount of ammonia gas was generated during ball milling process instead of the pure H_2 as stated in Xiong et al.'s study. The ammonia gas release and the presence of NaH were rationalized by the following dimerization reactions:[47]

$$NaNH_2BH_3 + NH_3BH_3 \rightarrow [Na(NH_3)][BH_3(NH_2)BH_3] \quad (2)$$

$$[Na(NH_3)][BH_3(NH_2)BH_3] \rightarrow Na^+[BH_3(NH_2)BH_3]^- + NH_3 \quad (3)$$

$$Na^+[BH_3(NH_2)BH_3]^- \rightarrow BH_3(NHNa)BH_2 + H_2 \quad (4)$$

Where the freshly formed $NaNH_2BH_3$ by the initial reaction between NaH and AB may continue to react with the unconverted AB thus forming a small amount of $BH_3(NHNa)BH_2$ species [termed sodium imidodiborane, which is analogous to the well know $BH_3(NH_2)BH_3$ imidodiborane] during milling. The reaction will thus leave some unreacted NaH in the sample despite the 1:1 initial ratio and expel ammonia gas during milling.

Initial efforts have also been made to extend the number of metal amibodoranes by ball milling hydrides and AB such as MgH_2/NH_3BH_3 and TiH_2/NH_3BH_3.[35] However, these hydrides are relatively weaker Lewis bases, and are not capable of acquiring the protic hydrogen from NH_3BH_3 to form amidoborane. Interestingly, even though no metal amidoboranes observed in the ball milled mixture of MgH_2/NH_3BH_3 and TiH_2/NH_3BH_3, noticeable improvement in dehydrogenation were observed in the mixtures compared to the pure AB in terms of hydrogen release temperatures and rates.

Besides the first two reported solution synthesis methods for $NaNH_2BH_3$ and $LiNH_2BH_3$,[38,39] solution chemical synthesis methods have also been further developed to prepare other new metal amidoboranes. By mixing and stirring a tetrahydrofuran (THF) solution of AB and a THF suspension of CaH_2 at low temperature, a crystalline THF adduct of calcium amidoborane, $[Ca(NH_2BH_3)_2(THF)_2]_n$ can be produced.[33] $[Ca(NH_2BH_3)_2(THF)_2]_n$ is a coordination polymer, which loses most of the coordinated THF under vacuum. The evacuated product still contains small amount of coordinated THF, which cannot reflect the real crystal structure and coordination environments of $Ca(NH_2BH_3)_2$. Thus, the solvent-free crystalline $Ca(NH_2BH_3)_2$ directly prepared by ball-milling was later used to successfully determine its crystal structure (Figure 2).[35] Similar solution method was also be applied to prepare $NaNH_2BH_3$ by reacting NaH/AB at low temperature or $NaNH_2$/AB at ambient temperature in THF.[48] Using solution chemical synthesis method, many new amidoborane compounds containing cations with relatively week ionicity that cannot be formed by direct ball milling have been prepared, including $Mg(NH_2BH_3)_2$, $Zn(NH_2BH_3)_2$, $Ti(NH_2BH_3)_4$, $Al(NH_2BH_3)_3$, $LiAl(NH_2BH_3)_4$, $Li_2Zn(NH_2BH_3)_4$ and $LiZn(NH_2BH_3)_3$.[11]

The dehydrogenation products of metal amidoboranes were reported to be amorphous,[33,34,35] which impedes structural analysis of intermediates and products. In order to overcome these difficulties, a hydrocarbon-soluble calcium hydride complex [{DIPP-nacnac)CaH(THF)}$_2$] has been developed to investigate the reaction between calcium hydride and AB at a molecular level.[40] By mixing the THF solution of AB with a toluene solution of [{DIPP-nacnac)CaH(THF)}$_2$], a $NH_2BH_3^-$ ligand-containing adduct [(DIPP-nacnac)$CaNH_2BH_3(thf)_2$] was formed. Interestingly, the dehydrogenation intermediates and products of such compounds are also crystalline, which allow the determination of their structures and understanding the dehydrogenation mechanisms. Although the compound is not a true metal amidoborane, its structure cannot really be correlated to the true crystal structure of $Ca(NH_2BH_3)_2$, and its dehydrogenation processes in solution may follow different reaction pathways compared to that of $Ca(NH_2BH_3)_2$ in the solid state, the insight into the dehydrogenation and the intermediates could be advantageous for useful investigations on the regeneration of hydrogen depleted products to amidoboranes and further development of high-performance solid state hydrogen materials.

Structures and Crystal Chemistry

1. Metal Amidoboranes

Spectroscopy techniques such as NMR are useful tools to probe the formation and local structures of metal amidoboranes. These techniques are not restricted to crystalline systems, therefore, they are especially helpful when the target compound is amorphous or nanocrystalline. In fact, the first reported metal amidoboranes, i.e. $LiNH_2BH_3$ and $NaNH_2BH_3$ formed in solutions, were detected using NMR. Of course, for crystalline compounds, these techniques cannot provide direct structural information. In some cases, the reported characteristic spectral modes can be quite different even for the nominally "same" compound. For example, ^{11}B resonance peak in $LiNH_2BH_3$ prepared by wet chemical method was -22 ppm,[39] which agrees with -22.8 ppm observed in $LiNH_2BH_3$ formed by ball milling,[36] whereas, this characteristic resonance was reported as -19.7 ppm in another ball milling study.[34]

To directly obtain the structure information of crystalline compounds, a crystallographic study based on high-quality diffraction data is required. However, crystal structural reports on compounds containing $[NH_2BH_3]^-$ are scarce. The difficulty of determining structures of metal amidoborane lies in several factors: the insensitivity of x-ray to light elements (e.g., H and Li), difficulties in making pure isotope enriched samples for neutron diffraction, possible orientational disorder of $–NH_3$ and $–BH_3$ groups at room temperature, various possible off-stoichiometric phases present in the samples, and poor crystallinity. Although many metal amidoborane complexes with a variety of metal cations have been prepared using wet chemical synthesis techniques, only three crystal structures of amidoborane, i.e. $LiNH_2BH_3$, $NaNH_2BH_3$ and $Ca(NH_2BH_3)_2$, have been reported.[35] Among them $LiNH_2BH_3$ and $NaNH_2BH_3$ are isostructural.

Figures 3 and 4 show the crystal structures for $LiNH_2BH_3$ and $Ca(NH_2BH_3)_2$ with charge density distribution indicated on each atom.[35] Particularly, the calculated Mulliken charges are +0.98 for Li and +1.67 for Ca cations, close to their formula valences.

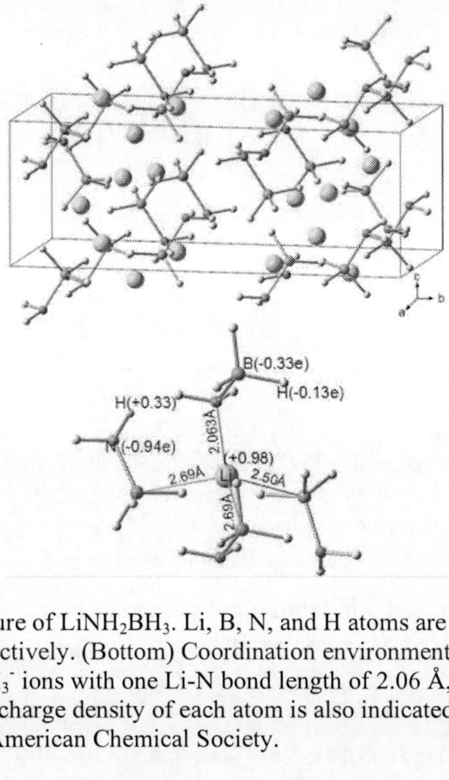

Figure 3. (Top) Crystal structure of $LiNH_2BH_3$. Li, B, N, and H atoms are represented by yellow, green, blue, and white spheres, respectively. (Bottom) Coordination environment of Li^+. Each Li^+ is coordinated with four $NH_2BH_3^-$ ions with one Li-N bond length of 2.06 Å, and three Li-B distances of 2.50 - 2.69 Å. The calculated charge density of each atom is also indicated. Adapted from [35] with permission. Copyright 2008 American Chemical Society.

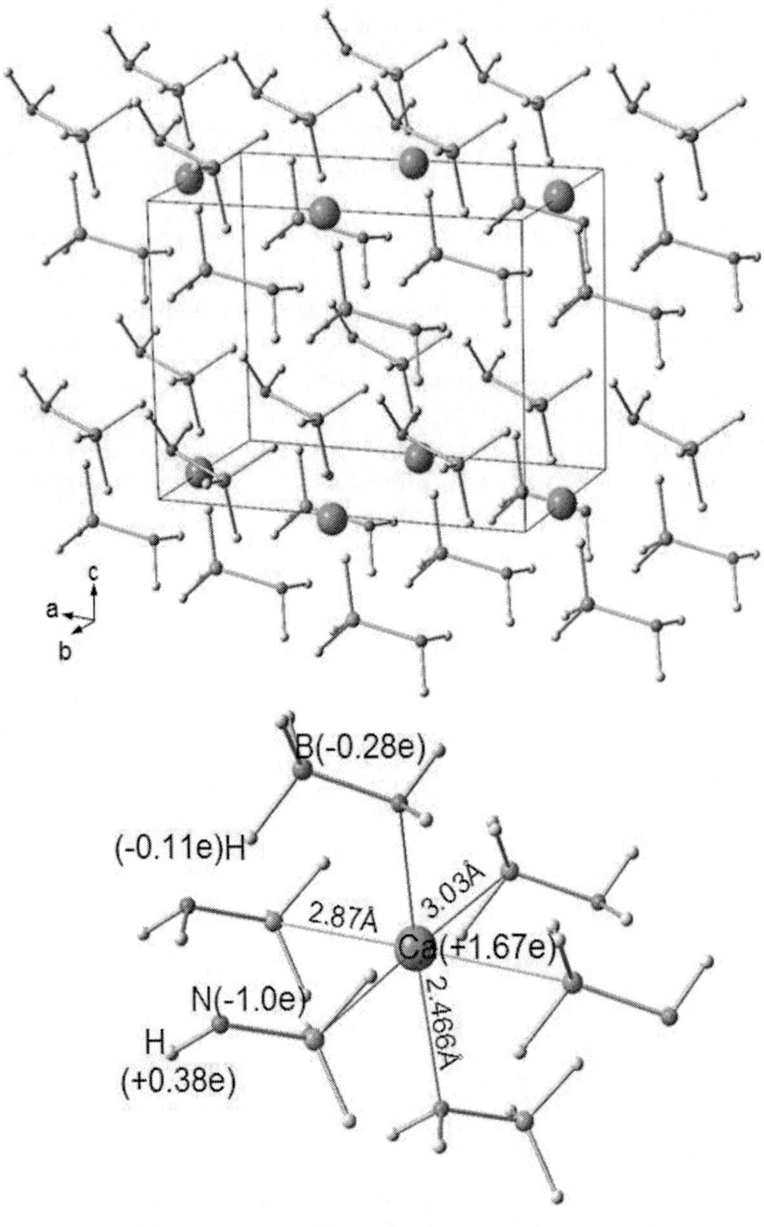

Figure 4. (Top) Crystal structure of $Ca(NH_2BH_3)_2$. Ca, B, N, and H atoms are represented by orange, green, blue, and white spheres, respectively. (Bottom) Coordination environment of Ca^{2+}. Each Ca^{2+} is coordinated with six $NH_2BH_3^-$ ions with two Li-N bond lengths of 2.46 Å and four Li-B distances of 2.87 Å and 3.03 Å. The calculated charge density of each atom is also indicated. Adapted from [35] with permission. Copyright 2008 American Chemical Society.

Therefore, unlike AB, metal amidoboranes are mainly ionic compounds with a significant ionic character between $NH_2BH_3^-$ anion and cations. From the refinement on the XRD data and the First-principles density functional theory (DFT) calculations, $LiNH_2BH_3$ forms an orthorhombic *Pbca* (No.61, Z=8) structure with a=7.1051(8) Å, b=13.930(1) Å and c=5.1477(7) Å, and $Ca(NH_2BH_3)_2$ crystallizes in a monoclinic *C2* (No. 5 Z=2) cell with a=9.100(2) Å, b=4.371(1) Å, c=6.441(2) Å, and β=93.19°,[35] which is different from the solvent-containing compound $Ca(NH_2BH_3)(THF)_2$.[33] The coordination preference and cation-anion distances in $LiNH_2BH_3$ and $Ca(NH_2BH_3)_2$ are consistent with other ionic Li- and Ca-complex hydrides. For example, in $LiNH_2BH_3$, the distance between Li^+ and N in the nearest $[NH_2BH_3]^-$ ion is 2.063 Å, similar to the Li-N distances (2.06-2.21Å) in $LiNH_2$.[48] In addition, each Li^+ is also surrounded by three other $[NH_2BH_3]^-$ ions with Li-B distances in the range of 2.50-2.69 Å, close to the Li-B distances (2.37-2.62 Å) in $LiBH_4$.[50] Therefore, each Li^+ is actually coordinated with four $[NH_2BH_3]^-$ groups, consistent with the Li (IV) tetrahedral coordination preferred in the commonly observed complex hydrides, e.g., $LiNH_2$,[49] $LiBH_4$,[50] Li_2BNH_6 [29]and $Li_4BN_3H_{10}$.[29] In $Ca(NH_2BH_4)_2$, each Ca^{2+} directly bonds with two $[NH_2BH_3]^-$ ions with the closest Ca-N distance ~2.466 Å, similar to those in $Ca(NH_2)_2$ (Ca-N=2.441-2.573Å).[51] Each Ca^{2+} is also coordinated with other four $[NH_2BH_3]^-$ group with Ca-B distance in the range of 2.87-3.03 Å, aligning with those distances (2.89-2.96 Å) in $Ca(BH_4)_2$.[52] Therefore, Ca^{2+} in $Ca(NH_2BH_3)_2$ also satisfies its octahedral coordination preference (VI) in the Ca-related complex hydrides, e.g., $Ca(NH_2)_2$,[51] CaNH,[53] and $Ca(BH_4)_2$.[52]

As for the $NH_2BH_3^-$ ion unit, the B-N bond lengths in $LiNH_2BH_3$ and $Ca(NH_2BH_3)_2$ are 1.547 Å and 1.546 Å,[35] respectively, both shorter than that (~1.58 Å) in solid NH_3BH_3,[12,54,55,56,57] which may indicate a stronger bonding between B and N in the metal amidoboranes (note: a shorter bond distance on its own cannot serve as proof for a stronger bond.[58]). The relative orientations of NH_2– and BH_3– (the dihedral angle H-N-B-H) in $NH_2BH_3^-$ units are different in $LiNH_2BH_3$ and $Ca(NH_2BH_3)_2$ with the dihedral angle close to 60° in $LiNH_2BH_3$ as it is in the ordered NH_3BH_3 (space group, $Pmn2_1$) and the angle approximately 24.4° in $Ca(NH_2BH_3)_2$. The B-N-Li and the B-N-Ca bond angles are 113° and 116° in $LiNH_2BH_3$ and $Ca(NH_2BH_3)_2$, respectively, similar to the H-N-B bond angles (109-114°) in NH_3BH_3. The DFT calculated B-H and N-H bond lengths after structural optimization are ~1.244 Å and 1.025 Å in $LiNH_2BH_3$, and ~1.241 Å and 1.025 Å in $Ca(NH_2BH_3)_2$, respectively.[35] The experimentally- determined N-H and B-H bonds are consistent with the calculations. The N-H bond in these metal amidoboranes are similar to the N-H bonds in solid NH_3BH_3 crystal structure, while the B-H bond lengths are longer than those in solid NH_3BH_3, indicating the similar N-H covalent bonds but the weakened covalent characteristics of B-H bonds. Finally, in both $LiNH_2BH_3$ and $Ca(NH_2BH_3)_2$ the calculated shortest BH···HN intermolecular distances (H-H distance) are longer than those in solid NH_3BH_3 (2.02 Å),[57] but still less than 2.4 Å, the van der Waals distance for the interaction constituting a dihydrogen bond.[2]

In addition to crystal structures, the electronic structures of metal amidoboranes also show interesting variations compared to AB. Analyzing the electronic density of states (DOS) of $LiNH_2BH_3$ and $Ca(NH_2BH_3)_2$ shows that hydrogen bonded with boron accounts for most of the states in the region close to the Fermi level (Figure 5), suggesting the H-B bond to be more reactive than H-N bond in a chemical reaction in general.

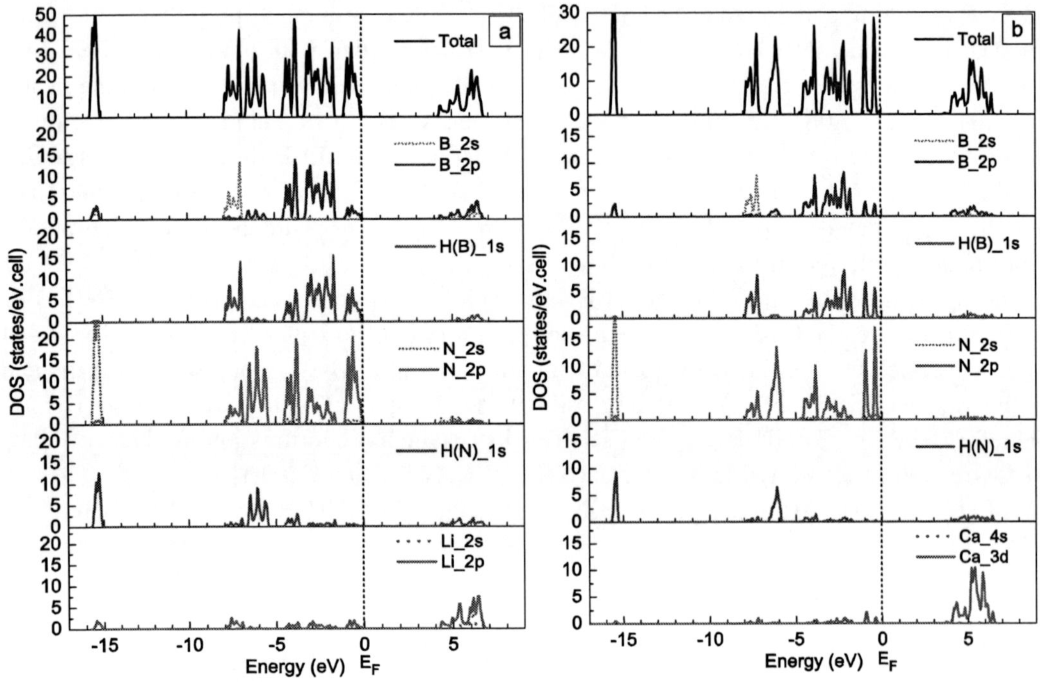

Figure 5. The total electron densities of states (DOS) and their projections around the different atomic sites for a) $LiNH_2BH_3$ and b) $Ca(NH_2BH_3)_2$. Fermi Level (E_f) is at zero. Adapted from [35] with permission. Copyright 2008 American Chemical Society.

The charge density calculations indicate a substantial increase in the hydridic nature of the boron hydrogens (-0.13 e) in $LiNH_2BH_3$ and (-0.11 e) $Ca(NH_2BH_3)_2$ compared to those in AB (-0.05 e).[35] These changes in bonding nature may be responsible for the different dehydrogenation properties between metal amidoboranes and AB. A recent study in $LiNH_2/NH_3BH_3$ mixture[37] showed the formation of an intermediate compound $LiNHBH_2NH_3$ and its rapid hydrogen release, indicating the promoted interaction between stronger hydridic boron hydrogens with the protic amine hydrogens. Further direct experimental evidence for a dehydrogenation mechanism in metal amidoborane is still needed.

2. Other Amidoborane Complexes

Besides the metal amidoborane salts ($M(NH_2BH_3)_n$), some amidoborane ligand $NH(R)BH_3$ (R= various organic linkers) containing metal organic complexes have also been reported recently and are worth mentioning. In these compounds, the local coordination environments of metal cations are not necessarily the same as in the solid metal amidoborane salts. While they all exhibit a common structural element of the chelating $NH(R)BH_3^-$ ligand, which binds to the metal through a M-N bond and a M-H-B bridge (Figure 6). The resulting planar MNBH ring has been observed as the most important feature in several main-group and transition-metal complexes.[40,42,43,,44,45,46] In these compounds, the B-N distances are similar to that of AB (~1.58 Å), while M-N bonds and M-B distances are slightly shorter than those of the corresponding $M(NH_2BH_3)_2$, e.g. Ca-complex[44] (Ca-N: 2.399-2.460 Å and

Ca-B: 2.570-2.867 Å with various R organic components) compared to $Ca(NH_2BH_3)_2$. These differences reflect the additional effects of other organic components in terms of altering the $NH(R)BH_3^-$ ligand molecular geometry, varying the metal cation coordination environments, and consequently tuning the bonding strength. The changes in the bonding strength may hold the promise for the future catalytic regeneration with H_2 back to the amidoborane anion ligands $N(R)HBH_3^-$.[40,44.]

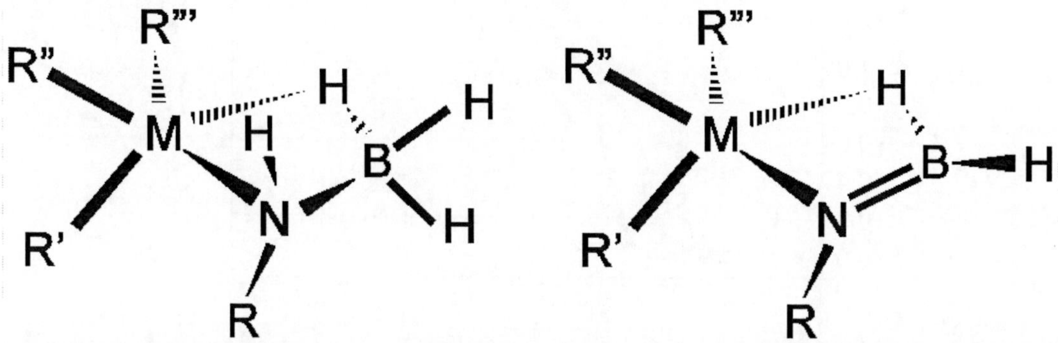

Figure 6. Schematic molecular structures of metalorganic complex containing H_3B-NH(R)- ligand (left) and its dehydrogenation product containing H_2B-=N(R) (right). (R= various organic linkers). The common feature that has been observed in such complexes with various metals is the edge-on coordination of the chelating $H_3BNH(R)$ ligand with metals. The $H_3BNH(R)$ ligand binds to the metal through an M-N bond and aan M-H-B bridge, and results in a planar MNBH ring.

Dehydrogenation Properties

Generally, metal amidoboranes have been shown to release hydrogen at lower temperatures, at higher rates and with the release of fewer volatile contaminates. They also appear to release hydrogen less exothermically than AB. Therefore, the superior dehydrogenation properties compared AB advance these materials as an attractive new class of compounds for hydrogen storage application. However, the reported information on dehydrogenation such as intermediate and final dehydrogenation products is very inadequate due to the limited number of amidoborane compounds studied so far, amorphous nature of dehydrogenation products, and complexity of dehydrogenation mechanisms in different systems. In addition, contradictions exist in the release gas species and types of solid products during processing and dehydrogenation, making it even more difficult to understand the formation and hydrogen storage mechanisms of these compounds. Below, we focus on the dehydrogenation properties of several representative metal amidoboranes.

1. Lithium Amidoborane

Dehydrogenation of $LiNH_2BH_3$ has been investigated and compared with AB in both thermodynamic and kinetic prospects. The solid state $LiNH_2BH_3$ decomposes in two distinct steps. From temperature programmed desorption (TPD) measurement (Figure 7 and 8), $LiNH_2BH_3$ first dehydrogenates at lower temperatures compared to pure AB, e.g., 86 or 92°C under different heating rate,[34,35,36] and releases rest of the hydrogen at 120°C with a final composition of "$LiNBH_{1.32}$" after TPD measurement.[35]

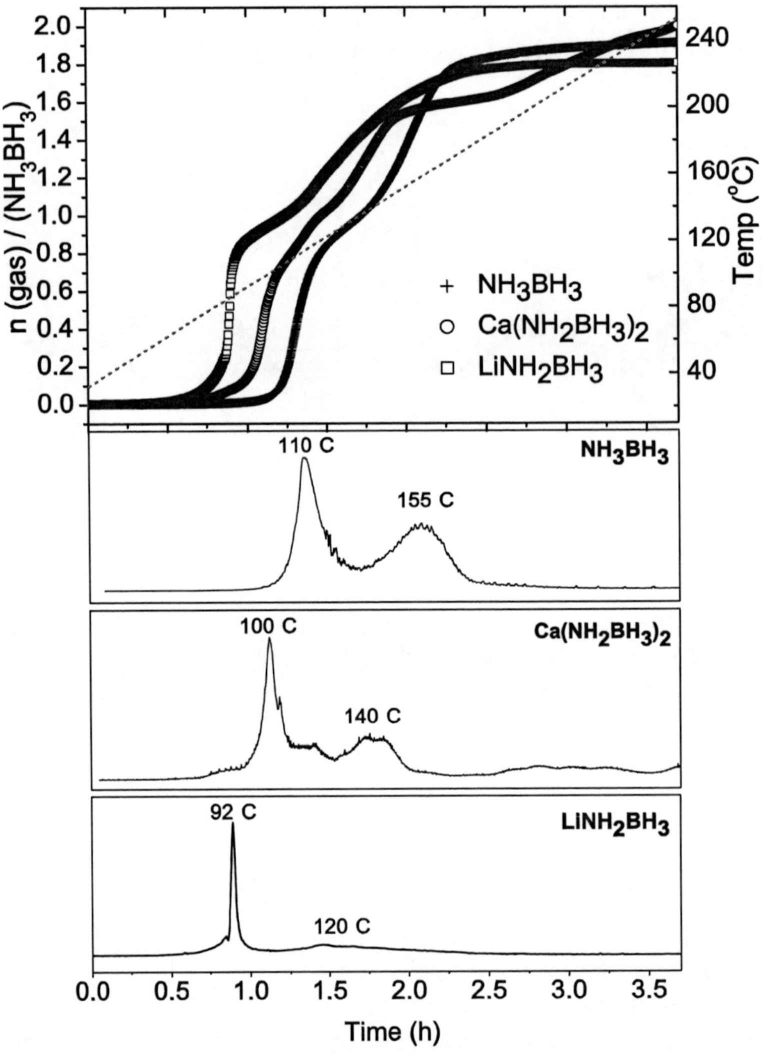

Figure 7. TPD results of hydrogen release for $LiNH_2BH_3$, $Ca(NH_2BH_3)_2$, and NH_3BH_3 with 1°C/min heating rate. Dehydrogenation of $LiNH_2BH_3$ and $Ca(NH_2BH_3)_2$ begins at lower temperatures than NH_3BH_3. The amount of hydrogen gas released has been normalized as n (H_2 gas)/ mol NH_3BH_3. Adapted from [35] with permission. Copyright 2008 American Chemical Society.

From an isotherm measurement, $LiNH_2BH_3$ also decomposes in a faster fashion, e.g. ~ 2.0 equivalent mole H_2 from $LiNH_2BH_3$ while only 0.82 equivalent mole H_2 from AB releases at 91°C within 20 h.[34] Such promoted dehydrogenation performance is similar to the observations in AB and ionic liquid solutions.[15]

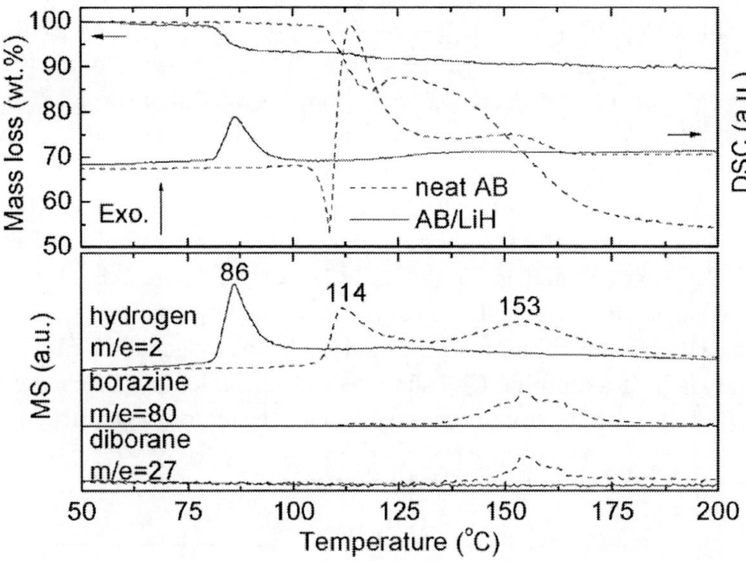

Figure 8. Thermal decomposition of $LiNH_2BH_3$ (solid line) and pure AB (dashed line) at 2°C min-1 heating rate. Thermogravimetric/differential scanning calorimetry (TGA/DSC) profiles (top), and synchronous MS profiles of H_2 (m/e=2), borazine (c-$(NHBH)_3$, m/e=80), and diborane (B_2H_6, m/e=27) (bottom). Reproduced from [36] with permission. Copyright Wiley-VCH Verlag GmbH & Co. KGaA.

Combined differential scanning calorimetry (DSC) and thermogravimetry measurements (Figure 8) indicated that dehydrogenation reaction enthalpy of $LiNH_2BH_3$ is -8 kJ mol^{-1} H_2[36] (-3 kJ mol^{-1} H_2 from another DSC measurement[34]), significantly less exothermic than that of the pure AB (-22 kJ mol^{-1} H_2[6]). Though dehydrogenated $LiNH_2BH_3$ cannot be rehydrogenated under hydrogen pressures duo to its exothermic desorption nature, it is known that the reaction enthalpy can be tuned through some chemical modifications. For example, the enthalpy of hydrogen release from AB in nanoporous-silica was measured as -1±1 kJ mol-1,[19] approximately 20 kJ mol^{-1} less exothermic than hydrogen loss from pure AB. A recent study indicated that rehydrogenation can be achieved in the nano-scaffolded $Li_3BN_2H_8$ system.[59] After desorption, the nano-sized $Li_3BN_2H_8$ showed a dramatic hydrogen uptake at relatively moderate conditions (300°C under 50 atm H_2) compared to the almost irreversible bulk $Li_3BN_2H_8$. It was believed that an increase in the surface area and the resulting increased surface energy will lead to a new van't Hoff relation and a significant reduction in the absorption heat for materials with sufficiently small particle sizes. This change with the reduced particle size is then likely to alter the thermodynamics of the system and to make the nano-sized material “rehydridable”. Therefore, rehydrogenation of the less exothermically dehydrogenated $LiNH_2BH_3$ under hydrogen pressures may be feasible if a nearly neutral free energy can be realized by inducing surface energy terms via tuning down particle sizes.

As for the kinetics for hydrogen release, many approaches have been found to be able to afford a much wider span of control over the rates and extent of hydrogen release from AB. For example, various transition metal complex catalysts have been reported to show a dramatic enhancement in the hydrogen release kinetics of AB.[16,17] In addition, nanostructured hydrides, with their increased surface energy associated with the very high surface area, excess enthalpy and strains at the grain boundary, and shorter transport path for hydrogen, have been shown to promote faster desorption rates and lower desorption temperatures as compared to the bulk hydrides.[59,60] A significantly smaller activation energy was observed in the nanoporous silica scaffolded AB (67±5 kJ mol^{-1})[19] through one or the combination of several ways: promotion of hydrogen loss rate from the confinement effect and the defect sites from microstructured silica framework; the catalytic effect and control of material distribution by the molecular structure of the substrate walls includes SiOH. For the metal amidoborane, from the Arrhenius dehydrogenation kinetics profiles of $LiNH_2BH_3$ and pure AB (Figure 9),[36] the apparent activation energies of H_2 release from $LiNH_2BH_3$ were calculated as 75 kJ mol^{-1}, which is substantially lower than pure AB (184 kJ mol^{-1}). The decreased energy barrier for $LiNH_2BH_3$ is responsible for the enhanced kinetics compared to AB. Apparently, with its already dramatically reduced dehydrogenation enthalpy, the desired hydrogen release kinetics within the DOE targeted operation temperature range may even be possible for $LiNH_2BH_3$ with the aid of the proper catalysts and/or the reduced particle size.

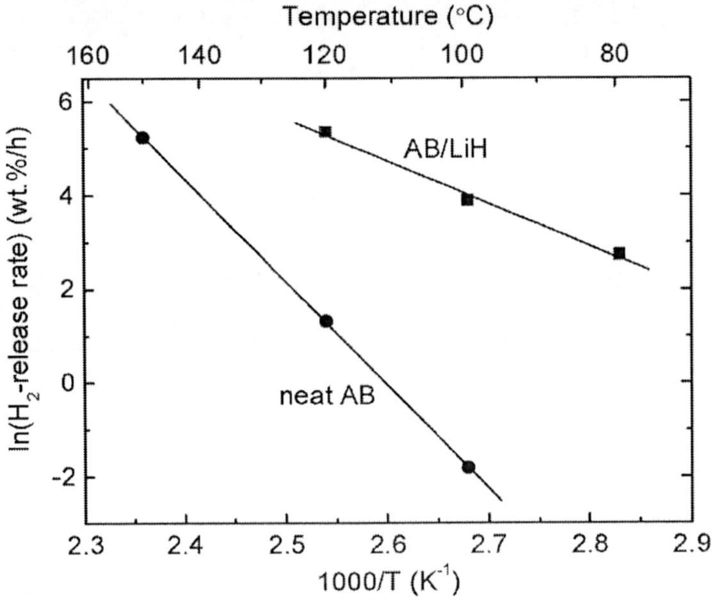

Figure 9. Arrhenius plots of the dehydrogenation kinetics of $LiNH_2BH_3$ and pure AB. The apparent activation energy can be deduced from the slope according to the Arrhenius equation: rate (wt % /h) = $k_0 \exp(-E_a/RT)$. Reproduced from [36] with permission. Copyright Wiley-VCH Verlag GmbH & Co. KGaA.

In addition to the improvement in dehydrogenation temperature and hydrogen loss rate, another remarkable advantage in dehydrogenation of $LiNH_2BH_3$ is the fewer volatile contaminants release such as borazine and diborane (Figure 8), as suggested in an early work.[34] This is especially important when considering use of these materials in conjunction with current fuel cell technology. However, a more recent report indicated that $LiNH_2BH_3$ releases more than ten times higher amount of ammonia (NH_3) compared to AB during decomposition,[61] which largely degraded its potential as a practical hydrogen storage material.

XRD on the products of $LiNH_2BH_3$ after desorption indicated formation of amorphous phases,[34,35] which prevents direct determination of their structures. Other characterization techniques such as NMR spectroscopy were used to probe and understand the atomic bonding and local environments in the non-crystalline decomposition products. Quadrupolar lineshaped peaks at about +25.1 ppm were observed in the ^{11}B NMR spectra collected on $LiNH_2BH_3$ after dehydrogenated up to 200 °C.[36] Similar shaped peaks at +22 ppm were observed in NH_3BH_3 dehydrogenated at the same conditions, and were ascribed to the polyiminoborane $[HNBH]_n$ previously.[4,5,6,7,9] This suggests that the decomposition products of $LiNH_2BH_3$ be polyiminoborane-like chain structure. Xiong at el found a singlet resonance peak at +29.8 ppm in ^{11}B NMR of $LiNH_2BH_3$ after dehydrogenated to 140 °C, comparable to the +26.5 ppm chemical shift which was observed in polyborazylene, and thus concluded that the decomposition product is a borazine-like or poly-borazine-like structure.[34] However, the +29.8 ppm peak is consistent with formation of internal B=N or terminal $B=NH_2$ units observed in the chain-like structure of the dehydrogenated NH_3BH_3.[15] Apparently, the structure of decomposed products has strong dependence on the dehydrogenation conditions (temperature, ramping rate, etc.), purity of the initial chemicals, and even characterization settings.

2. Sodium Amidoborane

$NaNH_2BH_3$ is isostructural to $LiNH_2BH_3$ with a smaller gravimetric hydrogen capacity (7.5 wt%). Similar to $LiNH_2BH_3$, it shows a depressed thermolysis onset temperature and an enhanced hydrogen loss rate compared to AB.[34] The dehydrogenation of $NaNH_2BH_3$ is also exothermic, -5 kJ mol^{-1} H_2,[34] but much less than that of pure AB. The purity of desorption gas and the dehydrogenation products of $NaNH_2BH_3$ seemed controversial. In the first study of dehydrogenation properties of $NaNH_2BH_3$,[34] Xiong et al. claimed that $NaNH_2BH_3$ releases as much as 7.4 wt% of hydrogen free of borazine at 91 °C, and produces amorphous $(NaNBH)_n$ with a similar structure as the dehydrogenated $LiNH_2BH_3$. In a subsequent study,[48] the same group reported evolution of pure hydrogen from $NaNH_2BH_3$, but the dehydrogenation occurs in two steps 4.4 wt% loss at 70-91 °C and 3.1 wt% at 90-200 °C, and finally results in NaH and amorphous BN as the dehydrogenation products. A more recent study conducted by a different group reported a substantial emission of ammonia gas during dehydrogenation of $NaNH_2BH_3$,[47] e.g. 1 mol NH_3 loss per 5.5 mol evolved H_2 during ball milling and 1 mol NH_3 loss per 10 mol H_2 during heating up to 110 °C, and revealed a third exothermic reaction at 55-80 °C (Figure 10). From detailed studies of gas species evolution using the combined TGA, DSC, mass spectrometers, FT-IR and ^{11}B and ^{1}H NMR techniques, Fijalkowski and Grochala suggested a more complex decomposition mechanism upon heating and ball milling, which involves the formation of an unstable intermediate $Na^+[BH_3(NHNa)BH_3]^-$ while evolving NH_3 (Figure 11).[47] Similar to the head-to-tail dimerization observed in AB, it was believed that $NaNH_2BH_3$ also underwent a head-to-tail dimerization during thermal decomposition at 55°C, leading to the formation of an ionic salt $[NH_3Na]^+[BH_3(HNNa)BH_3]^-$ (Figure 11). This ionic salt decomposes at 90-110 °C, resulting ammonia gas and the intermediate $Na^+[BH_3(NHNa)BH_3]^-$. The latter is unstable and supposed to spontaneously decompose into NaH and $BH_3(NHNa)BH_2$. $BH_3(NHNa)BH_2$ decomposes further at this temperature range with the release of H_2 (Figure 11). However, the mechanism of the major H_2 release process from $NaNH_2BH_3$ still remains unclear.

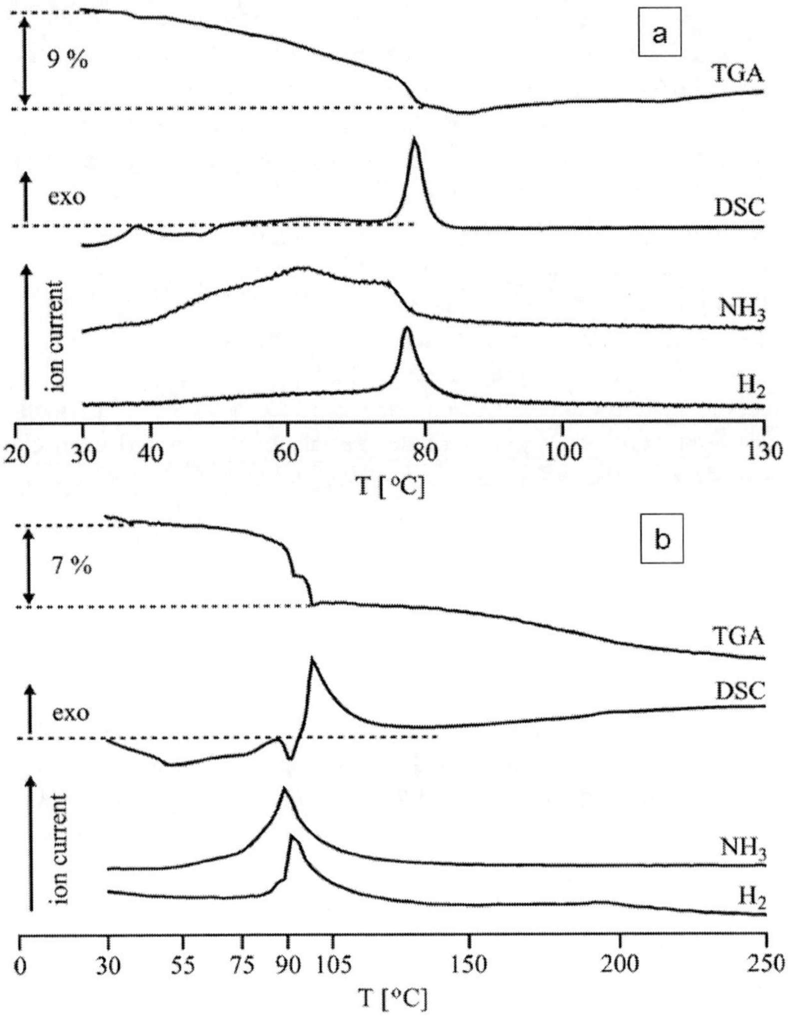

Figure 10. Thermal decomposition of $NaNH_2BH_3$ at a) 1 °C min^{-1} and b) 10 °C min^{-1} heating rates. The TGA, DSC, and evolved gas analysis (EGA) results are shown, respectively. Adapted from [47] by permission of The Royal Society of Chemistry.

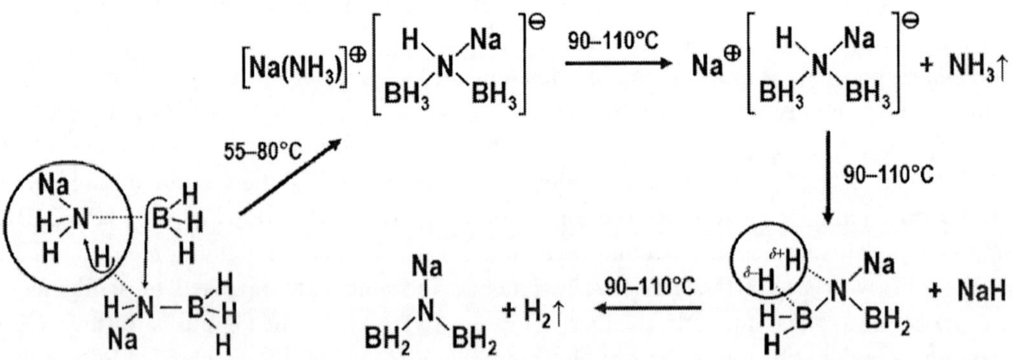

Figure 11. Proposed mechanism of secondary chemical transformations of $NaNH_2BH_3$ during its thermal decomposition. Reproduced from [47] by permission of The Royal Society of Chemistry.

In addition, contradict to the previous claim that $NaNH_2BH_3$ is non-flammable and non-explosive at the ambient condition,[34] It was found in a later study that $NaNH_2BH_3$ reacts slowly with atmosphere moisture, violently with water and leads to autoignition.[47] $NaNH_2BH_3$ is also flammable in air producing a steady orange flame. Therefore, the contaminated hydrogen release from the considerable amount of ammonia may disqualify this compound from on-board vehicle applications due to poisoning of low-temperature fuel cells with ammonia.

Although the application as a hydrogen storage material is not appealing for $NaNH_2BH_3$, it can be used as a precursor for preparing other metal amidoborane compounds such as $Mg(NH_2BH_3)_2$, $Zn(NH_2BH_3)_2$, and $Ti(NH_2BH_3)_4$ through exchange reactions by wet chemical synthesis techniques.[11]

$$MgCl_2 + 2NaNH_2BH_3 \rightarrow Mg(NH_2BH_3)_2 \quad (5)$$

$$ZnCl_2 + 2NaNH_2BH_3 \rightarrow Zn(NH_2BH_3)_2 \quad (6)$$

$$TiCl_2 + 4NaNH_2BH_3 \rightarrow Ti(NH_2BH_3)_4 \quad (7)$$

3. Calcium Amidoborane

Dehydrogenation properties of $Ca(NH_2BH_3)_2$ and its THF adduct $[Ca(NH_2BH_3)_2(THF)_2]_n$ have also been reported.[33,35] Pure $Ca(NH_2BH_3)_2$ starts to desorb hydrogen at about 80ºC with vigorous hydrogen release at approximately 100 and 140 ºC (Figure 7). Ramping the temperature to 170 ºC results in 3.6 equivalent of H_2 released (corresponding to 7.2 wt%) with <0.1% ammonia and borazine were detected. After the reaction finished at 250 ºC, about 4 equivalent of hydrogen was desorbed, resulting in a composition of "$CaB_2N_2H_{2.33}$" desorption product. Similar to $LiNH_2BH_3$ and $NaNH_2BH_3$, the decomposed product of $Ca(NH_2BH_3)_2$ is also amorphous as indicated by XRD. In contrast the exothermic hydrogen release of NH_3BH_3, $LiNH_2BH_3$, and $NaNH_2BH_3$, the hydrogen loss reaction from $Ca(NH_2BH_3)_2$ was first found to be endothermic with an enthalpy of 3.5 kJ mol^{-1}.[33] The endothermic dehydrogenation suggests that the rate and volume of hydrogen release from $Ca(NH_2BH_3)_2$ can be varied with temperature, and direct regeneration of hydride is possible through gas-solid reaction. However, the most recent report identified that all metal amidoboranes discovered so far including $LiNH_2BH_3$, $NaNH_2BH_3$, KNH_2BH_3, $Mg(NH_2BH_3)_2$, $Ca(NH_2BH_3)_2$ and $Al(NH_2BH_3)_3$ have exothermic hydrogen release in different degrees,[62] indicating that these compounds cannot be on-board regenerated under hydrogen pressure.

4. Structural Characteristics Responsible for the Dehydrogenation Properties

The improved dehydrogenation properties of metal amidoboranes have been ascribed to the presence of both positive and negative hydrogen.[34] However, the protic and hydridic hydrogen are actually present in the pure NH_3BH_3 as well. From the structural study and first-principles calculations,[35] the changes in the reaction kinetics were attributed to the different natures of hydrogen in the metal amidoboranes and in the pure solid NH_3BH_3. The charge density calculations indicate there is a considerable increase in the hydridic nature of B-H bond with more electrons being donated from metal to $[NH_2BH_3]^-$ ions. This effect is similar to the active intermediate BH_4^- induced by the ionic liquid. As noted previously in AB system, more reactive intermediates or transition states, such as the diammoniate of diborane $[(NH_3)_2BH_2]^+[BH_4]^-$ [8] and/or BH_4^- formed by the dehydrogenation of NH_3BH_3 in ionic liquids,[15] and the borenium cation $[H_2B\text{-}NH_3]^+$ produced in acid initiation of NH_3BH_3 dehydrogenation,[14] have been shown to effectively enhance dehydrogenation and reduce the borazine release. Therefore, the barrier of the reaction between $[NH_2BH_3]^-$ ions would be lower than that between two neutral NH_3BH_3 molecules. In addition, the charged $[NH_2BH_3]^-$ ion creates more polar surroundings compared to the symmetric NH_3BH_3 complex. As proposed for the enhanced dehydrogenation of NH_3BH_3 in the ionic liquids, the reaction environment controls the course of hydrogen release.[15] Compared to the reactions between NH_3BH_3 in solid AB, the polar environments of the $[NH_2BH_3]^-$ ions coupled with a change in the reactivity among these ions in the metal amidoborane compounds will facilitate the B-H···H-N interactions between the adjacent $[NH_2BH_3]^-$ ions. More recent experimental studies indeed observed the active intermediates, i.e. $LiNH_2BH_3NH_3$ in $LiNH_2$/AB[37], $[NH_3Na]^+[BH_3(NHNa)BH_3]^-$ and $Na^+[BH_3(NHNa)BH_3]^-$ in $NaNH_2BH_3$[47], during their dehydrogenation. Such approach to introducing polar species has also been used in improving the dehydrogenation properties of hydrazine borane N_2H_4—BH_3 (HB).[63] A 1:1 mixture of solid LiH and N_2H_4—BH_3 creates $Li^+(N_2H_3BH_3)^-$, which contains equal amount of protic and hydridic hydrogen and has been found to show significantly enhanced dehydrogenation rate compared to HB. The substantially increased the hydridic H(B) in these derivative compounds promotes stronger interaction with the protic H(N) in the rest dehydrogenation steps.

From the different released gas species and/or dehydrogenation products, the reaction pathways for various metal amidoboranes may not necessarily be the same. Thus far there is no sufficient information to determine the reaction pathway and the dehydrogenation products for the known metal amidoboranes. Nonetheless, the observed improved hydrogen release process/kinetics and the decreased dehydrogenation temperatures of metal amidoborane compared to solid AB is more likely related to the changes in their crystal structures and the resulting altering in the nature and reactivity of hydrogens in $[NH_2BH_3]^-$.

5. Other Amidoborane Complexes

Hydrogen release in the bulk amidoborane complexes has also been examined. Studies in calcium amidoborane complexes, (DIPP-nacnac)CaNH(iPr)BH_3·(THF) and (DIPP-nacnac)CaNH(DIPP)BH_3·(THF), DIPP= 2, 6-$i\text{Pr}_2C_6H_3$ and DIPP-nacnac = CH{(CMe)(2,6-$i\text{Pr}_2C_6H_3N$)}$_2$, found that during dehydrogenation the *i*Pr-containing complex will form dimeric product and NBNB chain, whereas the DIPP substituted complex is bulky enough to prevent such dimerization.[44] Decomposition of DIPP-containing complex resulted in the monomeric product with formation of H_2B^-=(R)N that is edge-on coordinated to Ca^{2+} as in the hydrogenated H_3B-(R)NH ligand (Figure 6). The *p*-type lone pair on the N atom is supposed to be shielded by the *i*Pr-substituents in the *ortho*-position. This shielding is responsible for the stability of the monomeric decomposition product of DIPP-complex. The bulky groups on N also prevent the dimerization of *i*PrN=BH_2.

The anionic BN-fragment in such monomeric decomposition product could be seen as a donor/acceptor pair, which is believed to be able to activate molecular hydrogen, similar to a frustrated Lewis pair,[64] A recent theoretical study shows that hydrogenation of $R_2NBR'_2$ to $R_2HNBHR'_2$ is favored for substrates with electron-rich N and electron-poor B atoms,[65] thus hydrogenation of $H_2B(R)N^-$ might thermodynamically be feasible. However, after desorption these compounds cannot be directly hydrogenated to restore the initial amidoborane complexes under hydrogen pressure. This might be due to the strong B=N π-bond or sluggish kinetics.

Conclusions

Metal amidoboranes are a new class of materials that provide great potentials for hydrogen storage applications. Like many complex metal hydrides, these ionic compounds contain various metal cations and $[NH_2BH_3]^-$ anions. With electrons transferred from cations to anions, the H(B) in the $[NH_2BH_3]^-$ anions becomes significantly more hydridic, and the charged $[NH_2BH_3]^-$ ion creates more polar surroundings compared to the symmetric NH_3BH_3 complex, therefore, unlike NH_3BH_3, metal amidoboranes have been shown to release hydrogen at a moderate temperature range but at higher rates. Some of them release fewer volatile contaminants during dehydrogenation. Particularly, several metal amidoboranes also appear to release hydrogen less exothermically than AB. With proper catalytic modification and/or particle size control, the dehydrogenation of these materials can be tuned to be nearly thermally neutral, which may lead to a possible energy-efficient regeneration. Metal amidoborane for hydrogen storage is still a relatively emergent research field. Further investigations are still needed to understand the dehydrogenation mechanism and pathways, to explore the role of catalysts, and to achieve facile reversibility. Considering the large number of metal elements in the periodic table that may form metal amidoboranes, the ongoing research in this field will definitely be fruitful.

Acknowledgment

T.Y. acknowledges partial support from the DOE BES through Grant DE-FG02-08ER46522.

References

[1] Züttel, A.; Schlapbach, L. *Nature* 2001, *414*, 353–358
[2] Stephens, F.; Pons, V.; Baker, R.T. *Dalton Transaction* 2007, 2613-2626
[3] Hamilton, C. W.; Baker, R. T.; Staubitz, A.; Manners, I. *Chem. Soc. Rev.* 2009, *38*, 279-293.
[4] Hu, M.G.; Geanangel, R.A.; Wendlandt, W.W. *Thermochim. Acta* 1978, *23*, 249-255.
[5] Baitalow, F.; Baumann, J.; Wolf, G.; Jaenicke-Rossler, K.; Leitner, G. *Thermochim. Acta* 2002, *391*, 159-168.

[6] Wolf, G.; Baumann, J.; Baitalow, F.; Hoffmann, F.P. *Thermochim. Acta* 2000, *343*, 19-25.
[7] Sit, V.; Geanangel, R. A.; Wendlandt, W.W. *Thermochim. Acta* 1987, *113*, 379-382.
[8] Stowe, A.C.; Shaw, W.J.; Linehan, J.C.; Schmid, B.; Autrey, T. *Phys. Chem. Chem. Phys*. 2007, *9*, 1831-1836.
[9] Baumann, J.; Baitalow, F.; Wolf, G. *Thermochim Acta* 2005, *430*, 9-14.
[10] Damle, A. S. (2006). *Development of Regenerable High Capacity Boron Nitrogen Hydrides as Hydrogen Storage Materials, DOE Hydrogen Program FY 2006 Annual Progress Report.* http://www.hydrogen.energy.gov/pdfs/review06/iv_b_5_damle.pdf
[11] Ott, K. C.; Baker, R. T.; Burrell,A.; Davis, B.; Diyabalanage, H.; Gordon, J. C.; Hamilton, C.; Henson, N.; Inbody, M.; John, K.D.; Semelsberger, T.; Shrestha, R.; Stephens, F. (2008). *Chemical hydrogen storage R&D at Los Alamos National Laboratory, DOE Hydrogen Program FY 2008 Annual Progress Report.* http://www.hydrogen. energy.gov/pdfs/review08/iv_b_1f_ott.pdf
[12] Miranda, C. R.; Ceder, G. *J. Chem. Phys.* 2007, *126*, 184703.
[13] Diwan, M.; Diakov, V.; Shafirovich, E.; Varma, A. *Int. J. Hydrogen Energy* 2008, *33*, 1135.
[14] Stephens, F. H.; Baker, R. T.; Matus, M. H.; Grant, D.J.; Dixon, D. A. *Angew. Chem. Int. Ed.* 2007, *46*, 746-749.
[15] Bluhm, M. E.; Bradley, M. G.; Butterick, R.; Kusari, U., Sneddon, L.G. *J. Am. Chem. Soc*. 2006, *128*, 7748-7749.
[16] Jaska, C.A.; Temple, K., Lough, A.J.; Manners, I. *J. Am. Chem. Soc*. 2003, *125*, 9424-9434.
[17] Denney, M. C.; Pons, V., Hebden, T. J.; Heinekey, M.; Goldberg, K. I. *J. Am. Chem. Soc.* 2006, *128*, 12048-12049.
[18] Keaton, R. J.; Blacquiere, J. M.; Baker, R.T. *J. Am. Chem. Soc.* 2007, *129*, 1844-1845.
[19] Gutowska, A.; Li, L.; Shin, Y.; Wang, C. M.; Li, X. S.; Linehan, J. C.; Smith, R. S.; Kay, B.D.; Schmid, B.; Shaw, W.; Gutowski, M. Autrey, T. *Angew. Chem. Int. Ed.* 2005, *44*, 3578-3582.
[20] Feaver, A; Sepehri, S.; Shamberger, P.; Stowe, A.; Autrey, T.; Cao, G. *J. Phys. Chem. B* 2007, *111*, 7469-7472.
[21] Tumas, W.; Baker, R. T.; Burrell, A.; Thorn, D. (2006). Chemical Hydrogen Storage Research at Los Alamos National Laboratory, DOE Hydrogen Program FY 2006 Annual Progress Report. http://www.hydrogen.energy.gov/pdfs/progress06/iv_b_4_tumas.pdf.
[22] Sneddon, L. G. (2006). *New Methods for Promoting Amineborane Dehydrogenation /Regeneration Reactions, DOE Hydrogen Program FY 2006 Annual Progress Report.* http://www.hydrogen.energy.gov/pdfs/progress06/iv_b_4g_sneddon.pdf.
[23] Sneddon, L. G. (2007). Amineborane Based Chemical Hydrogen Storage, DOE Hydrogen Annual Merit Review. http://www.hydrogen.energy.gov/pdfs/review07/st_27_ sneddon.pdf
[24] Hausdorf, S.; Baitalow, F.; Wolf, G.; Mertens, F. O. R. L. *Int. J. Hydrogen Energy*, 2008, *33*, 608.
[25] Vajo, J. J.; Skeith, S. L.; Mertens, F. *J. Phys. Chem. B* 2005, *109*, 3719.
[26] Pinkerton, F. E.; Meyer, M. S.; Meisner, G. P.; Balogh, M. P.; Vajo, J. J. *J. Phys. Chem. C* 2007, *111*, 12881-5.
[27] Pinkerton, F. E.; Meisner, G. P.; Meyer, M. S.; Balogh, M. P.; Kundrat, M. D. *J. Phys. Chem. B* 2005, *109*, 6.
[28] Meisner, G. P.; Scullin, M. L.; Balogh, M. P.; Pinkerton, F. E.; Meyer, M. S. *J. Phys. Chem. B* 2006, *110*, 4186.

[29] Wu, H.; Zhou, W.; Udovic, T. J.; Rush, J. J.; Yildirim, T. *Chem. Mater.* 2008, *20*, 1245-1247.
[30] Chater, P.A.; David, W. I. F.; Johnson, S. R.; Edwards, P. P.; Anderson, P.A., Chem Commun., 2006, *23*, 2439.
[31] Filinchuk, Y. E.; Yvon, K.; Meisner, G.P.; Pinkerton, F.E.; Balogh, M. P., *Inorg. Chem.*, 2006, *45*, 1433.
[32] US Patent Application Publication No. US 2006/0097221 A1
[33] Diyabalanage, H.V.K.; Shrestha, R. P.; Semelsberger, T. A.; Scott, B. L.; Bowden, M. E.; Davis, B. L.; Burrell, A. K. *Angew. Chem. Int. Ed.* 2007, *46*, 8995-8997.
[34] Xiong, Z.; Yong, C. K.; Wu, G.; Chen, P.; Shaw, W.; Karkamkar, A.; Autrey, T.; Jones, M. O.; Johnson, S. R.; Edwards, P. P.; David, W. I. F. *Nature Mater.* 2008, *7*, 138-141.
[35] Wu, H.; Zhou, W.; Yildirim, T. *J. Am. Chem. Soc.* 2008, *130*, 14834–14839.
[36] Kang, X. D.; Fang, Z. Z.; Kong, L. Y.; Cheng, H. M.; Yao, X. D.; Lu, G. Q.; Wang, P. *Adv. Mater.* 2008, *20*, 2756–2759.
[37] Graham, K.R.; Kemmitt, T.; Bowden, M.E. *Energy Environ. Sci.* 2009, *2*, 706-710.
[38] Schlesinger, H. I.; Burg, A. B. *J. Am. Chem. Soc.* 1938, *60*, 290-299.
[39] Myers, A. G.; Yang, B. H.; Kopecky, D. J. *Tetrahedron Lett.* 1996, *37*, 3623-3626.
[40] Spielmann, J.; Jansen, G.; Bandmann, H.; Harder, S. *Angew. Chem. Int. Ed.* 2008, *47*, 6290-6295.
[41] Hillier, A. C., Fox, T., Schmalle, H. W. and Berke, H. *J. Organomet. Chem.* 2003, *669*, 14–24.
[42] Binger, P., Sandmeyer, F. and Krger, C. *Organometallics* 1995, *14*, 2969–2976.
[43] Fryzuk, M. D., MacKay, B. A., Johnson, S. A. and Patrick, B. O. *Angew. Chem., Int. Ed.* 2002, *41*, 3709–3712.
[44] Spielmann, J. and Harder, S. *J. Am. Chem. Soc.* 2009, *131*, 5064–5065.
[45] Zimmerman, P. M., Paul, A., Zhang, Z. and Musgrave, C. B. *Angew. Chem., Int. Ed.* 2009, *48*, 2201–2205.
[46] Forster, T.D.; Tuononen, H. M.; Parvez, M.; Roesler, R. *J. Am. Chem. Soc.* 2009, *131*, 6689-6691.
[47] Fijalkowski, K. J.; Grochala, W. *J. Mater. Chem.* 2009, *19*, 2043-2050.
[48] Xiong, Z.; Wu, G.; Chua, Y. S.; Hu, J.; He, T.; Xu, W.; Chen, P. *Energy Environ. Sci.* 2008, *1*, 360-363.
[49] Nagib, M.; Jacob, H. *Z. Anorg. Allge. Chem.* 1972, *391*, 271.
[50] Hartman, M. R.; Rush, J. J.; Udovic, T. J.; Bowman, Jr., R. C.; Hwang, S. J. *J. Solid State Chem.* 2007, *180*, 1298.
[51] Senker, J.; Muller, M.; Press, W.; Muller, P.; Mayer, H. M.; Ibberson, R. M. *J. Phys. Chem. B* 1998, *102*, 931-940.
[52] Buchter, F.; Lodziana, Z.; Remhof, A.; Friedrichs, O.; Borgschulte, A.; Mauron, Ph.; Züttel, A.; Sheptyakov, D.; Barkhordarian, G.; Bormann, R.; Chlopek, K.; Fichtner, M.; Sørby, M.; Riktor, M.; Hauback, B.; Orimo, S. *J. Phys. Chem. B* 2008, *112(27)*; 8042-8048.
[53] Sichla, T.; Jacobs, H. *Zeitschrift fuer Anorganische und Allgemeine Chemie* 1996, *622*, 2079-2082.
[54] Hughes, E.W. *J. Am. Chem. Soc.* 1956, *78*, 502.
[55] Hoon, C. F.; Reynhardt, E. C. *J. Phys. C.* 1983, *16*, 6129.
[56] Buehl, M.; Steinke, T.; von Rague Schleyer, P.; Boese, R. *Angew. Chem. Int. Ed.* 1991, *30*, 1160.
[57] Klooster, W. T.; Koetzle, T. F.; Siegbahn, P. E. M.; Richardson, T. B.; Crabtree, R. H. *J. Am. Chem. Soc.* 1999, *121*, 6337.
[58] Bessac, F.; Frenking, G. *Inorg. Chem.* 2006, *45*, 6956.

[59] Wu, H.; Zhou, W.; Wang, K.; Udovic, T. J.; Rush, J. J.; Yildirim, T.; Bendersky, L. A.; Gross, A. F.; Atta, S. L. V.; Vajo, J. J.; Pinkerton, F. E.; Meyer, M. S. *Nanotechnology* 2009, *20*, 204002.
[60] Gross, A. F.; Vajo, J. J.; Atta, S. L.; Olsen, G. L. *J. Phys. Chem C* 2008, *112(4)*, 5651-5657.
[61] Autrey, T. (2009). PNNL *Progress as Part of the Chemical Hydrogen Storage Center of Excellence, DOE Hydrogen Annual Merit Review.* http://www.hydrogen.energy.gov/pdfs/review09/st_18_ autrey.pdf
[62] Burrell, A. (2009). *Chemical Hydrogen Storage R&D at Los Alamos National Laboratory, DOE Hydrogen Annual Merit Review.* http://www.hydrogen.energy.gov/pdfs/review09/st_17_ burrell.pdf
[63] Hugle, T.; Kuhnel, M. F.; Lentz, D. *J. Am. Chem. Soc.* 2009, *131*, 7444-7446.
[64] Welch, G. C.; San Juan, R. R.; Masuda, J. D.; Stephan, D. W. *Science* 2006, *314*, 1124–1126.
[65] Staubitz, A., Besora, M., Harvey, J. N. and Manners, I. *Inorg. Chem.* 2008, *47*, 5910–5918.

In: Boron Hydrides High Potential Hydrogen Storage Material ISBN:978-1-61668-361-0
Editors: Umit B. Demirci and Philippe Miele ©2011 Nova Science Publishers, Inc.

Chapter 10

CONCLUSION AND OUTLOOK: WHICH FUTURE FOR BORON HYDRIDES?

***Umit B. Demirci*[1] *and Philippe Miele*[2]**
Laboratoire des Multimatériaux et Interfaces, UMR 5615 CNRS Université Lyon 1, Université de Lyon, 43 boulevard du 11 Novembre 1918, F-69622 Villeurbanne, France.

ABSTRACT

Since about ten years, boron hydrides, and especially sodium borohydride, lithium borohydride and ammonia borane (and derivatives such as amidoboranes), are hydrogen storage materials that attract much attention by virtue of their high theoretical gravimetric hydrogen storage capacities. This has been the main subject of the present book. From Chapter 1 to Chapter 9, their state-of-the-art and their potential (especially for on-board automotive applications) have been reviewed, discussed and analyzed. The present chapter aims to conclude all of these chapters and tries to answer, or at least to give some elements of an answer, to the following question: which future for boron hydrides?

Boron hydrides are widely recognized as being promising hydrogen storage materials but none has met the technical criteria set for automotive applications yet. They all suffer from low effective (gravimetric and volumetric) hydrogen storage capacities (i.e. < 9 wt% at 85 °C) and storage irreversibility (at best they are off-board regenerable). These are the main issues. There are other important issues to be addressed, which are e.g. the hydrogen release kinetics and the hydride cost. The issues list is, unfortunately, not exhaustive. However, it is also widely recognized that research on these hydrides is at its early beginnings. In other words, there is still an ample room for improvement, especially in the case of materials like ammonia borane.

[1]Umit.Demirci@univ-lyon1.fr, Tel.: +33 (0)4 72 44 84 03; +33 (0)4 72 43 12 05, Fax:+33 (0)4 72 44 06 18
[2]Philippe.Miele@univ-lyon1.fr, Tel.: +33 (0)4 72 44 84 03; +33 (0)4 72 43 12 05, Fax:+33 (0)4 72 44 06 18

1. Introduction

"Hydrogen storage is the key technology for the introduction of the hydrogen society". This is the idea on which Mori and Hirose [1] emphasized in a recent paper dealing with the "recent challenges of hydrogen storage for fuel cell vehicles". This sentence reports none other than the general feeling researchers involved in hydrogen storage have. In other words, the hydrogen economy (or hydrogen society) is harshly hindered by many issues, the most significant one being certainly hydrogen storage. Today, intense (political, financial, technical, scientific and human) efforts are devoted to address this huge issue. For example, the number of publications about 'hydrogen storage' has exponentially increased since 1989 (Figure 1). It was of 18246 the 10th of October 2009 and more than 91 % of the articles have been published since 1989 (8.6 % in 1962-1988) [2].

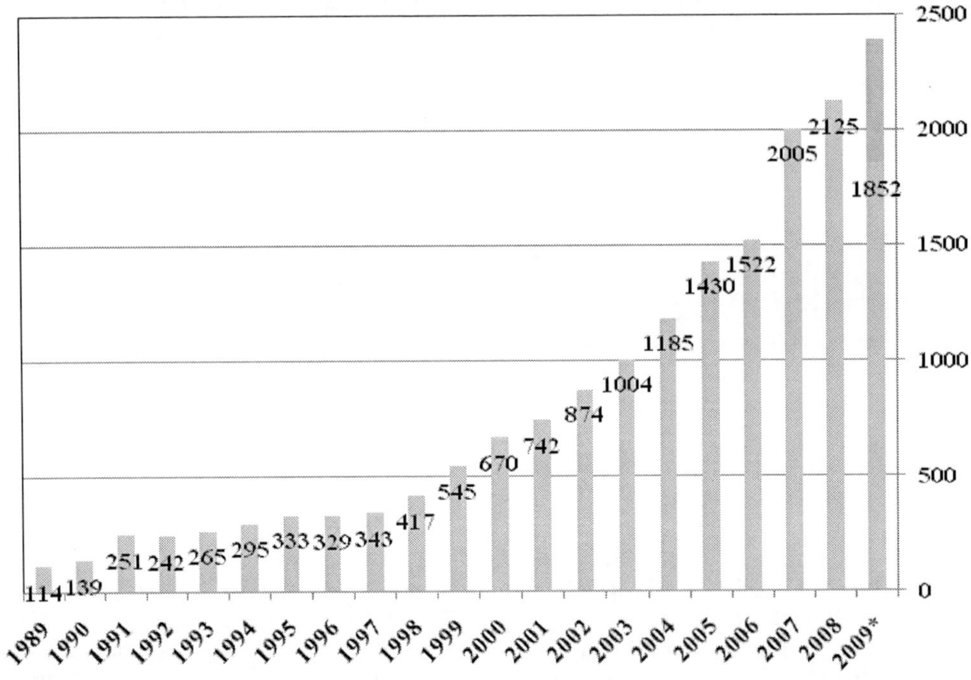

Figure 1.Number of publications dealing with hydrogen storage since 1989; note that for 2009, the search has been limited to the 10th of October; the grey 'rod' for 2009 is a projection for the 31st of December [2].

Several solutions are today under investigation. Many studies are in progress. As well, reported by Sartbaeva et al. [3], the accepted target for automotive applications is a reliable, safe and economical method to store 4-5 kg of hydrogen whilst minimizing volume, weight, storage energy, cost, refueling time and in providing prompt hydrogen release on demand. However, currently no candidate meets all of the requirements (safety, performance, cost, technical adaptation for the infrastructure, and scalability) [1]. Even though it has been discussed in Chapter 1, it is important to notice here, one more time, that the automotive industry is certainly the keenest sector to see the development of the hydrogen economy and that severe technical targets have been set for on-board hydrogen storage systems [4]. Chapter 1 has discussed the hydrogen storage issue while listing exhaustively the various storage solutions that have been suggested throughout the open literature. It has been showned that solid-state hydrogen storage is potentially the most convenient and the safest method [5], and that among the various possibilities of solid materials that could be used complex hydrides show the highest gravimetric hydrogen storage capacities (as well as volumetric hydrogen storage capacities).

Complex hydrides are in fact a large family of materials but one has to recognize that boron hydrides are often introduced as being the most attractive materials in term of hydrogen density [6,7]. This kind of hydrides has been the core of the present book. Introduced in Chapter 1, boron hydrides have been listed in Chapter 2 and the most promising (that are besides the most investigated), i.e. borohydrides $M(BH_4)_n$, ammonia borane NH_3BH_3, and amidoboranes MNH_2BH_3, have been reviewed in Chapters 3 to 9. Actually lithium borohydride $LiBH_4$, sodium borohydride $NaBH_4$ and ammonia borane NH_3BH_3 are by far the most studied boron hydrides. Few borohydrides such as calcium borohydride $Ca(BH_4)_2$ and amidoboranes such as sodium amidoborane $NaBH_2BH_3$ are emerging materials. Chapters 3 to 9 have reported the state-of-the-art for each of these classes of boron hydrides and show that even if e.g. sodium borohydride attracts attention since the early 2000s research on it and on each of the boron hydrides introduced here is still at the early beginning. In other words, they are still far from being in a technological development stage.

Recently, a timely paper [1] gave an updated description of the material and system development of three storage solutions, i.e. high pressure tank system, liquid hydrogen tank and hydrogen adsorbing alloy tank. Even though this paper does not really tackle boron hydrides, it has two merits. First, it shows that there are many large-scale automotive demonstrations of the storage ways the paper introduce. Unfortunately, it is not the case for the complex hydrides [8]. Second, none of the studied storage ways (high pressure, liquid hydrogen, hydrogen-adsorbing alloy) meets the application targets (in terms of gravimetric and volumetric hydrogen storage capacities). This second remark asks then the following questions. What about the complex hydrides, and more specifically, the boron hydrides? Can one expect that they will meet the targets in a near-future?

The questions asked above may actually be merged into one question, that is: which future can one expect for boron hydrides? After a short summary of the state-of-the-art of the boron hydrides (developed in Chapters 3 to 9), the present chapter concludes the present book and attempts to give some elements of answer to our last question.

2. Current Potentials of the Boron Hydrides

2.1 Borohydrides

The best known boron hydride is without doubt sodium borohydride (or sodium tetrahydroborate, $NaBH_4$). The research devoted to sodium borohydride is still today active and vivid [9]. The first recent studies date back the late 1990s (Figure 2) and the starting point may be ascribed to the Amendola et al.'s works [10,11]. This recent period 1999-2010, for which research is catalyzed by societal challenges that are energetic, environmental, civilian, social, and political has to be distinguished from the first period, i.e. the 1950s, which is the period of its discovery and when the challenges underlying its development were mainly military [9]. Sodium borohydride is a hydrogen carrier (i.e. a hydrogen storage material) as well as an electron carrier (i.e. a fuel for direct liquid-fed fuel cell). As that has been relevantly reported by Aardahl and Rassat [8], sodium borohydride is one of the only two (chemical) hydrides that have permitted to develop a chemical storage system prototype to be interfaced with vehicles (large-scale automotive demonstrations). However, as that has been reviewed in Chapters 4 and 5, technologies based on sodium borohydride, i.e. both solid-state storage system (releasing hydrogen on demand) and fuel of direct borohydride fuel cell, are far from being commercially developed. Many issues hinder the development of sodium borohydride. Actually, despite a ten-year-old research, studies may be considered as being at their early beginnings yet. Moreover, the situation is even much more critical for sodium borohydride as hydrogen carrier since a critical publication of the U.S. Department of Energy [12]. To briefly report this publication, one can remark that it was concluded that sodium borohydride has no potential for automotive applications due to low effective storage capacities, storage irreversibility and hydride cost. Accordingly, it has been recommended a no-go for it. Note that a prototype vehicle using the sodium borohydride-based storage system has been used but, as relevantly reported by Eberle et al. [13], the practical experience was not sufficiently convincing to perform a second prototype test. Nevertheless, as well remarked by Umegaki et al. [14], hydrogen release by hydrolysis of sodium borohydride is a good choice for one-time application (i.e. portable applications). Niche applications are also an alternative way of valorizing sodium borohydride [13]. With respect to sodium borohydride as fuel of direct borohydride fuel cell, it is still in its initial development phase [15].

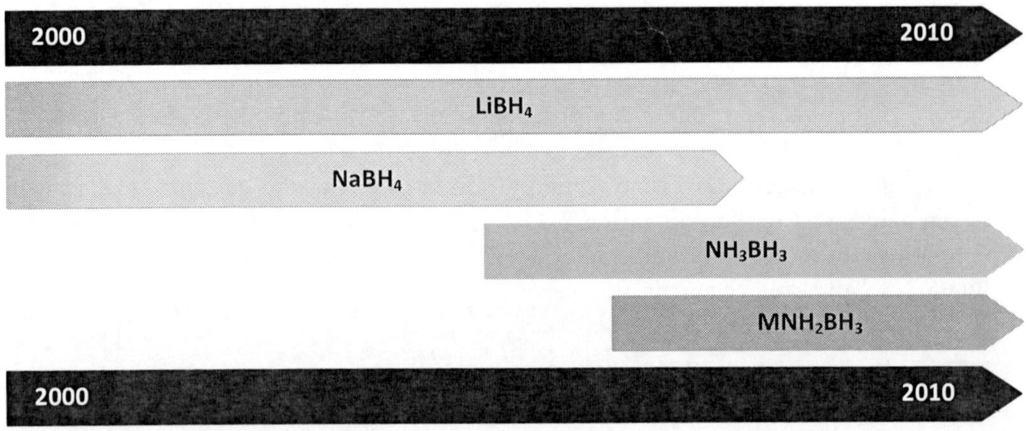

Figure 2. From borohydrides to boranes in less than 10 years.

Among the great number of borohydride compounds that have been listed in Chapters 2 to 6, lithium borohydride (or lithium tetrahydroborate, $LiBH_4$) is one of the most known materials [16] whereas the studies focusing the others are still rather marginal. Its theoretical gravimetric and volumetric hydrogen storage capacities are among the highest ones since they are higher than 18 wt% and 120 kg m^{-3}, respectively [14]. With lithium borohydride, stored hydrogen is recovered only by thermal decomposition (thermolysis). The first thermal desorption of hydrogen was reported in 1964 [17]. Note that its hydrolysis is strictly avoided because it is highly exothermic and the quick hydrogen release can be hazardous. Thermal destabilization (with hydrogen release) of neat lithium borohydride requires too high temperatures because it has a large decomposition enthalpy (for example, 1 wt% released at > 400 °C) [14,18,19]. Today, the objective is to decrease the hydrogen release temperature while increasing the release kinetics in order to reach the specifications set for automotive applications (e.g. equivalent of 9 wt% of pure hydrogen releases at 85°C in 2015) [4]. Despite the addition of destabilizing agents (e.g. silica SiO_2 or metal oxides M_xO_y [20], or metal halides M_xX_y [21]) into the lithium borohydride matrix or the synthesis of $LiBH_4$ nanoparticles [22], the onset dehydrogenation temperature of lithium borohydride is higher than the target 85°C. In other words, it is too high in the view of portable and mobile applications [14]. It is nevertheless important to report that decomposition at low temperatures, i.e. 60°C, were reached for titanium chloride-modified lithium borohydride via the formation of titanium borohydride $Ti(BH_4)_3$ [21]. However, in that case, the storage is fully irreversible, which makes lithium borohydride much less attractive, given the decomposition products rehydrogenation is still limited [14]. Despite increasing efforts since the 1960s, it has not been possible to overcome the thermodynamic, kinetic and reversibility challenges that would enable $LiBH_4$ to be employed as a practical hydrogen storage material [16].

As well as both sodium borohydride and lithium borohydride, no other borohydride compound has been the object of attention in hydrogen storage (and even in fuel cell). Today, few borohydrides like calcium borohydride $Ca(BH_4)_2$ [23] arouse interest but, in our opinion, it does not appear to be relevant to talk about their respective state-of-the-art. Several of those borohydrides have attractive theoretical gravimetric hydrogen storage capacities (see Chapter 2) but their effective gravimetric hydrogen storage capacities have still to be experimentally demonstrated. According to Yang and Hirano [7], it is challenging to use borohydrides alone for low temperature reversible hydrogen storage. And, the state-of-the-art (see Chapters 2, 3 and 4) confirms.

2.2 Ammonia Borane and Amidoboranes

While the U.S. Department of Energy [12] recommended a no-go for sodium borohydride for on-board vehicular hydrogen storage, it strongly suggested to work on ammonia borane (or borazane, NH_3BH_3), a novel (even if first synthesized in 1955 [24]) highly promising boron hydride that stores more than 19 wt% of hydrogen. Since this recommendation, i.e. November 2007, one may assert that ammonia borane (Figure 2) is become the most 'popular' boron hydride [14,25,26,27]. On the one hand, it has been considered as hydrogen storage material and, on the other hand, it has been envisaged as fuel (as an aqueous solution) of direct liquid-fed fuel cell [15]. Furthermore, its safety features are often showed as being important advantages [16]. Chapters 7 and 8 have reviewed these potential applications respectively. In both cases, the studies are in their early beginnings and the progresses vary from one application to the other. Chapter 7 has showed that ammonia borane thermolysis is a route that enables to reach most of the U.S. Department of Energy targets [4], with e.g. hydrogen yields of about 7 wt% at 85°C, if ammonia borane is dispersed in ionic liquid solvents [28], incorporated into a scaffold such as mesoporous silica [29], and doped with transition metal complex-based catalysts [30]. Hydrolysis of ammonia borane is another way for releasing hydrogen [14] but the formation of ammonium ion NH_4^+ and the need of an excess of water drastically decrease the effective gravimetric storage capacities of the system ammonia borane-water, what makes Demirci and Miele conclude that ammonia borane hydrolysis is much more problematic than sodium borohydride hydrolysis [15]. With respect to the utilization of ammonia borane as fuel of fuel cell, it is concluded in Chapter 8 as well as in reference [15] that the direct ammonia borane fuel cell is less effective, less practical (because of NH_4^+) and more complex than the direct borohydride fuel cell. To conclude, one can today assert that ammonia borane has a high potential if its thermolysis is only considered. However, another significant issue, that is storage reversibility, has to be absolutely addressed. The ammonia borane regeneration is not at all straightforward [31]. Eberle et al. [13] emphasized the exothermicity of the decomposition steps and the fact that ammonia borane is not useable for reversible on-board hydrogen storage. As that has been concluded in Chapter 7, it has to be stressed that the regeneration of spent fuel is a crucial open point.

Chapters 1, 2, and 9 have reported that another interesting strategy for improving the potential of ammonia borane in hydrogen storage is its chemical activation via substitution of the protonic hydrogen in the hydride H–NH_2BH_3 by a metal atom M. By this way, metal amidoboranes MNH_2BH_3 have been obtained. The metal M decreases the enthalpy of hydrogen loss, eliminates the formation of volatile by-products and helps the control of the hydrogen release. Several metal amidoboranes have been synthesized and the most attractive is lithium amidoborane $LiNH_2BH_3$, which has a gravimetric hydrogen storage capacity of 13.6 wt% [32]. Actually, amidoboranes are a new class of materials in hydrogen storage (Figure 2). Very few studies have been published and very few metals have been used. Accordingly the range of metals used in metal amidoborane complexes can be expanded [27,33,34]. Compared to ammonia borane, the first results reported so far for amidoboranes are quite promising with decreased hydrogen release temperatures, less exothermic dehydrogenation and no formation of volatile by-products [35]. For example, Xiong et al. [33] reported releases of 10.9 and 7.5 wt% of hydrogen for $LiNH_2BH_3$ and $NaNH_2BH_3$, respectively, at about 90 °C. Nevertheless, one has no information concerning e.g. their stability, toxicity, and storage reversibility (or recyclability) yet.

2.3 Summary

From a survey of the open literature and from the reviews the present book proposes, five main conclusions emerge. The first conclusion is related to ammonia borane. Today it seems to be the only boron hydride that has a real potential for automotive applications despite the severe specifications that have been set for [4,27]. The second and third conclusions concern sodium borohydride. As a hydrogen storage material it still has a potential for small portable, mobile applications [15]. This is the second conclusion. And as a fuel of fuel cell it is most attractive than ammonia borane [15]. The fourth conclusion concerns metal amidoboranes. Very few studies have been currently published and it is too soon to critically discuss their technological potential. Nevertheless they represent a class of boron hydrides which potential is perhaps among the most promising. Furthermore, it seems to be opportunities to develop amidoborane-based new hydrides. For example, very recently, calcium amidoborane ammoniate $Ca(NH_2BH_3)_2{\cdot}2NH_3$ with a storage capacity of 12 wt% has been synthesized. It releases more than 8 wt% hydrogen at 150 °C (without gaseous by-product emission) [36]. Finally, the fifth conclusion is about one of the most important criteria that hydrogen storage materials must meet, that is, storage reversibility. The state-of-the-art clearly reveals that today none of the investigated boron hydrides is a reversible storage material. At most, few can be partly off-board regenerated.

3. Always the Same Challenges

Many (maybe too many) challenges have to be addressed before realistically expecting the development of a hydrogen economy. Recently, several discussion papers stressing on the challenges have been published. Sartbaeva et al. [3] critically analyzed the ability of hydrogen in replacing carbon-based fuels in the future. They discussed "hydrogen nexus" that has been defined as follows: "hydrogen may fulfill part of the central role currently occupied by carbon, and provide the means of connection and link to a sustainable energy economy." The state-of-the-arts of hydrogen production, distribution, storage were briefly surveyed in order to highlight the scientific, technological and socio-economic challenges the hydrogen economy is facing. Unfortunately, it stands out that there are many challenges and that all are significant. The first (pessimistic) feeling that such discussion may leave is that nothing has changed despite years of efforts in research and development.

3.1 Hydrogen Production and Distribution

With respect to hydrogen production, the challenges are still the same. Actually there are two main challenges. The first one is the hydrogen production cost [3,37,38]. The second one concerns a sustainable and carbon dioxide-free production route (e.g. solar methods) [3,38]. Concerning the hydrogen delivery and distribution, it is still expected a safe and efficient infrastructure to be developed [3].

3.2 Hydrogen Storage

Like hydrogen production and also like hydrogen distribution, hydrogen storage faces the challenges it was facing ten years ago [1]. Today solid-state hydrogen storage system using hydrides seems to be the most promising storage alternative [1]. According to Sartbaeva et al. [3], it may represent perhaps the only viable method of achieving a high weight percent and high volume density of stored hydrogen (Figure 3).

Analyzing the potential of a solid-state storage system can only be done on the basis of the specifications set for vehicular applications since to our knowledge no criterion set for applications other than these ones is as important and recurrent throughout the open literature [15]. The technical, technological, and economic requirements can be summarized as being as follows: safety, performance in terms of charge/discharge and weight/volume, cost, technical adaptation for the infrastructure, and scalability (applicability to small and large vehicles) [1].

In other words, it is expected the solid-state storage system is compact, low-weight, low-cost, safe and efficient and it operates at or near room temperature and low pressures [3]. Note that Wenger et al. [39] have proposed detailed information about the requirements by a typical 100 kW fuel cell vehicle. However, currently no candidate, even the hydrides like e.g. ammonia borane, meets all of these critical requirements. This agrees many authors' conclusion [1,3,13,40].

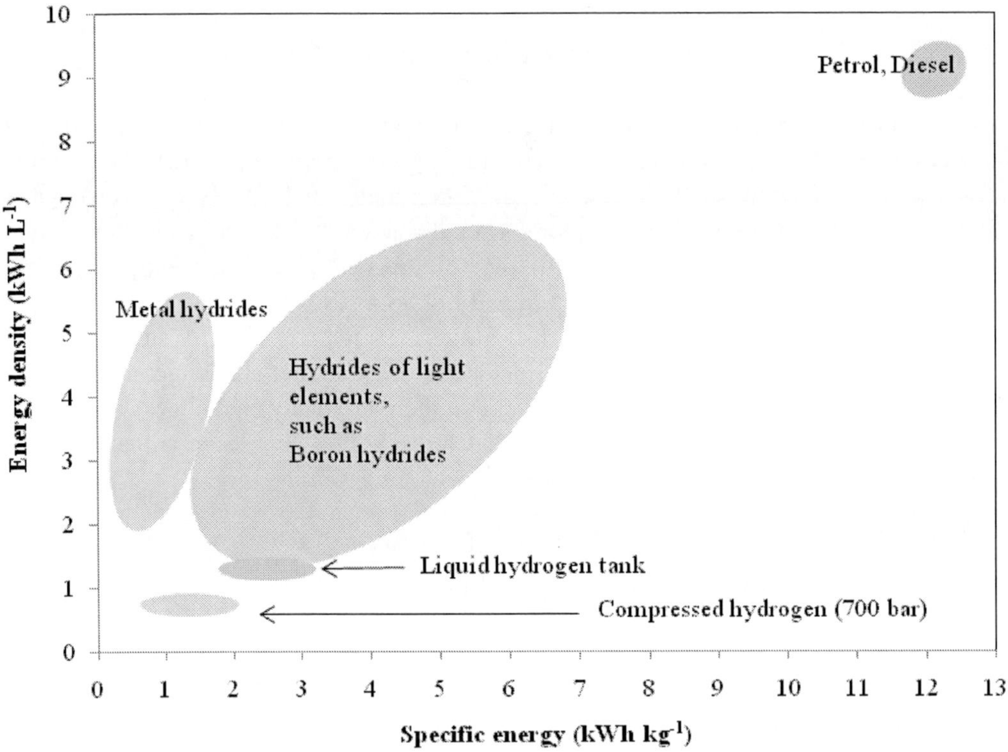

Figure 3. Energy densities of various energy storage materials and technologies; adapted from ref. [3].

Recently, researchers at Daimler AG, Chrysler LLC and TU München have published a technical paper to emphasize four important, but neglected, issues in solid-state hydrogen storage systems: i.e. volumetric storage capacity, heat transfer for desorption, recharging at low temperatures, and cold start of the vehicle [40]. For example, it has been showed that the volumetric hydrogen storage capacity of the solid-state storage system decreases despite the increasing gravimetric hydrogen storage capacity of the material. This makes the authors suggest placing more emphasis on the volumetric density. One has here to recognize that they are true. Even in the present book, through Chapters 1 to 9, the gravimetric hydrogen storage capacity of the boron hydrides is the main tool used to compare the storage materials each others. Wenger et al. [40] insistently suggested that the four issues they discussed should be taken into account to adequately explore the limitations and capabilities of the storage technologies. This is never done. One never does it whereas one is searching for new boron hydrides. Wenger et al.'s timely article asks actually one question: are complex hydrides, and especially boron hydrides, mature enough to be developed in the view of automotive applications? From the surveys in the present book, they appear not to be. However, and, fortunately, there is still an ample room for improvement, especially in the case of materials like ammonia borane.

3.3 Summary

From a general point of view, the drawn picture is quite pessimistic with respect to the level of performance the complex hydrides, and so the boron hydrides, have reached. None of these materials obviously have met the technical criteria for on-board vehicular applications (Figure 4). This has been very recently reported by Eberle et al. [13] while they were surveying, exhaustively, the chemical and physical solutions for hydrogen storage. The challenges are still the same than those listed years ago. All of the challenges discussed above are important but four of them are often showed as being the most significant ones. The first one is the effective hydrogen storage capacity of the storage system though the gravimetric hydrogen storage capacity is much more 'popular' than the volumetric one. The second one is the storage reversibility, or maybe one should rather say irreversibility, which is very problematic for automotive applications. Ideally, it is expected that one can refuel the hydrogen storage system as one uses to refuel the current petrol or diesel tanks. The third one is cost. The boron hydrides are today expensive and, within the current context, they are not competitive at all [31,35,41]. Nevertheless, this aspect is generally briefly announced. Hydrogen storage cost is not as informed as hydrogen production cost yet. Actually discussing the hydrides cost seems to be irrelevant as these materials are still in a research stage. The fourth challenge is the purity of released hydrogen, which is critical for fuel cell applications. Nevertheless it does not appear to be as important as the first challenge. This is actually a priority concern; namely, once the effective storage capacity improved, one should be able to find a (technical at least) solution to the hydrogen purity.

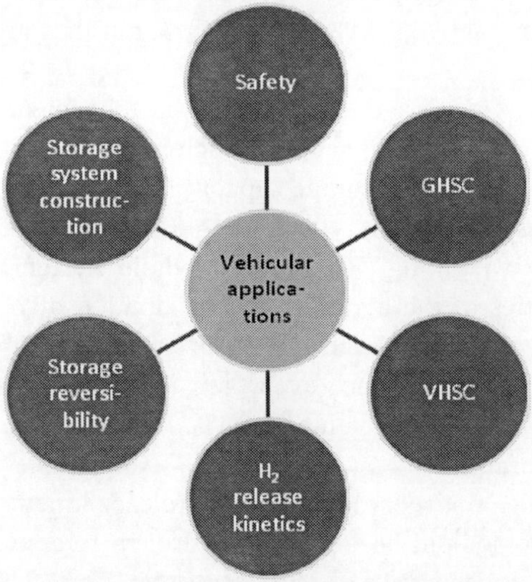

Figure 4. Technical criteria for on-board vehicular applications that have to be fulfilled, according to references [1,3,39]; with GHSC and VHSC as gravimetric/volumetric hydrogen storage capacity.

Many efforts are now focused on optimizing the effective storage capacities (first issue) and making the storage reversible (second issue). In such a context, decreasing the cost appears to be a secondary issue. For this reason, this third issue (as well as the fourth one as specified above) will not be discussed in the next section that deals with the most critical challenges facing boron hydrides.

4. The Most Critical Challenges Facing Boron Hydrides

As it has been several times remarked throughout the present book, it is today widely expected from chemical hydrides, and especially from boron hydrides, that they provide in a near-future a solution to overcome the hydrogen storage problem encountered by on-board vehicular applications. They theoretically have such a potential, especially in terms of gravimetric and volumetric hydrogen storage capacities. It is much less obvious regarding the storage reversibility. Both aspects, which are the most critical challenges the boron hydrides presently discussed face, are discussed hereafter.

4.1 Effective Storage Capacity

Hydrogen has to be supplied at a sufficiently high amount to meet the targets for on-board vehicular applications (e.g. 9 wt% for 2015). Theoretically, all of the boron hydrides discussed here should meet such targets. However, their effective storage capacities are well below the theoretical value. To be brief, none of them has reached the 2015 target yet (specifically discussed in the next paragraph). In this context, sodium borohydride seems to be definitely dismissed [4]. Lithium borohydride and ammonia borane are still far from releasing 9 wt% of their hydrogen at 85 °C. Regarding amidoboranes, the first results are rather attractive with values above 9 wt% at 90 °C [33] but they are not enough in the view of applications.

Regarding the effective storage capacities, we, the scientists involved in hydrogen storage, often make a mistake when discussing such parameters. Eberle et al. [13] relevantly stressed this mistake. To calculate a storage capacity, the storage system has to be considered as a whole as it is the case in all energy technologies. If in a first approximation the storage system is accounted as weighting 50% of the whole system, the effective gravimetric hydrogen storage capacities of the hydrides will be divided by a factor 2 [15]. Accordingly none of the hydrides would today reach the 2015 target of 9 w% (Figure 5). Furthermore, most of them should be dismissed since even their theoretical gravimetric hydrogen storage capacities would be below of the value 9 wt%. It would be the case of e.g. sodium borohydride and amidoboranes.

The present section has not regarded hydrogen release kinetics as a challenge yet. But in fact, it is. With respect to sodium borohydride, hydrogen release rate is not a problem since with the addition of a good catalyst very high rates can be reached [15]. One may advance that in this case this is not an issue anymore. Chapter 4 has illustrated this. Concerning the hydrides destabilized by thermolysis, hydrogen release rate is critical but it may have a dependence on the hydrogen release temperature. In other words, the decrease of the dehydrogenation temperature of the hydride may be related to an increase in the hydrogen release rate. Actually it is expected that the hydride destabilization impact both the dehydrogenation temperature and the hydrogen release rate [7] due to lowered apparent activation energy of the rate determining steps and more favorable hydrogen reaction thermodynamics, respectively. The former feature is all the same much more studied. The objective in both cases is to reach temperatures close to fuel cell working temperature range [7], namely around 85 °C.

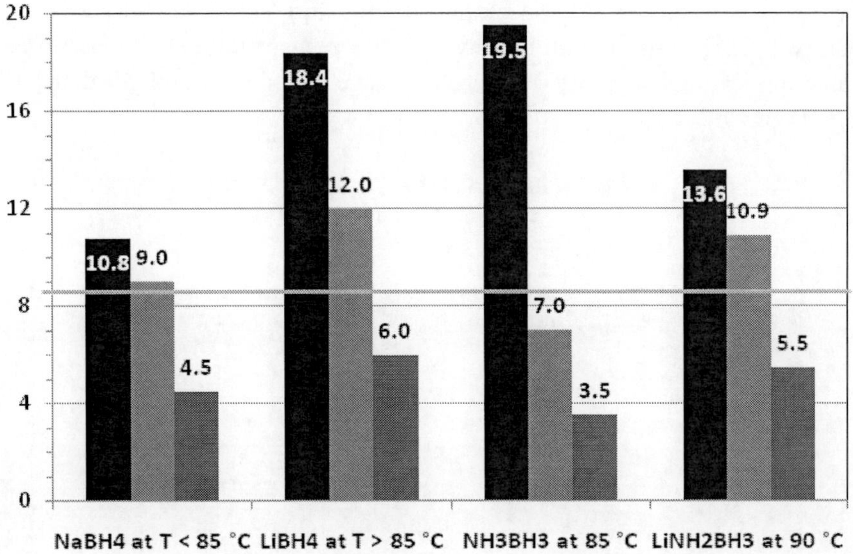

Figure 5. Gravimetric hydrogen storage capacity (GHSC) of the storage system taken as a whole; the black rods for the theoretical GHSC, the blue ones for the effective GHSC, and the brown ones for the effective GHSC when the storage system weight is taken into account (50 %); the yellow line is the 2015 target for automotive applications (effective GHSC of 9 wt%).

To summarize the present subject, two observations stand out. First, the effective hydrogen storage capacities that have been reached are far from the theoretical values and the targets but marked improvements have been showed in more than a decade of sustained efforts. Second, when discussing the effective hydrogen storage capacities of a hydride one has to keep in mind that the storage system has to be taken into account as a whole. In many scientists' opinion, there is still an ample room of improvement. To illustrate that let cite Eberle et al.'s opinion [13]. "Advances of solid hydrogen storage materials based on hydrides are difficult to predict. Chemically, it appears that higher storage capacities can be achieved. The thermodynamics of these systems are not known at present, and the same holds for the kinetics."

4.2 Storage Reversibility

Storage reversibility in boron hydrides, i.e. in sodium borohydride, lithium borohydride, ammonia borane and amidoboranes, remains a significant challenge that has to be necessarily overcome. Here will be distinguished storage reversibility with sodium borohydride from that with the other boron hydrides because of the difference in the hydrogen release reaction.

Sodium borohydride is today exclusively dehydrogenated through the catalytic hydrolysis with the formation of four hydrogen molecules and of a solid by-product that is the hydrated sodium metaborate [15]. Such a reaction and such a by-product do not enable sodium borohydride to be a material storing reversibly hydrogen. It has been attempted to recycle (hydrated) sodium metaborate back to sodium borohydride but, according to the U.S. Department of Energy [12], the attempts failed in proposing viable, efficient and cost-effective recycling routes. Frankly speaking, the literature survey effectively shows that the attempts failed [15]. Today, sodium borohydride has perhaps no potential as storage material for vehicular applications anymore. Hydrogen storage in sodium borohydride is, thus, clearly irreversible (Figure 6).

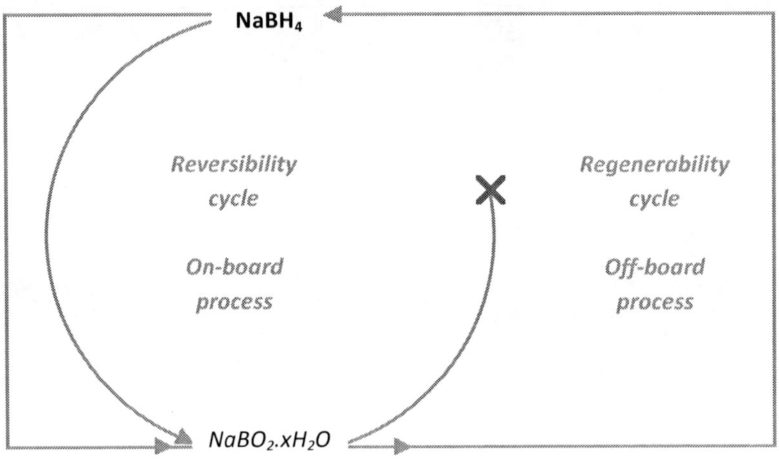

Figure 6. Sodium borohydride, a non-reversible hydrogen storage material.

Thermolysis is a dehydrogenation reaction that leads to decomposition products that may offer hydrogenation possibilities. Numerous attempts have been reported throughout the open literature and, evidently, this research area still today is mobilizing many research groups and considerable efforts. Thermolysis is the main reaction applied to lithium borohydride, ammonia borane and amidoboranes.

Lithium borohydride does not store hydrogen reversibly. Wang and Kang [35] recently reviewed the huge literature dealing with dehydrogenation as well as hydrogenation of this boron hydride. To improve the storage reversibility, a strategy consisting in destabilizing lithium borohydride while stabilizing the solid thermal decomposition product with the aid a third body (metal hydride, scaffold, cation/anion substitution and catalyst) has mainly been used. Despite improvements (i.e. partial hydrogenation), hydrogen storage in lithium borohydride is irreversible or, to be more optimistic, partly reversible but it is not enough for the expected technological (automotive) applications.

Thermal decomposition of ammonia borane with hydrogen release is exothermic and leads to polymeric solid products (i.e. BNH_x). Both aspects do not make storage reversibility easy. Ammonia borane has been showed as being the most promising boron hydride but its storage irreversibility is detrimental to its development [27,31]. According to Mandal and

Gregory [16], ammonia borane and amidoborane systems offer several advantages but the most serious drawback is the lack of facile reversibility which makes them unsuitable for on-board storage vehicular applications. These last years, significant progress has been achieved but the efficient off-board regeneration (indirect reversibility [41]) of spent ammonia borane remains a grand scientific and technical challenge [35]. Among the recent progresses, one study is worth being cited. Davis et al. [42] have shown that one form of spent ammonia borane, i.e. polyborazylene BNH_x, can be regenerated efficiently in a one-pot process by the stepwise addition of appropriate digesting (1,2-benzenedithiol, $C_6H_4(SH)_2$) and reducing (tributylstannane, Bu_3SnH) agents (Figure 7). By this regeneration route, an overall yield of 67% isolated ammonia borane would have been obtained. Given the U.S. Department of Energy target of 60% thermodynamic efficiency for off-board regeneration [4], this is, as well remarked by Hamilton et al. [27], "an intriguing result". According to Davis et al. [42], optimization studies are still being developed and should be soon reported.

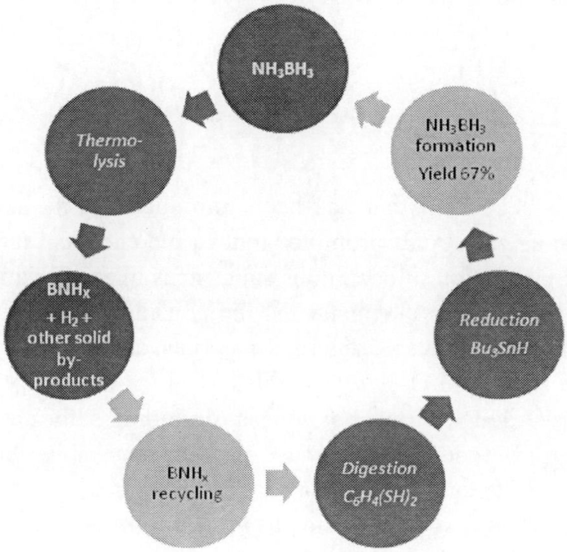

Figure 7. Off-board regeneration of one form of spent ammonia borane, namely BNHx [42].

As a conclusion to this sub-section, one may cite Crabtree's critical analysis on hydrogen storage [31]. It is relevantly noticed that boron is not a renewable element and that boron hydride compounds should be seen as depleting resources. Accordingly, it is crucial to develop either boron hydrides reversibly storing hydrogen or boron hydrides which solid product(s) would be efficiently recycled back to the 'mother' compound. In other words, (direct or indirect) reversibility of storage is essential in terms of application and sustainability (greenness).

4.3 Summary

Effective hydrogen storage capacity and storage reversibility are as important as they were more than a decade ago when hydrogen became one of the most promising energy carriers to replace in near-future depleting fossil fuels. Years of efforts have been fruitful in improving the features of the hydrides already known as well as in discovering new types of boron hydrides. Some of them showed storage capacities meeting the 2007 target (i.e. 4.5 wt% of hydrogen). Few of them would be able to meet the 2010 target of 6 wt%. However, none of them currently meets the severe target of 2015, which is of 9 wt%. Actually the targets evolve more quickly than do the effective storage capacities. Besides, none of the hydrides discussed here is able to store hydrogen reversibly. To summarize, "there remain immense challenges ahead before we enter into a genuine hydrogen energy area" [16].

5. Greenness Considerations

Sustainability and greenness are concepts that emerged in the 1990s. Since, they have become important, especially in chemistry. Both concepts take their strength in the fact that they are seen as powerful tools to construct a future more respectful of humans, environment and Earth.

5.1 Green Chemistry

Green chemistry is a concept that has been introduced in the early 1990s by the US Environmental Protection Agency. It promotes innovative chemical technologies that reduce or eliminate the use or generation of hazardous substances in the design, manufacture and use of chemical products [43]. Green chemistry can be considered as a series of reductions (in risk and hazards, waste, non-renewable, energy, materials, and cost) [44,45]. In 1998, Anastas and Warner [46] have stated the twelve principles of green chemistry: prevention of waste, atom economy, less hazardous chemical synthesis, designing safer chemicals, safer solvents and auxiliaries, design for energy efficiency, use of renewable feedstock, reduction of derivatives, catalysis utilization, design for degradation, real-time analysis for pollution prevention, and safer chemistry for accident prevention. Green chemistry provides an enormous number of challenges to chemists and is an opportunity to enhance the much-tarnished image of chemistry that is seen as being the cause of most of the environmental problems [44].

Green chemistry is a concept that addresses hazards, whether physical (flammability, explosivity), toxicological (carcinogenicity, endocrine disruption), or global (ozone depletion, climate change) as an inherent property of a molecule [45-48]. According to Anastas [47], green chemistry is engaged in addressing energy needs through the development of more sustainable energy technologies like hydrogen and fuel cells.

5.2 Greenness of the Boron Hydrides

According to Horváth and Anastas [49], the materials that will be required for alternative energy futures (e.g. biofuels, hydrogen or other alternatives) should be designed in such a way that the new materials do not introduce unacceptable problems of toxicity, persistence, bioaccumulation, and depletion of finite resources.

Boron hydrides are chemical compounds and in that sense they may be hazardous towards human and environment. Sodium borohydride, lithium borohydride and ammonia borane are often reported as being safe and stable compounds at room temperature if they are stored in closed vessel and anhydrous medium [15]. Regarding amidoboranes, there is no specific study tackling safety problems that could be encountered with such materials. Hence, these materials will not be further discussed here but one has to keep in mind that there is no information concerning e.g. their stability and toxicity yet and this should be quickly made good.

With respect to sodium borohydride, recent papers dealing with direct liquid-fed fuel cells [50,51], and especially with direct borohydride fuel cell using sodium borohydride as fuel, reported that this hydride is not as safe as often claimed. Its handling as well as that of its by-product sodium metaborate is hazardous and may be risky. Safety is important for the chemicals intended to large-scale utilizations. Safety is also important from the greenness standpoint. Information on safety is generally available in the chemicals material safety data sheet. The few ones about ammonia borane are incomplete (see *e.g.* reference [52]). The ones on sodium borohydride are much more complete [53]. Lane [54] discussed the safety aspects relating to ammonia borane. Sodium borohydride and ammonia borane are white solids. Both are moisture sensitive but only the former hydrolyses and generates hydrogen in the presence of water. Ammonia borane is rather stable in aqueous solution, hydrolysis being obtained only in the presence of a catalyst (homogeneous catalysts as acids or heterogeneous catalysts as metals). When handling or storing them, oxidising agents and acids must be avoided. To be stable and safe, ammonia borane must be of high purity. The main difference between sodium borohydride and ammonia borane is the thermal stability. The latter decomposes at very low temperatures in relation to the former. In summary, it is emphasized that if the handling and storage recommendations are scrupulously respected the risk should be highly minimized [15,50,51]. Since green chemistry simply advocates safe chemicals and safe chemistry, the relative safety (if handling and storing recommendations are fully respected) of both sodium borohydride and ammonia borane may conform to the green chemistry concept [15].

With respect to lithium borohydride, the material safety data sheets report hazards that are similar to those reported for sodium borohydride (see e.g. ref. [55]). It would be as health-hazardous, flammable and reactive. One has nevertheless to add that lithium borohydride is much more water-reactive. For example, Tanaka et al. [56] assessed potential hazards of several complex hydrides, which were subjected to standardized safety tests. Lithium borohydride as well as the other hydrides were judged flammable, pyrophoric and water-reactive.

Green chemistry is also engaged in sustainability [45]. It intensely supports the utilisation of renewable feedstocks. Boron sources and, accordingly, sodium borohydride, lithium borohydride and ammonia borane, are not renewable [31]. Nevertheless, sodium metaborate stemmed from the hydrolysis of sodium borohydride and polymeric BNH_x products stemmed from the thermal decomposition of ammonia borane can be recycled back to the hydrides (even if the processes are still too expensive and are not efficient yet). Spent lithium borohydride can be partially hydrogenated. From this point of view, sodium borohydride, lithium borohydride and ammonia borane may be regarded as being relatively green [15], with maybe lithium borohydride less green than the other hydrides.

5.3 Summary

To summarize, the safety aspects and the possible regeneration of the spent fuels are in favour of these hydrides. In other words, they may be seen as being relatively green (Figure 8).

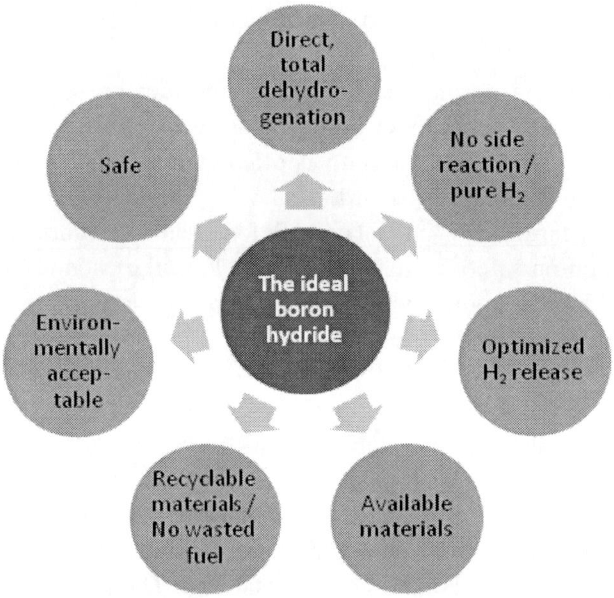

Figure 8. The ideal boron hydride.

CONCLUSION AND OUTLOOK

In 2009, several reviews discussing solid-state hydrogen storage have been published. Each surveyed the open literature, exhaustively for some of them, partly for the others. Further, while some especially focused on hydrides, the other ones considered all of the materials for solid-state hydrogen storage. And interestingly the conclusions regarding boron hydrides are systematically the same. Namely, boron hydrides are promising storage materials for on-board vehicular applications but several significant issues have to be addressed, the issues being related to the followings features: effective (gravimetric and volumetric) hydrogen storage capacities, optimized hydrogen release kinetics, storage reversibility, hydrogen with high purity, and low cost. Few of the reviews relevantly emphasize the fact that the effective storage capacities have to be absolutely calculated while considering the storage system as a whole, namely taking into account the weight of each component (tank, pipes, valves and so on). Further, as reported by Graetz [18], the utility of a hydride will ultimately be determined by its system performance.

Eberle et al. [13] have very recently proposed an 'outlook' section as a last section of their survey. On the basis of the state-of-the-art of each storage solution, it has been suggested that liquid hydrogen is the most promising option for large volume storage and pressure storage is the best solution for smaller-scale applications (e.g. cars). Nevertheless, it has been remarked that this may change with possible advances in other technologies. Eberle et al. have also discussed the outlook of solid-state storage materials. They have recognized that advances on hydrides are difficult to predict because the thermodynamics and the kinetics are not well known yet. Their potential is high and they might outperform the high pressure storage as well as the low temperature storage (liquid hydrogen). But they might outperform in near-future if improvements linearly evolve.

Research on boron hydrides, and maybe also on all of the hydrogen storage solutions, is still novel. Borohydrides are investigated since about ten years. Ammonia borane is more recent and amidoboranes are much more recent (Figure 2). They appears to be highly promising materials and solid-state hydrogen storage is certainly the best storage solution as it is commonly admitted throughout the open literature. However, it is extremely difficult to predict today if they will be able to meet the criteria set for e.g. the automotive applications. Eberle et al. [13] concluded: "a sound prediction of the best future system is very difficult, as 'best' may depend on progress in different technology sectors". New approaches or new materials could be solutions to further improve performance of hydrogen storage materials. To conclude the present book, we would like to join van den Berg and Areán [40] when they write "we are eager to see fruitful developments coming out soon", especially in the area devoted to boron hydrides (Figure 9).

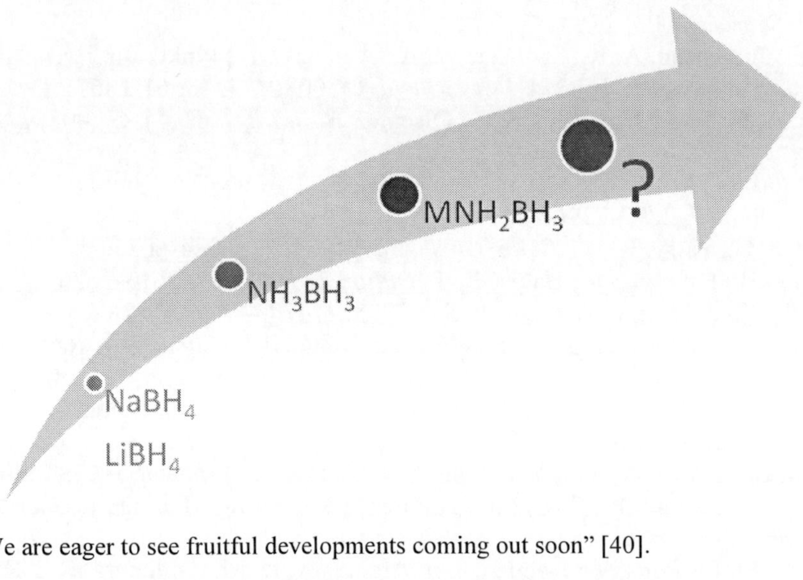

Figure 9. "We are eager to see fruitful developments coming out soon" [40].

REFERENCES

[1] Mori, D.; Hirose, K. *Int. J. Hydrogen Energy* 2009, *34*, 4569-4574.

[2] ISI Web of Knowledge. http://www.isiknowledge.com. *Search using 'hydrogen storage' as topic criterion.* Searched the 2009, 10th October.

[3] Sartbaeva, A.; Kuznetsov, V. L.; Wells, S. A.; Edwards, P. P. *Energy Environ. Sci.* 2008, *1*, 79-85.

[4] U.S. Department of Energy, Energy Efficiency and Renewable Energy (2007). Hydrogen storage. http://www1.eere.energy.gov/hydrogenandfuelcells/mypp/pdfs/storage.pdf.

[5] Mao, S. S.; Chen, X. *Int. J. Energy Res.* 2007, *31*, 619-636.
[6] Schlapbach, L.; Züttel, A. *Nature* 2001, *414*, 353-358.
[7] Yang, J.; Hirano, S. *Adv. Mater.* 2009, *21*, 1-6.
[8] Aardahl, C. L.; Rassat, S. D. *Int. J. Hydrogen Energy*, 2009, *24*, 6676-6683.
[9] Demirci U. B.; Miele, P. *C. R. Chim.* 2009, *12*, 943-950.
[10] Amendola, S. C.; Onnerud, P.; Kelly, M. T.; Petillo, P. J.; Sharp-Goldman, S. L.; Binder, M. *J. Power Sources* 1999, *84*, 130-133.
[11] Amendola, S. C.; Sharp-Goldman, S. L.; Janjua, M. S.; Spencer, N. C.; Kelly, M. T.; Petillo, P. J.; Binder, M. *Int. J. Hydrogen Energy* 2000, *25*, 969-975.
[12] U.S. Department of Energy (2007). Independent review: go/no-go recommendation for sodium borohydride for on-board vehicular hydrogen storage. NREL/MP-150-42220.
[13] Eberle, U.; Felderhoff, M.; Schüth, F. *Angew. Chem. Int. Ed.* 2009, *48*, 2-25.
[14] Umegaki, T.; Yan, J.M.; Zhang, X.B.; Shioyama, H.; Kuriyama, N.; Xu, Q. *Int. J. Hydrogen Energy* 2009, *34*, 2303-2311.
[15] Demirci U. B.; Miele, P. *Energy Environ. Sci.* 2009, *2*, 627-637.
[16] Mandal, T. K.; Gregory, D. H. *Annu. Rep. Prog. Chem., Sect. A* 2009, *105*, 21-54.
[17] Fedneva, E. M.; Alpatova, V. L.; Mikheeva, V. I. *Russ. J. Inorg. Chem.* 1964, *9*, 826-827.
[18] Graetz, J. *Chem. Soc. Rev.* 2009, *38*, 73-82.
[19] Stasinevich, D. S.; Egorenko, G. A. *Russ. J. Inorg. Chem.*. 1968, *13*, 341-343.
[20] Züttel, A.; Wenger, P.; Rentsch, S.; Sudan, P.; Mauron, Ph.; Ernmenegger, Ch. *J. Power Sources* 2003, *118*, 1-7.
[21] Au, M.; Jurgensen, A. R.; Spencer, W. A.; Anton, D. L.; Pinkerton, F. E.; Hwang, S. J.; Kim, C.; Bowman Jr., R. C. *J. Phys. Chem. C* 2008, *112*, 18661-18671.
[22] Gross, A. F.; Vajo, J. J.; Atta, S. L.; Olsen, G. L. *J. Phys. Chem. C* 2008, *112*, 5651-5657.
[23] Nakamori, Y.; Li, H. W.; Matsuo, M.; Miwa, S.; Towata, S.; Orimo, S. *J. Phys. Chem. Solids* 2008, *69*, 2292-2296.
[24] Shore, S. G.; Parry, R. W. *J. Am. Chem. Soc.* 1955, *77*, 6084-6085.
[25] Stephens, F. H.; Pons, V.; Baker, R. T. *Dalton Trans.* 2007, 2613-2626.
[26] Peng, B.; Chen, J. *Energy Environ. Sci.* 2008, *1*, 479-483.
[27] Hamilton, C. W.; Baker, R. T.; Staubitz, A.; Manners, I. *Chem. Soc. Rev.* 2009, *38*, 279-293.
[28] Bluhm, M. E.; Bradley, M. G.; Butterick III, R.; Kusari, U.; Sneddon, L. G. *J. Am. Chem. Soc.* 2006, *128*, 7748-7749.
[29] Gutowska, A.; Li, L.; Shin, Y.; Wang, C. M.; Li, X. S.; Linehan, J. C.; Smith, R. S.; Kay, B. D.; Schmid, B.; Shaw, W.; Gutowski, M.; Autrey, T. *Angew. Chem. Int. Ed.* 2005, *44*, 3578-3582.
[30] Denney, M. C.; Pons, V.; Hebden, T. J.; Heinekey, D. M.; Goldberg, K. I. *J. Am. Chem. Soc.* 2006, *128*, 12048-12049.
[31] Crabtree, R.H. *Energy Environ. Sci.* 2008, *1*, 134-138.
[32] Wu, H.; Zhou, W.; Yildirim, T. *J. Am. Chem. Soc.* 2008, *130*, 14834-14839.
[33] Xiong, Z.; Wu, G.; Chua, Y. S.; Hu, J.; He, T.; Xu, W.; Chen, P. *Energy Environ. Sci.* 2008, *1*, 360-363.
[34] Xiong, Z.; Yong, C. K.; Wu, G.; Chen, P.; Shaw, W.; Karkamkar, A.; Autrey, T.; Jones, M. O.; Johnson, S. R.; Edwards, P. P.; David, W. I. F. *Nature Mater.* 2008, *7*, 138-141.
[35] Wang, P.; Kang, X. D. *Dalton Trans.* 2008, 5400-5413.
[36] Chua, Y. S.; Wu, G.; Xiong, Z.; He, T.; Chen, P. *Chem. Mater.* 2009, *21*, 4899-4904.
[37] Moriarty, P.; Honnery, D. *Int. J. Hydrogen Energy* 2009, *34*, 31-39.
[38] Bockris, J. O'M. *Int. J. Hydrogen Energy* 2008, *33*, 2129-2131.

[39] Wenger, D.; Polifke, W.; Schmidt-Ihn, E.; Abdel-Baset, T.; Maus, S. *Int. J. Hydrogen Energy* 2009, doi:10.1016/j.ijhydene.2009.05.072.
[40] van den Berg, A. W. C.; Areán, C. O. *Chem. Commun.* 2008, 668-681.
[41] Guo, Z. X.; Shang, C.; Aguey-Zinsou, K. F. *J. Eur. Ceram. Soc.* 2008, *28*, 1467-1473.
[42] Davis, B. L.; Dixon, D. A.; Garner, E. B.; Gordon, J. C.; Matus, M. H.; Scott, B.; Stephens, F. H. *Angew. Chem. Int. Ed.* 2009, *48*, 6812-6816.
[43] Lancaster, M. In *Green chemistry: an introductory text*; RSC, Cambridge: RSC Paperbacks: 2002.
[44] Clark, J. H. In *Part 1 Green chemistry for sustainable development;* in Green separation Processes; Weinheim; Wiley-VCH Verlag GmbH & Co.: 2005.
[45] Beach, E. S.; Cui, Z.; Anastas, P. T. *Energy Environ. Sci.* 2009, *2*, 1038-1049.
[46] Anastas P. C.; Warner, J. C. In *Green chemistry: Theory and practice*. New York: Oxford, University Press: 1998.
[47] Anastas, P. T. *Green Chem.* 2003, G29-G34.
[48] Warner, J. C.; Cannon, A. S.; Dye, K. M. *Environ. Impact Assess. Rev.* 2004, *24*, 775-799.
[49] Horváth, I. T.; Anastas, P. T. *Chem. Rev.* 2007, *107*, 2169-2173.
[50] Demirci U. B. *J. Power Sources* 2007, *169*, 239-246.
[51] Demirci U. B. Environ. Int. 2009, *35*, 628-631.
[52] Sigma Aldrich, *Material safety data sheet of ammonia borane,* Product number 287717.
[53] Sigma Aldrich, *Material safety data sheet of sodium borohydride*, Product number 452173.
[54] Lane, C. F. *Ammonia-borane and related N-B-H compounds and materials: safety aspects, properties and applications. U.S. Department of Energy Chemical Hydrogen Storage Center.* http://www1.eere.energy.gov/hydrogenandfuelcells/pdfs/nbh_h2_storage_survey.pdf.
[55] Sigma Aldrich, Material safety data sheet of lithium borohydride, Product number 62460.
[56] Tanaka, H.; Tokoyoda, K.; Matsumoto, M.; Suzuku, Y.; Kiyobayashi, T.; Kuriyama, N. *Int. J. Hydrogen Energy* 2009, *34*, 3210-3218.

INDEX

C

D

E

F

G

H

I

K

L

M

N

O

P

R

S

Y

Z